Developments in Emerging ar

Antimicrobial Stewardship

Developments in Emerging and Existing Infectious Diseases

Antimicrobial Stewardship

Edited by

Céline Pulcini
Professor of Infectious Diseases,
University of Lorraine,
Nancy University Hospital, Nancy, France

Önder Ergönül
Professor of Infectious Diseases, Koc University,
School of Medicine, Istanbul, Turkey

Füsun Can
Professor of Microbiology, Koc University,
School of Medicine, Istanbul, Turkey

Bojana Beović
Professor of Infectious Diseases,
Department of Infectious Diseases,
University Medical Centre Ljubljana and
Faculty of Medicine, University of Ljubljana, Slovenia

Academic Press is an imprint of Elsevier
125 London Wall, London EC2Y 5AS, United Kingdom
525 B Street, Suite 1800, San Diego, CA 92101-4495, United States
50 Hampshire Street, 5th Floor, Cambridge, MA 02139, United States
The Boulevard, Langford Lane, Kidlington, Oxford OX5 1GB, United Kingdom

Library of Congress Cataloging-in-Publication Data
A catalog record for this book is available from the Library of Congress

British Library Cataloguing-in-Publication Data
A catalogue record for this book is available from the British Library

ISBN: 978-0-12-810477-4
ISSN: 2214-4129

For information on all Academic Press publications
visit our website at https://www.elsevier.com/books-and-journals

Publisher: Sara Tenney
Acquisition Editor: Linda Versteeg-buschman
Editorial Project Manager: Tracy Tufaga
Senior Production Project Manager: Priya Kumaraguruparan
Cover Designer: Vicky Pearson Esser

Typeset by SPi Global, India

Contents

Contributors

Numbers in Parentheses indicate the pages on which the author's contributions begin.

Murat Akova (205, 331), Hacettepe University; Hacettepe University School of Medicine, Ankara, Turkey

Attila Altiner (167), Rostock University Medical Center, Rostock, Germany

Arjana T. Andrašević (263), University Hospital for Infectious Diseases, Zagreb, Croatia

Ozlem K. Azap (147), Başkent University, Ankara, Turkey

Luis Bavestrello (243), Clínica Reñaca, Viña del Mar, Chile

Bojana Beović (55, 235, 239, 305), University Medical Centre Ljubljana; Faculty of Medicine, University of Ljubljana, Ljubljana, Slovenia

Femke Böhmer (167), Rostock University Medical Center, Rostock, Germany

Guillaume Béraud (271), CHU de Poitiers, Poitiers, France

Kirsty Buising (247), University of Melbourne, Melbourne, VIC, Australia

Füsun Can (3, 115, 331), Koç University School of Medicine, Istanbul, Turkey

José L. Castro (243), Pan American Health Organization/World Health Organization, Washington, DC, United States

Milan Čižman (305), University Medical Centre Ljubljana; Faculty of Medicine, University of Ljubljana; University Medical Centre Ljubljana, Ljubljana, Slovenia

Sara E. Cosgrove (13, 235, 335), Johns Hopkins University School of Medicine, Baltimore, MD, United States

Menino O. Cotta (85), The University of Queensland; Royal Brisbane and Women's Hospital, Brisbane, QLD, Australia

Marlieke E.A. de Kraker (341), Infection Control Program, Geneva University Hospitals and Faculty of Medicine, Geneva, Switzerland

Katja de With (275), University Hospital Carl-Gustav-Carus at the TU Dresden, Germany

Jan De Waele (193), Ghent University Hospital, Ghent, Belgium

Oliver J. Dyar (55, 139), Karolinska Institutet, Stockholm, Sweden

Önder Ergönül (147, 331), Koç University, Istanbul, Turkey

Mats Erntell (321), Strama Halland, Halmstad, Sweden

Abdul Ghafur (285), Coordinator, Chennai Declaration, India

Inge C. Gyssens (29, 299), Radboud University Medical Center, Nijmegen, The Netherlands; Hasselt University, Hasselt, Belgium

Stephan Harbarth (341), Infection Control Program, Geneva University Hospitals and Faculty of Medicine, Geneva, Switzerland

Katarina Hedin (321), Region Kronoberg, Växjö, Sweden

Cristhian Hernndez-Gómez (243), Centro Internacional de Entrenamiento e Investigaciones Médicas (CIDEIM), Cali, Colombia

Philip Howard (129, 267), Leeds Teaching Hospitals NHS Trust, Leeds, United Kingdom

Marlies E.J.L. Hulscher (41), Radboud University Medical Center, Nijmegen, The Netherlands

Onur Karatuna (115), Acibadem University School of Medicine, Istanbul, Turkey

Winfried V. Kern (275), University Hospital and Medical Center, and Faculty of Medicine, Albert-Ludwigs-University Freiburg, Germany

Emma Keuleyan (259), Medical Institute of the Ministry of Interior, Sofia, Bulgaria

Diamantis P. Kofteridis (281), University Hospital of Heraklion, Crete, Greece

Tomislav Kostyanev (3, 259), University of Antwerp, Antwerp, Belgium

Patrick Lacor (255), UZ Brussel, Brussel, Belgium

Gabriel Levy-Hara (243, 313), Hospital Carlos G Durand, Buenos Aires, Argentina

David X. Li (13), Johns Hopkins University School of Medicine, Baltimore, MD, United States

Theodore I. Markou (335), Johns Hopkins University School of Medicine, Baltimore, MD, United States

Marc Mendelson (309), University of Cape Town, Groote Schuur Hospital, Cape Town, South Africa

Peter Messiaen (255), Jessaziekenhuis, Hasselt, Belgium

Dilip Nathwani (301), Ninewells Hospital and Medical School, Dundee, United Kingdom

Leonardo Pagani (293), Bolzano Central Hospital, Bolzano, Italy

José R. Paño Pardo (69, 317), Hospital Clínico Universitario "Lozano Blesa", Instituto de Investigación Sanitaria Aragón, Zaragoza, Spain

Catherine Pluss-Suard (329), Lausanne University Hospital, Lausanne, Switzerland

Jan Prins (299), Academic Medical Center, Amsterdam, the Netherlands

Céline Pulcini (139, 185, 271), Université de Lorraine and Nancy University Hospital, Nancy, France

Pilar Ramón-Pardo (243), Pan American Health Organization/World Health Organization, Washington, DC, United States

Jason A. Roberts (85), The University of Queensland; Royal Brisbane and Women's Hospital, Brisbane, QLD, Australia

Jesús Rodríguez-Baño (317), Unidad Clínica de Enfermedades Infecciosas y Microbiología, Hospital Universitario Virgen Macarena and Departamento de Medicina, Universidad de Sevilla - Instituto de Biomedicina de Sevilla (IBiS), Seville, Spain

Jeroen Schouten (41, 193, 299), Radboud University Medical Center, and Canisius Wilhelmina Ziekenhuis; Canisius Wilhelmina Hospital, Nijmegen, The Netherlands

Mitchell J. Schwaber (289), National Center for Infection Control, Ministry of Health, Tel Aviv, Israel

Laurence Senn (329), Lausanne University Hospital, Lausanne, Switzerland

Keigo Shibayama (297), National Institute of Infectious Diseases, Tokyo, Japan

Mahipal G. Sinnollareddy (85), University of South Australia, Adelaide, SA; The Canberra Hospital, Canberra, ACT, Australia

Jacqueline Sneddon (301), Scottish Antimicrobial Prescribing Group, Glasgow, United Kingdom

Cecilia Stålsby Lundborg (321), Karolinska Institutet, Stockholm, Sweden

Evelina Tacconelli (239), Medical University Hospital Tübingen, Tübingen, Germany

Karin Thursky (99, 247), University of Melbourne; Peter MacCallum Cancer Centre, Melbourne, VIC, Australia

Antonis Valachis (281), University Hospital of Heraklion, Crete, Greece

Pierluigi Viale (293), University of Bologna, Bologna, Italy

María V. Villegas (243), Centro Internacional de Entrenamiento e Investigaciones Médicas (CIDEIM), Cali, Colombia

Vera Vlahović-Palčevski (29, 175, 263), University Hospital Rijeka; University of Rijeka Medical Faculty, Rijeka, Croatia

Agnes Wechsler-Fördös (251), Krankenanstalt Rudolfstiftung, Vienna, Austria

Anja Wollny (167), Rostock University Medical Center, Rostock, Germany

Giorgio Zanetti (329), Lausanne University Hospital, Lausanne, Switzerland

Foreword by Prof. Mario Poljak

This book promises to become a state-of-the-art reference in the field of antimicrobial stewardship, which is defined as a coordinated program promoting the appropriate use of antimicrobial agents in order to improve patient outcomes, reduce antimicrobial resistance, and decrease the spread of infections caused by multidrug-resistant organisms. The book "Antimicrobial Stewardship" was initiated and coordinated by the European Society of Clinical Microbiology and Infectious Diseases (ESCMID) Study Group for Antibiotic Policies (ESGAP) and benefited from the knowledge and long-standing practical experience of a team of more than 40 leading experts from all over the world. ESGAP is organized under the auspices of ESCMID, a non-profit society with more than 8,000 members, which has been committed to improving the diagnosis, treatment, and prevention of infection-related diseases since its foundation in 1983. Antimicrobial resistance has been at the core of ESCMID's activities—at its study groups, committees, courses, and conferences—over the past years. Its experts have been committed to developing hands-on solutions by supporting and promoting evidence-based research and professional training programs to tackle the problem around the world. ESGAP has been most intensively engaged in advancing scientific knowledge and disseminating professional guidelines in the field of antimicrobial stewardship, one of the key strategies to tackle antimicrobial resistance. As current ESCMID President, I am proud to support ESGAP's activities, especially the publication of a book, which perfectly exemplifies how ESCMID strives to accomplish its mission. The book is not only designed to provide guidance for young health professionals starting their training in this fascinating area of medicine, it also acts as a reference for experts seeking information on the current state of science and practice. I believe that this publication will help expand existing knowledge and increase awareness and understanding of antimicrobial stewardship. It will contribute to the body of evidence required to successfully reduce the misuse and overuse of antimicrobials and thus control antimicrobial resistance—one of the world's most pressing public health problems.

Prof. Mario Poljak, MD, PhD
ESCMID President

Introduction by Murat Akova

"The times they are a-changin": nothing describes better than this lyric song by the latest Nobel laureate, Bob Dylan, the current situation of antimicrobial resistance has become one of the worst nightmares of humanity. Antibiotics have saved lives and contributed significantly to the duration of life expectancy. We have come a long way from the discovery of penicillin to modern-day antibiotics against all of which microbes have been able to find ways to resist. So, the era of one of the most important successful discoveries in human medicine has come to an end. Well, there is a great concern about that, but not a definite answer beyond some pessimistic predictions. What we know best, in simple terms, is that, if one uses antibiotics prudently, then he/she may expect less resistance to develop.

A more complex strategy is offered by "antimicrobial stewardship." For that, you would need to know how to implement usage of new and old antibiotics into the current practice, better and faster diagnostics, coupled them with infection control practices, other preventive measures that may well include vaccination strategies, and above all a comprehensive educational program for both students and healthcare professionals. Although huge amount of data including several guidelines has been available in the literature, a textbook of comprehensive guidance about antimicrobial stewardship was lacking, so far. Thus, this new book by ESCMID Study Group for Antibiotic Policies (ESGAP) is timely and expected to fill in this gap. As the past president of ESCMID, I wholeheartedly congratulate the editors, Profs. Beović, Can, Ergönül, and Pulcini for their immeasurable efforts. I have no doubts that this textbook will be of great help for those who would like to be updated on antimicrobial stewardship.

Murat Akova
Past President and Communications Officer, ESCMID

Introduction by Jesús Rodríguez-Baño

Antimicrobial resistance is one of the more important and complex challenges of medicine at the present time. Among the actions to be undertaken to fight against resistance, one of the most urgent is improving the use of antibiotics. This is far from easy. The usual figures show that one out of every three prescriptions of antibiotics is inappropriate, which is probably related to intrinsic complexities of antibiotic use and also to the fact that antibiotics can be prescribed by all physicians in very different circumstances. Well-designed antibiotic stewardship programs carried out by specialized multidisciplinary teams are needed to significantly improve the use of antibiotics. However, their implementation is frequently limited by a lack of resources and institutional support and also by a scarcity of experts in the field. Antibiotic stewardship has been frequently neglected in many training programs for infectious diseases, microbiology, or hospital pharmacy.

Despite the body of knowledge related to antibiotic stewardship rapidly increasing, the textbooks compiling it are still rare. This is therefore the perfect time for a book that is comprehensive enough and orientated to the practice of professionals working in antibiotic teams. At the European Society of Clinical Microbiology and Infectious Diseases (ESCMID), we are proud of the leadership and work of the Editors and the whole European Study Group for Antibiotic Policies (ESGAP) for driving this initiative.

Jesús Rodríguez-Baño
President-Elect, ESCMID

Introduction by Evelina Tacconelli

The rate of infections caused by antimicrobial-resistant microorganisms is seen increasingly by the public and healthcare inspection organizations as an indicator of quality of healthcare and patient safety and a major threat to public health. For these reasons, the European Society of Clinical Microbiology and Infectious Diseases (ESCMID) and ESCMID Study Group for Antibiotic Policy (ESGAP) together with other major international stakeholders have been working restless, since many years, to develop new tools improving antibiotic usage and understanding major limitations and gaps in prescribing. Notably, ESGAP's activity spans a wide range of antibiotic research fields and is internationally recognized for setting the agenda in policy and stewardship. This extensive experience and international collaborations provided ESGAP with a unique opportunity to commit on this book scientists and doctors every day involved in implementation of antibiotic stewardship in hospitals and communities.

The book represents a practical aid for hospitals and healthcare institutions, in putting in place locally tailored stewardship measures taking also in consideration costs and adverse events. Step by step, the authors not only deliver the information you need to plan and implement in your own antimicrobial stewardship program but also specify quality and quantity metrics of antibiotic usage as well instructions on how to organize educational events for healthcare workers. In accordance with the purpose of this project, the book provides a valuable framework for the important collegial relationship between different specialists that must be involved in any stewardship action (infectious disease physicians, microbiologists, pharmacists, surveillance scientists, and infection prevention and control nurses) to have success. What I personally like more in this book is the evidence-based approach applied to provide suggestions and recommendations to be used in "real life." Of even more importance, the authors did not opt for a generic approach but instead clearly focused on different components of the stewardship as location (i.e., long-term care facilities or outpatient settings) or case-mix of patients (i.e., ICU or hematological patients).

I believe that this book represents a valuable resource and will improve your patients' quality of care. It will definitively help in intensifying the global efforts that, together with enhanced infection control and revitalized antibiotic R&D, are needed to reduce the burden of antibiotic resistance.

Evelina Tacconelli
ESCMID Education Officer

Preface

Antimicrobial resistance has developed like any other ecological catastrophe: human enthusiasm in controlling nature was followed by a rebound of uncountable numbers of microorganisms that have been adapting to life on earth for much longer than *Homo sapiens*. As mankind is intensively looking for new sources of sustainable and green energy, the efforts are parallel with the efforts for more responsible use of energy for transport, industry, heating, and other activities. We do need new antimicrobial drugs, but the lesson we have learned observing the rise of antimicrobial resistance taught us that prudent and responsible use of antimicrobials is necessary in all cases. We have also learned that defining and promoting prudent use based on drug pharmacodynamics and pharmacokinetics and development of practice guidelines is not enough. Antimicrobial stewardship has emerged as a set of interventions that brings the principles of prudent use to every healthcare professional and, most importantly, to every patient. The book provides a snapshot about where we are in antimicrobial stewardship and the motivation to challenge such a big public health problem.

This book summarizes the current 'state of the art' on antimicrobial stewardship in a short and practical way in order for it to be a 'hands-on' tool for students, trainees, physicians, microbiologists, pharmacists, nurses, or anyone involved in antimicrobial stewardship. The first part gives us an insight into the scientific background of antimicrobial stewardship and describes the current global situation regarding antimicrobial use and resistance. The antimicrobial strategies described in the second part are followed by a part on antimicrobial stewardship in special situations. In the last part, experts from around the globe describe their efforts, success stories, and problems related to the local situations in different parts of the world, together with the most important international governmental and nongovernmental initiatives. Unmet needs and challenges in research on antimicrobial stewardship are addressed in the last chapter.

Most of the book's authors are members of the European Society for Clinical Microbiology and Infectious Diseases (ESCMID) Study Group for Antibiotic Policies (ESGAP), which has almost two decades of experience in educational courses on antimicrobial stewardship. As ESGAP members and as healthcare professionals working in antimicrobial stewardship, we strongly believe that the book will find its place in the libraries and on the desks of antimicrobial stewardship teams. We hope that you will find it useful and that you will take as much pleasure in reading it as we did in writing it. We hope the book will give you the necessary background for motivation to change.

The editors of the book are

Prof. Bojana Beović (Slovenia), ESGAP chair
Prof. Önder Ergönül (Turkey)
Prof. Füsun Can (Turkey)
Prof. Céline Pulcini (France), ESGAP secretary

Acknowledgments

We are grateful to the members of the Scientific Committee for their insightful input and careful review of the chapters: Murat Akova, Sara Cosgrove, Stephan Harbarth, Dilip Nathwani, Mical Paul, Jesus Rodriguez-Bano, and Karin Thursky. We also thank Marlieke De Kraker for her help in the reviewing process.

Section A

The Global Picture of Antimicrobial Use and Resistance

Chapter 1

The Global Crisis of Antimicrobial Resistance

Tomislav Kostyanev* and Füsun Can**
*University of Antwerp, Antwerp, Belgium
**Koç University School of Medicine, Istanbul, Turkey

INTRODUCTION

Since the introduction of the first antibiotic in clinical practice for the treatment of infections in the 1940s, antibiotic resistance appeared and evolved against all classes of antibiotics. Nowadays, antimicrobial resistance (AMR) has been identified as a global public health threat, leading to at least 700,000 mortality cases every year around the world. It has been estimated that, if no measures are taken, this menace would cause the death of 10 million people in 2050 [1]. As stated in O'Neill's report, AMR is a truly global problem that "should concern every country irrespective of its level of income."

MECHANISMS OF AMR

There are many mechanisms by which bacteria might exhibit resistance to antibiotics. Bacteria can be intrinsically resistant to antibiotics or can also acquire resistance via spontaneous mutations in chromosomally located genes and by the horizontal acquisition of new genes.

Intrinsic Resistance

Intrinsic resistance is found within the genome of bacterial species and gives the bacteria an ability to resist the activity of a particular antimicrobial agent. It is independent of antibiotic selective pressure and horizontal gene transfer. Intrinsic resistance may be due to:

- a lack of affinity of the drug for the bacterial target
- inaccessibility of the drug into the bacterial cell
- extrusion of the drug by chromosomally encoded efflux pumps
- presence of drug-degrading enzymes

Antimicrobial Stewardship. http://dx.doi.org/10.1016/B978-0-12-810477-4.00001-5

In practice, knowledge of the intrinsic resistance of a pathogen is essential to avoid inappropriate antimicrobial therapy and to decrease the risk of acquired resistance. Recent studies using high-throughput screens of genome mutant libraries have shown various genes that are responsible for intrinsic resistance to several antibiotics, such as β-lactams, fluoroquinolones, and aminoglycosides. These studies have also provided possible synergistic drug combinations that can be used for treatment of some clinically important pathogens, including *Acinetobacter baumannii* and *Pseudomonas aeruginosa.* In the future, these data can be used for the development of new antibiotic combinations that have an expanded spectrum of activity against these problematic pathogens [2].

Acquired Resistance

In addition to intrinsic resistance, bacteria can obtain the ability to resist the activity of an antimicrobial agent that was previously effective. Unlike intrinsic resistance, acquired resistance develops in some strains or subpopulations of each particular bacterial species. Acquired resistance arises from gene change and/or exchange by mutation of a particular gene or horizontal gene transfer via transformation, conjugation, or transduction. These genetic changes can facilitate different resistance mechanisms that are classified into three main groups:

- Prevention of access to target due to low intracellular concentrations of the antibiotic by reduced intake or increased efflux activity
- Target modification of the antibiotic by genetic mutation or posttranslational modification
- Inactivation of the antibiotic by hydrolysis or modification of binding site

Prevention of Access to Target

Reduced Permeability

A Gram-negative bacterial cell wall is less permeable to antibiotics because of porin proteins embedded into the outer membrane. Porins create a size-selective channel for antibiotics and control the rate of diffusion of large antibiotics. The levels of porins in the bacterial cell can increase up to 10^6 copies per cell [3]. In some cases, mutations that cause the loss, downregulation, or replacement of porins limit the diffusion rate of antibiotics. In general, the loss of any particular porin initiates low-level resistance. However, the accumulation of independent mutational events can cause high-level resistance. Some clinically important Gram-negative bacteria like *A. baumannii*, Enterobacteriaceae species, and *P. aeruginosa* become resistant to key antibiotics such as carbapenems or cephalosporins by alterations in porin proteins. The mutations of CarO porins in *A. baumannii*, OmpK36 in *Klebsiella pneumoniae*, and OprD in *P. aeruginosa* cause carbapenem resistance [4–6].

Increased Efflux Pumps

Efflux pumps are one of the major contributors of resistance in Gram-negative bacteria and are encoded in the chromosome. Efflux pumps can either have narrow substrate specificity and only export one molecule, or they can be more broadly active and transport structurally dissimilar substrates. Overexpression of efflux pumps can result in moderate- to high-level resistance to antibiotics. There are five main classes of efflux proteins at the bacterial membrane: the ATP binding cassette (ABC), the major facilitator, the multidrug and toxic-compound efflux, the small multidrug resistance, and the resistance-nodulation-division (RND) family [7]. The resistance-nodulation-division family is known as MDR (multidrug resistant) efflux pumps and is largely responsible for antibiotic resistance in Gram-negative bacteria. The overexpression of efflux genes usually originates from mutations in the efflux-pump controlling networks. In *Neisseria gonorrhoeae,* a single-base-pair mutation that changes promoter causes an overexpression of the efflux pump and multidrug resistance [8]. In *Escherichia coli*, the *AcrAB-TolC* resistance-nodulation-division efflux system mediates the resistance of this organism to a broad range of antibiotics, such as the tetracyclines, fluoroquinolones, ß-lactams, and the macrolides [7]. Mutations in the AcrAB efflux system in combination with porin loss may function as stepping stones toward high-level carbapenem resistance without the existence of carbapenemases [9]. Recently, an INCH1 plasmid carrying both resistance-nodulation-division pump and New Delhi metallo-β-lactamase 1 (NDM-1) has been isolated from *Citrobacter freundii* [10]. This finding has indicated the presence of mobilized efflux pump genes that can be rapidly disseminated between clinically relevant pathogens.

Development of molecules that prevent efflux pump activation and use of these agents in combination with antibiotics could be a new therapeutic approach against MDR bacteria.

Target Modification of the Antibiotic by Genetic Mutation or Posttranslational Modification

Some bacteria develop a resistance to antimicrobials by altering the target structure to avoid recognition. Therefore, in spite of the presence of antimicrobials at the site, the target is still able to maintain its normal function.

Some clinically important examples of bacterial resistance related to target site modification are:

- *Alteration in penicillin-binding protein (PBPs) results in reduced affinity of β-lactam antibiotics* [11]: Acquisition of the staphylococcal cassette chromosome *mec* (SCC*mec*) element, which carries the *mecA* gene, encoding PBP2a causes resistance to β-lactam antibiotics in Staphylococci. In *Streptococcus pneumoniae,* mutations in PBPs decrease the affinity of β-lactams. In *N. gonorrhoeae,* the mosaic structure of the *penA* gene (PBP encoding gene) is associated with high-level resistance to extended spectrum cephalosporins.

- *Changes in cell wall thickness reduces the activity of vancomycin:* A thickened cell wall in vancomycin intermediate *Staphylococcus aureus* (VISA) prevents the diffusion of vancomycin into the cell and causes resistance to all glycopeptide antibiotics.
- *Changes in vancomycin precursors inhibit the activity of vancomycin:* Synthesis of abnormal peptidoglycan precursors terminating in D-Ala–D-lactate instead of D-Ala–D-Ala prevents the binding of vancomycin to its target.
- *Alterations in subunits of DNA gyrase and topoisomerase IV reduce the activity of fluoroquinolones:* Resistance to fluoroquinolones mainly arises from mutations in DNA gyrase in Gram-negative bacteria or mutations in topoisomerase IV in Gram-positive bacteria. The plasmid encoded *qnr*-type genes are also associated with quinolone resistance and protect topoisomerase IV and DNA gyrase from the lethal action of quinolones.
- *Alteration in negative charge of cell membrane reduces the activity of polymyxins:* Colistin resistance in *K. pneumoniae* is associated with mutations in the *mgrB* gene or genes encoding the *PhoPQ* two-component system. These alterations change the negative charge of the Lipopolysaccharide (LPS) through the addition of phosphoethanolamine to lipid A and reduces the ability of colistin to bind to LPS. Recently, the first plasmid-mediated polymyxin resistance gene, *MCR-1*, in Enterobacteriaceae was discovered in China. *MCR-1* is a member of the phosphoethanolamine transferase enzyme family and is responsible for the addition of phosphoethanolamine to lipid A [12].
- *Alterations in binding sites of daptomycin confer daptomycin resistance:* Daptomycin resistance in *S. aureus*, is acquired by a single nucleotide polymorphisms in the multipeptide resistance factor gene (mprF: a membrane lysylphosphatidylglycerol synthetase), the yycFG (a histidine kinase), *rpoC*, and *rpoB* (subunits of RNA polymerase) [13].

Direct Modification of Antibiotics

As well as preventing antibiotics from diffusing into the cell or altering their targets, bacteria can destroy or modify antibiotics.

Degradation of Antibiotics by Enzyme Hydrolysis

Bacteria have developed several enzymatic hydrolysis mechanisms to inactivate antibiotics. A classic example of enzymatic degradation is the production of β- lactamases that hydrolyze the β-lactam ring of penicillins. These enzymes have evolved to hydrolyze a broad range of extended spectrum cephalosporins for years and are called extended spectrum β-lactamases, or

ESBLs. The TEM, SHV, and CTX-M ESBLs are now known to be produced by many Gram-negative bacteria.

Carbapenems are frequently used for the treatment of bacterial infections as a last choice of therapy. Carbapenem resistance is usually associated with the acquisition of β-lactamases with carbapenem-hydrolyzing activity (known as carbapenemases). Carbapenemases were first identified on the chromosomes of single species; however, various types of plasmid-mediated carbapenemases have now been reported in Enterobacteriaceae, *P. aeruginosa,* and *A. baumannii* [11]. The most common carbapenemases are KPC, VIM, IMP, NDM, and OXA-48 types. The KPC carbapenemases were first isolated from *K. pneumoniae* and quickly spread across a wide range of Gram-negative bacteria and are no longer limited to *K pneumoniae*. OXA carbapenemases are widely dispersed in *P. aeruginosa*, *Klebsiella pneumoniae,* and in *A. baumannii*. The VIM carbapenemases have been identified in *Pseudomonas spp.* and rarely in members of *Enterobacteriaceae* and *Acinetobacter spp.* The NDM carbapenemases are detected in Gram-negative pathogens such as *A. baumannii*, *K. pneumoniae* and *E. coli* and confer resistance to all β-lactams except aztreonam [14].

Like β-lactamases, bacteria produce macrolide esterases and fosfomycin epoxidases enzymes. Macrolide esterases inactivate erythromycin A and oleandomycin by hydrolyzing the lactone ring [15].

Inactivation of Antibiotics by Transfer of a Chemical Group

Some bacteria become resistant to antibiotics through the addition of chemical groups to active sites of the antibiotics, which prevents binding of the antibiotic to its target protein. The aminoglycoside-modifying enzymes (acetyltransferases, phosphotransferases, and nucleotidyltransferases) catalyze the modification at different −OH or −NH2 groups of the 2-deoxystreptamine nucleus or the sugar moieties and cause high levels of resistance to aminoglycosides [11].

AMR IN GRAM-POSITIVES

Some of the most clinically significant resistant Gram-positive bacteria are methicillin-resistant *S. aureus* (MRSA) and vancomycin-resistant enterococci (VRE).

The clonal spread of MRSA is of particular concern in many hospital settings, causing outbreaks of hospital-associated MRSA. The transmission of these MDR bacteria between patients is facilitated by poor hospital hygiene and a lack of infection control measures. MRSA also causes problems in the community; community-associated MRSA is one of the leading pathogens causing difficult-to-treat invasive infections, such as pneumonia, endovascular infections, sepsis, etc. However, the transfer of healthcare-associated MRSA clones into the community is another underlying reason for community-onset infections caused by MRSA clones. According to the ECDC AMR surveillance report for 2014 [16], there is a decreasing MRSA trend in Europe, which is less substantial, however, compared with the one for the period 2009–2012. A total of 6 out of 29 countries (Italy, Cyprus, Greece, Malta, Portugal, Romania) reported invasive MRSA isolates above 30%.

In contrast to the overall rate of MRSA in Europe, vancomycin resistance in enterococci is on the rise, much more pronounced in *Enterococcus faecium* than in *E. faecalis*. Invasive VRE isolates have spread clonally, and now, more than a third of the countries reporting to ECDC show an increasing trend for the period 2011–2014 [16]. Ireland, Cyprus, Greece, and Romania particularly report rates of vancomycin-resistant *E. faecium* higher than 25%.

The ECDC report also reports large intercountry variations in terms of invasive penicillin-nonsusceptible *S. pneumoniae* (PNSP) isolates. Despite the high rate of PNSP in some countries, such as Bulgaria, Croatia, Spain, and Romania (reporting more than 25%), the invasive pneumococcal disease seems to be well contained due to the implemented routine immunization programs for children with multivalent pneumococcal conjugate vaccines.

Overall, there has been success in many countries in terms of limiting the initially increasing trends of MDR Gram-positives. This could be mainly due to numerous infection control measures and successful interventions targeting these problematic microorganisms [17]. Hand hygiene campaigns, improvements of contact isolation and environmental control with reduced cross transmissions, robust screening for MRSA carriage, antibiotic policies with reduced antibiotic selective pressure and natural fluctuation, and change in the prevalence of certain clones are some of the most plausible reasons for overall success in the battle with AMR in Gram-positives.

AMR IN GRAM-NEGATIVES

In contrast to the resistance in Gram-positives, AMR in Gram-negatives has been booming, with certain clones spreading very quickly, leading to outbreaks of epidemic proportions in some countries. Resistance to antibacterial agents, such as carbapenems and colistin, has made some Gram-negatives a substantial threat and a challenge to overcome. Over the years, this problem has been aggravated, and certain countries (such as Greece, Italy, Israel, India, and others) became endemic for several MDR Gram-negatives.

Being much more promiscuous than Gram-positives, Gram-negative bacteria are genetically more flexible and frequently acquire genetic material from other bacteria. The successful transfer of resistant genes amongst Gram-negatives and the establishment of certain high-risk clones have made these bacteria persist and spread in various environments. These clones play the role of a source for the further spread of genetic components of AMR (i.e., genes, integrons, transposons, and plasmids) [18]. ESBL-producing Gram-negatives (such as TEM, SHV, CTX-M, and others) appeared gradually in the 1980s and 1990s and spread widely in many countries, some of which became endemic (i.e., Southern and some Eastern European countries, India, and many others) [16]. The resistance to several beta-lactams conferred by these enzymes is frequently combined with resistance to quinolones and aminoglycosides, which makes these Gram-negatives more problematic and difficult to treat. Some of the last-resort antibiotics against ESBL infections, the carbapenems, are extensively targeted by other beta-lactamases, called carbapenemases. Since the identification of the first carbapenemase producer in *Enterobacteriaceae* in the 1990s [19], many carbapenemase genes have been found and classified into three classes of β-lactamases: the Ambler class A, B, and D β-lactamases. The main representative and most commonly found class A carbapenemase, the Klebsiella pneumoniae carbapenemase (KPC), has spread globally and has been causing outbreaks in many countries such as the United States, Puerto Rico, Colombia, Greece, Israel, Italy, and the People's Republic of China [20]. One of the main reasons for the spread of carbapenem resistance is the successful dissemination of *K. pneumoniae* isolates members of the clonal complex 258 (CC258), particularly sequence type (ST) 258, usually harboring KPC variants. Class B metallo-β-lactamases (MBLs) are represented mainly by the Verona integron–encoded metallo-β-lactamase (VIM), IMP carbapenemases, and the NDM-1 type. Countries such as Greece, Taiwan, and Japan have been found to be endemic for VIM- and IMP-type enzymes; however, these types have been detected in many other countries [20]. NDM-1 producers have been reported from almost all around the world, with India, the Middle East, and the Balkan countries being the main reservoirs [21]. The most important Class D carbapenemase in Enterobacteriaceae, OXA-48, first isolated in Turkey in 2003, has spread ever since in several countries in Europe, in the southern and eastern part of the Mediterranean Sea, and Africa [20,22].

The rates of carbapenem resistance have been alarmingly increasing, reaching very high levels in many countries. There is a substantially increasing trend among the countries reporting to ECDC in 2014, with three countries (Greece, Italy, Romania) showing carbapenem resistance rates in invasive *K. pneumoniae* isolates higher than any other country in Europe (62.3%, 32.9%, and 31.5 %, respectively) [16].

Associated with higher mortality rates [23], the nosocomial infections caused by MDR Gram-negatives are left with less therapeutic options. Three of the last-resort antibiotics in such cases are colistin, fosfomycin, and tigecycline [24]. Colistin, a polymyxin antibiotic, one of the first antibiotics, discovered in the 1940s, was left aside because of its nephrotoxicity and neurotoxicity. With the rise of MDR Gram-negatives and its more frequent use as the drug-of-choice, especially in carbapenemase-endemic countries, unfortunately, colistin resistance did not lag behind. The reemergence of colistin as an antimicrobial therapy and its broad use in the veterinary sector have logically led to a rise in the colistin resistance rates [25]. More alarmingly, it has been found that apart from the chromosomal resistance, there is also a plasmid-mediated polymyxin resistance mechanism, MCR-1, in Enterobacteriaceae [13], which is highly transferrable and might have the potential to become the next menace in the field of AMR. However, from the many scientific publications that appeared after the first report of MCR-1, it was made clear that this was not a new problem [26] and that the gene has been present but remained undetected for a long time. The oldest isolate, harboring MCR-1, was reported in chickens in China dating back to the 1980s. In Europe, that was found to be *E. coli* from a diarrheic veal calf isolated in France in 2005. The earliest reported isolate from humans, *Shigella sonnei,* originated in Vietnam in 2008.

MDR Gram-negatives have effectively spread in many countries, and some of the numerous reasons for this, apart from the pandemicity of certain clones, plasmids and genes, could be the lack of rapid diagnostic methods locally combined with a lack of screening and an overall failure to detect resistant isolates in time or at all. Ineffective hand hygiene and increased antibiotic selective pressure add to this problem. Transmission of MDR pathogens can also persist despite some antimicrobial stewardship (AMS) and infection control measures.

It becomes more and more challenging to overcome the increasing rates of MDR Gram-negatives worldwide with very limited therapeutic options. As summarized by Roca *et al.* [27], there are five key aspects that need to be tackled to more effectively combat AMR, which include:(i) limiting the emergence and spread of resistant bacteria in the animal sector, (ii) undertaking AMS measures in the community, (iii) undertaking AMS measures in the healthcare settings, (iv) improving AMR diagnostics, and (v) fueling the antibiotic pipeline.

While the first four have been extensively addressed within the past 20 years, the latter has slowed to an unacceptable level with no new classes of antibiotics within the past two decades, and the antibiotic pipeline is running dry. There is an unmet public health need for new antibacterial products that can boost the armamentarium against MDR infections. Due to several, mainly economical, drawbacks, many pharmaceutical companies have invested less in antibiotic drug discovery and development in the past two decades. One of the most ambitious and substantial investment of resources in this field of was made by the Innovative Medicines Initiative (IMI) program, a public–private partnership for moving drug development forward [28]. One of IMI's programs, the New Drugs for Bad Bugs (ND4BB) initiative has a budget 660 million Euros and targets both drug discovery and development and also addresses improvement of clinical trial designs and responsible use of antibiotics. Collaborative funding into the research and development of antibiotics is very much needed and represents a substantial long-term investment in public health globally. It will overcome the inequity created by the business model dependent on sales volume, which has made R&D of antibiotics less attractive for the pharmaceutical industry compared to medicines for chronic diseases.

REFERENCES

[1] O'Neill J. Antimicrobial resistance: tackling a crisis for the health and wealth of nations, The review on antimicrobial resistance; 2014, http://amr-review.org/sites/default/files/AMR%20Review%20Paper%20%20Tackling%20a%20crisis%20for%20the%20health%20and%20wealth%20of%20nations_1.pdf.

[2] Principe L, Capone A, Mazzarelli A, D'Arezzo S, Bordi E, Di Caro A, *et al.* In vitro activity of doripenem in combination with various antimicrobials against multidrug-resistant Acinetobacter baumannii: possible options for the treatment of complicated infection. Microb Drug Resist 2013;19(5):407–14.

[3] Fernandez L, Hancock RE. Adaptive and mutational resistance: role of porins and efflux pumps in drug resistance. Clin Microbiol Rev 2012;25(4):661–81.

[4] Fernandez-Cuenca F, Tomas M, Caballero-Moyano FJ, Bou G, Martinez-Martinez L, Vila J, *et al.* Reduced susceptibility to biocides in Acinetobacter baumannii: association with resistance to antimicrobials, epidemiological behaviour, biological cost and effect on the expression of genes encoding porins and efflux pumps. J Antimicrob Chemother 2015;70(12):3222–9.

[5] Shen J, Pan Y, Fang Y. Role of the outer membrane protein OprD2 in carbapenem-resistance mechanisms of *Pseudomonas aeruginosa*. PLoS ONE 2015;10(10):e0139995.

[6] Wassef M, Abdelhaleim M, AbdulRahman E, Ghaith D. The role of OmpK35, OmpK36 porins, and production of beta-lactamases on imipenem susceptibility in *Klebsiella pneumoniae* Clinical Isolates, Cairo, Egypt. Microb Drug Resist 2015;21(6):577–80.

[7] Piddock LJ. Multidrug-resistance efflux pumps—not just for resistance. Nat Rev Microbiol 2006;4(8):629–36.

[8] Demczuk W, Martin I, Peterson S, Bharat A, Van Domselaar G, Graham M, *et al.* Genomic epidemiology and molecular resistance mechanisms of azithromycin-resistant *Neisseria gonorrhoeae* in Canada from 1997 to 2014. J Clin Microbiol 2016;54(5):1304–13.

[9] Adler M, Anjum M, Andersson DI, Sandegren L. Combinations of mutations in envZ, ftsI, mrdA, acrB and acrR can cause high-level carbapenem resistance in *Escherichia coli.* J Antimicrob Chemother 2016;71(5):1188–98.

[10] Dolejska M, Villa L, Poirel L, Nordmann P, Carattoli A. Complete sequencing of an IncHI1 plasmid encoding the carbapenemase NDM-1, the ArmA 16S RNA methylase and a resistance-nodulation-cell division/multidrug efflux pump. J Antimicrob Chemother 2013;68(1):34–9.

[11] Blair JM, Webber MA, Baylay AJ, Ogbolu DO, Piddock LJ. Molecular mechanisms of antibiotic resistance. Nat Rev Microbiol 2015;13(1):42–51.

[12] Liu YY, Wang Y, Walsh TR, Yi LX, Zhang R, Spencer J, *et al.* Emergence of plasmid-mediated colistin resistance mechanism MCR-1 in animals and human beings in China: a microbiological and molecular biological study. Lancet Infect Dis 2016;16(2):161–8.

[13] Montero CI, Stock F, Murray PR. Mechanisms of resistance to daptomycin in *Enterococcus faecium.* Antimicrob Agents Chemother 2008;52(3):1167–70.

[14] Nordmann P, Poirel L. The difficult-to-control spread of carbapenemase producers among Enterobacteriaceae worldwide. Clin Microbiol Infect 2014;20(9):821–30.

[15] Morar M, Pengelly K, Koteva K, Wright GD. Mechanism and diversity of the erythromycin esterase family of enzymes. Biochemistry 2012;51(8):1740–51.

[16] ECDC. Antimicrobial resistance surveillance in Europe. Annual report of the European Antimicrobial Resistance Surveillance Network (EARS-Net), http://ecdc.europa.eu/en/publications/Publications/antimicrobial-resistance-europe-2014.pdf; 2014.

[17] Eggimann P, Pittet D. Nonantibibiotic measures for the prevention of Gram-positive infections. Clin Microbiol Infect 2001;7(Suppl 4):91–9.

[18] Pitout JD, Nordmann P, Poirel L. Carbapenemase-producing *Klebsiella pneumoniae*, a key pathogen set for global nosocomial dominance. Antimicrob Agents Chemother 2015;59(10): 5873–84. http://dx.doi.org/10.1128/AAC.01019-15.

[19] Naas T. Nordmann P Analysis of a carbapenem-hydrolyzing class A β-lactamase from *Enterobacter cloacae* and of its LysR-type regulatory protein. Proc Natl Acad Sci U S A 1994;91:7693–7. http://dx.doi.org/10.1073/pnas.91.16.7693.

[20] Nordmann P, Naas T, Poirel L. Global spread of carbapenemase-producing Enterobacteriaceae. Emerg Infect Dis 2011;17(10):1791–8. http://dx.doi.org/10.3201/eid1710.110655.

[21] Nordmann P, Poirel L, Toleman MA. Walsh TR Does broad-spectrum β-lactam resistance due to NDM-1 herald the end of the antibiotic era for treatment of infections caused by Gram-negative bacteria? J Antimicrob Chemother 2011;66:689–92. http://dx.doi.org/10.1093/jac/dkq520.

[22] Albiger B, Glasner C, Struelens MJ, Grundmann H, Monnet DL. the European Survey of Carbapenemase-Producing Enterobacteriaceae (EuSCAPE) working group. Carbapenemase-producing Enterobacteriaceae in Europe: assessment by national experts from 38 countries, May 2015. Euro Surveill 2015;20(45). http://dx.doi.org/10.2807/1560-7917.ES.2015.20.45.30062. pii: 30062.

[23] Tzouvelekis LS, Markogiannakis A, Psichogiou M. Carbapenemases in *Klebsiella pneumoniae* and other Enterobacteriaceae: an evolving crisis of global dimensions. Clin Microbiol Rev 2012;25:682–707.

[24] Karaiskos I, Giamarellou H. Multidrug-resistant and extensively drug-resistant Gram-negative pathogens: current and emerging therapeutic approaches. Expert Opin Pharmacother 2014;15(10):1351–70. http://dx.doi.org/10.1517/14656566.2014.914172.

[25] Lim L, Ly N, Anderson D, Yang JC, Macander L, Jarkowski J, *et al.* Resurgence of colistin: a review of resistance, toxicity, pharmacodynamics, and dosing. Pharmacotherapy 2010;30(12):1279–91. http://dx.doi.org/10.1592/phco.30.12.1279.

[26] Skov R, Monnet D. Plasmid-mediated colistin resistance (mcr-1 gene): three months later, the story unfolds. Euro Surveill 2016;21(9). http://dx.doi.org/10.2807/1560-7917.ES.2016.21.9.30155. pii: 30155.

[27] Roca I, Akova M, Baquero F, Carlet J, Cavaleri M, Coenen S, *et al.* The global threat of antimicrobial resistance: science for intervention. New Microbes New Infect 2015;6:22–9. http://dx.doi.org/10.1016/j.nmni.2015.02.007.

[28] Kostyanev T, Bonten MJM, O'Brien S, Steel H, Ross S, François B, *et al.* The Innovative Medicines Initiative's New Drugs for Bad Bugs programme: European public–private partnerships for the development of new strategies to tackle antibiotic resistance. J Antimicrob Chemother 2016;71:290–5. http://dx.doi.org/10.1093/jac/dkv339.

Chapter 2

Antimicrobial Stewardship: Efficacy and Implementation of Strategies to Address Antimicrobial Overuse and Resistance

David X. Li and Sara E. Cosgrove
Johns Hopkins University School of Medicine, Baltimore, MD, United States

KEY POINTS

- Antibiotic use has been linked to the emergence of Gram-negative and Gram-positive bacterial resistance
- Antimicrobial stewardship programs (ASPs) utilize a broad range of clinical interventions to reduce unnecessary antibiotic use with the goals of curbing the emergence of resistance, improving patient safety and outcomes, and reducing hospital and patient costs
- Successful ASPs require the cooperation and active participation of a diverse group of key players, including physicians, pharmacists, infection control, and hospital administration

INTRODUCTION

Infections due to multidrug-resistant bacteria are increasing in incidence and have been associated with poor patient outcomes and increased healthcare costs [1]. The overuse of antibiotics, particularly broad-spectrum agents, has contributed significantly to the emergence of increasingly resistant bacteria [2]. Considering the slow pace of development of novel antimicrobial agents, reducing the emergence of drug resistance and conserving the active agents that are currently available are primary goals of antimicrobial stewardship programs (ASPs). Other important ASP goals include improving patient safety and outcomes, reducing the harms associated with excessive antibiotic

Antimicrobial Stewardship. http://dx.doi.org/10.1016/B978-0-12-810477-4.00002-7

use, and reducing antibiotic and general healthcare expenditures. In this chapter, we describe a number of antimicrobial stewardship strategies and review the scientific evidence supporting their efficacy. We further describe the key parties involved in the implementation of an effective ASP and offer suggestions on the most effective strategies to empower providers, administrators, and patients to become practicing antimicrobial stewards.

THE RELATIONSHIP BETWEEN ANTIBIOTIC USE AND DRUG RESISTANCE

Selection pressure exerted by exposure to antimicrobials may lead to the selective survival and dissemination of drug-resistant bacterial strains. Both in vitro and clinical studies have established a direct link between antibiotic use and drug resistance in Gram-positive organisms. A meta-analysis of 76 clinical studies found a relative risk of 1.8 for MRSA among patients with prior antibiotic exposure compared to unexposed patients [3]. On a population level, a recent time series analysis in Scotland identified thresholds for population consumption of certain antibiotics, including fluoroquinolones, macrolides, and clindamycin, above which the emergence of community-acquired MRSA was favored [4]. In a matched case–control study, antecedent treatment with various antibiotics, including third-generation cephalosporins, fluoroquinolones, and metronidazole, was associated with the emergence of vancomycin-resistant *Enterococcus* infection or colonization [5].

In recent years, infections due to multidrug-resistant Gram-negative bacteria have emerged as a growing public health threat. Two randomized trials have shown that shorter courses of antibiotic therapy for ventilator-associated pneumonia are associated with reductions in the subsequent development of antibiotic resistance while having no deleterious impact on patient outcomes [6,7]. A multinational pooled analysis showed that recent antibiotic use was significantly associated with an odds ratio of 1.8 for the development of community-acquired extended-spectrum β-lactamase-producing Enterobacteriaceae infection [8].

THE EVIDENCE SUPPORTING ANTIMICROBIAL STEWARDSHIP STRATEGIES

Antimicrobial stewardship encompasses a broad range of strategies and interventions. Although the specific components that make up an ASP vary by institution, a number of commonly implemented strategies have demonstrated efficacy in reducing unnecessary antibiotic use and improving patient outcomes [9]. Some strategies have further demonstrated efficacy in reducing resistance [10–13]. These are summarized below and in Table 1; a more in-depth discussion of these various strategies can be found in subsequent chapters.

TABLE 1 Overview of the characteristics and efficacy of select antimicrobial stewardship strategies

	Clinical practice guidelines [14]	Preprescription approval [10,11,15]	Postprescription review [12,16–18]	Computer-based interventions [13,19]	Syndrome-specific interventions [20–25]
Characteristics of the intervention					
Advantages	Institution-specific Typically well-received by other providers	Limits initiation of unnecessary antibiotics Optimizes use of broad-spectrum empiric therapy	More clinical data available at 48–72 h to make clinical decisions Can be performed less than daily if resources are limited Builds collegial relationships	Automated, less resource-intensive Flexible design adaptable to the institution's needs	Identifies specific targets and goals for prescribers to meet Sustained impacts on antibiotic use
Barriers to implementation	Dissemination of knowledge does not guarantee behavior change	Most resource-intensive, requiring a stewardship specialist to be on-call Greater focus on initial empiric antibiotic use than on downstream use	Labor-intensive, requiring stewardship specialists to review antibiotic regimens Stewardship may not have the authority to make changes to ordered antibiotics	Requires new software to be developed and validated Software prompts can be overridden by the provider	Resource-intensive Multicomponent approach required for efficacy
Requires changes to the electronic record system	No	No	No	Yes	No

Continued

TABLE 1 Overview of the characteristics and efficacy of select antimicrobial stewardship strategies—Cont'd

	Clinical practice guidelines [14]	Preprescription approval [10,11,15]	Postprescription review [12,16–18]	Computer-based interventions [13,19]	Syndrome-specific interventions [20–25]
Evidence of efficacy					
Effect on antibiotic use	–	↓24% antibiotic doses per patient [PPS]	↓37% days of unnecessary antibiotics [RCT] ↓22% broad-spectrum use [PPS]	↓94% antibiotic-susceptibility mismatches [PPS] ↓52% antibiotic duration [PPS]	↓Duration of therapy (e.g., 10 to 7 days for CAP) [PPS] ↓Days of unnecessary antibiotics [PPS]
Effect on drug resistance	–	↓Gram-negative resistance [PPS] ↓MRSA and *Stenotrophomonas* colonization or infection [PPS]	↓*Enterobacteriaceae* resistance in subsequent infection [PPS] ↓Carbapenem resistance in ICU [PPS]	↓Gram-negative resistance [PPS]	–
Effect on patient outcomes	↓14% (absolute) mortality [PPS]	↓50% *C. difficile* infection [PPS]	↓31% (95% CI 16.1-42.3) hospital length of stay [RCT] ↓44-62% *C. difficile* infections [PPS, RCT]	↓45% length of ICU stay [PPS] ↓22% hospital length of stay [PPS]	↓76% mortality [PPS] ↓24% hospital length of stay [PPS]

Note: CI, confidence interval; PPS, pre–post study; RCT, randomized controlled trial.

Education should be a fundamental component of all stewardship interventions and includes a broad range of passive interventions designed to improve prescribers' familiarity with the indications and optimal regimens for common infectious syndromes. Educational activities may consist of lectures, interactive modules, or distributed reading materials. However, education alone may not lead to sustained changes in antibiotic use and should be used in conjunction with other stewardship interventions [9]. In one study, a marketing campaign targeting perioperative antibiotic prophylaxis was effective in reducing antibiotic use in the immediate period, but the changes were not sustained at 12 months [26].

Clinical practice guidelines may be developed and published by individual institutions to standardize prescribing practices across the institution [9]. Such guidelines have three advantages: (a) they are typically institution-specific, incorporating both national guidelines as well as local epidemiology in microbiology and resistance patterns; (b) if revised and republished on a regular basis, they allow new clinical research findings related to the management of infectious syndromes to be disseminated to the broader population of hospital providers; and (c) institutional thought leaders outside of the ASP can be engaged in the development, updating, and dissemination of the recommendations to enhance the scope of stewardship across the institution. A study among patients in a surgical intensive care unit showed that the implementation of clinical practice guidelines for six common infectious diagnoses, including intra-abdominal infection and wound infection, led to a 14% absolute reduction in mortality ($P=0.02$) and a nearly 80% decrease in antibiotic costs, as well as a nonsignificant reduction in antibiotic resistance among *Escherichia coli* and *Pseudomonas aeruginosa* isolates [14]. It is unclear if these changes in prescribing behavior are durable over the long term; in a study of intensive care unit patients with pneumonia, dissemination of a pneumonia clinical pathway guideline led to sustained changes in antibiotic prescribing at 3-year follow-up [27]. In another study, however, discontinuation of measures to enhance guideline adherence led to increased rates of unnecessary antibiotic use [28].

In *preprescription approval*, the prescription of certain antibiotics is restricted and requires approval by a member of the ASP, such as a clinical pharmacist or physician. In a 3.5-year prospective study, implementation of a preprescription approval policy was associated with a 24% reduction in the mean number of antibiotic doses per patient and a 32% reduction in antibiotic costs [15]. Another study found that preprescription approval was associated with a 32% reduction in antibiotic costs and increased susceptibility of Gram-negatives to various antibiotics: the proportion of *Pseudomonas aeruginosa* isolates susceptible to aztreonam increased from 70% to 88%, and the proportion of isolates susceptible to ceftazidime increased from 76% to 92% [10]. Other studies have similarly shown that preprescription approval reduces antibiotic costs and may also decrease the incidence of resistant organisms, such as methicillin-resistant *Staphylococcus aureus* (MRSA) and *Stenotrophomonas* [11]. In a meta-analysis of 16 studies with nearly 450,000 patients, preprescription approval was associated with a risk ratio of 0.50 for *Clostridium difficile* infection ($P < 0.0001$) [29].

Preprescription approval strategies have demonstrated efficacy in reducing unnecessary antibiotic use and impacting patient outcomes [9]. However, these strategies are resource-intensive as they require a member of the stewardship team to be on-call to respond in real time to requests. Such strategies are also dependent on the ability of the ASP to gauge the clinical situation and determine antibiotic appropriateness as well as the ability of the provider seeking antibiotic approval to accurately communicate and justify the need for antibiotics; failure in either of these realms has been associated with higher rates of inappropriate antibiotic use [30,31]. Furthermore, preprescription approval focuses on optimizing initial or empiric antibiotic use, with a potentially smaller effect on downstream use. If not employed as a primary stewardship strategy, restriction of high-cost agents may still be considered.

In *postprescription review*, antibiotic regimens are reviewed for appropriateness by the antimicrobial stewardship team at 48–72 h after microbiological results and additional clinical data have returned. Postprescription review strategies have been shown to be efficacious in reducing unnecessary antibiotic use, promoting earlier de-escalation, and improving patient outcomes. In one randomized trial, patients in the postprescription review arm had a 3.3-day decrease in hospital length of stay ($P < 0.001$), a 6% decrease in mortality, and an over $6000 decrease in hospital costs compared to patients who underwent no stewardship intervention [16]. In a randomized controlled trial involving levofloxacin and ceftazidime, implementation of a postprescription review intervention led to a 37% reduction in days of unnecessary therapy [17]. In intensive care units, postprescription review has been associated with reduced Gram-negative resistance and earlier de-escalation of broad-spectrum empiric therapy [12,32]. The changes in antibiotic prescribing observed during a postprescription review intervention appear to be durable, with reductions in broad-spectrum antibiotic use and *C. difficile* infection sustained at 7-year follow-up in one prospective longitudinal study [18].

Postprescription review has a number of advantages: (a) at 48–72 h, the ASP has more data related to microbiological results and the patient's clinical course with which to make an informed decision about management; (b) postprescription review focuses on de-escalation of existing therapy, potentially leading to earlier discontinuation of broad-spectrum or intravenous therapy and optimization of duration of therapy; and (c) as opposed to a hard restriction on ordering an antibiotic, postprescription review utilizes a more personalized, conversational approach with the treating provider, facilitating one-on-one feedback and teaching. [9] However, postprescription review strategies are resource-intensive, requiring ASPs to regularly review antibiotic regimens. Furthermore, at most institutions, the ASP does not have the authority to make changes to antibiotic regimens that have already been ordered, meaning that any changes to the patient's antibiotics require the cooperation of the primary team.

Prescriber-led interventions are a subset of postprescription review in which prescribers are encouraged to regularly assess antibiotic regimens, often in the absence of direct input from an ASP. Such interventions include time-out tools, checklists, and stop orders and may be particularly beneficial in nonacute care settings that lack the resources required to maintain a rigorous ASP [9]. In one pre–post study, a twice-weekly antibiotic timeout tool and checklist was implemented in the absence of an ASP. The intervention was initially associated with changes in the antibiotic regimens of 15% of audited patients and reductions in overall antibiotic use; however, these improvements degraded over the 18-month study period [33]. In another study, a stop order was implemented requiring prescribers to review vancomycin prescriptions 72 h after antibiotic initiation. This intervention led to a halving of the rate of continued vancomycin therapy in the absence of documented Gram-positive infection [34]. Overall, these data suggest that prescriber-led interventions can be effective in improving antibiotic use in some circumstances, although regular prompting of prescribers and oversight by the ASP are likely needed. Additional studies are needed in this area.

Computer-based interventions refer to the use of automated, computerized prompts to provide real-time decision support for antibiotic prescribers either at the time of initial prescribing or later in the antibiotic course. Although these interventions are resource-intensive to develop and implement initially, requiring extensive informatics support and an existing electronic medical record system, they do not require direct ASP involvement to be maintained on a regular basis and can be widely and quickly disseminated across an entire health system. In particular, such interventions show promise in ambulatory settings in which a traditional ASP may not be feasible [9]. In one meta-analysis of six studies of computer-based decision support in the outpatient setting, four studies showed reductions in antibiotic prescribing, but none showed any improvements in patient outcomes at return clinic visits [35]. In an inpatient study, a computer program was designed to integrate patient data, including vital signs and microbiological results, with the institutional antibiogram and offer clinical decision support to providers. Implementation of this program led to reductions in the use of contraindicated antibiotics, such as agents to which the patient had a reported allergy and agents to which the infective organism was resistant [19]. In the intensive care setting, implementation of a computerized decision support system led to a significant and sustained decrease in Gram-negative resistance, including a 9.2%/year reduction in imipenem resistance among *Pseudomonas* isolates [13].

Syndrome-specific interventions utilize multicomponent strategies to target specific infectious syndromes rather than broadly targeting antibiotic prescription. These interventions are recommended as they have the advantage of offering a targeted message and specific recommendations on optimizing antibiotic regimens for specific infectious syndromes, and their effects appear to be sustained even after discontinuation of the intervention [9]. One study implemented a multicomponent intervention consisting of provider education and postprescription review and feedback to reduce the overtreatment of community-acquired pneumonia. This led to a reduction in the median duration of therapy from 10 to 7 days, and effects were sustained 3 years after the intervention [20,21]. In another study, a multicomponent intervention consisting of education and provider feedback reduced the rate of treatment of asymptomatic bacteriuria by over 60%; changes in prescribing were sustained at 1-year follow-up [22]. An intervention targeting inpatient skin and soft tissue infections led to a reduction in the median duration of therapy from 13 to 10 days; these changes were sustained 4 years after the intervention [23,24]. In a retrospective cohort study, patients with Gram-negative bacteremia who underwent active stewardship alerting and intervention were initiated on appropriate therapy 6 h earlier ($P=0.014$) and had a 24% reduction in length of stay and a 76% reduction in infection-related mortality compared to nonintervention patients [25].

CHOOSING A STEWARDSHIP STRATEGY

Although the above strategies form the toolbox from which most hospital ASPs are constructed, the evidence supporting one strategy over another remains sparse. A Cochrane review of 52 interrupted time series studies comparing preprescription authorization to postprescription review showed that preprescription authorization had a greater impact on antibiotic prescribing and patient outcomes in the short term, but the two strategies were equivalent at 12 and 24 months [36]. In a quasiexperimental study at a tertiary care center, a change from preprescription approval to postprescription review was associated with an increase in broad-spectrum antigram-negative antimicrobial use of 4.80 (95% CI 2.46–7.14) days of therapy per 1000 patient-days [37]. Rather than selecting a single optimal strategy, most ASPs utilize a combination of multiple strategies based on local resources and needs. An in-depth discussion of the institution-specific factors involved in selecting and successfully implementing these various strategies can be found in subsequent chapters.

IMPLEMENTATION OF AN ASP IN NONTERTIARY CARE SETTINGS

Studies have shown that ASPs can be effectively tailored to healthcare settings with fewer resources, such as community hospitals, long-term care facilities, and ambulatory clinics. For instance, a study conducted in the internal medicine department of an 80-bed community hospital found that education and weekly postprescription review during ward rounds led to a 36% reduction in antibiotic use ($P = 0.001$) and a 53% reduction in drug costs [38]. Of note, this hospital had no infectious diseases specialists on staff, and the intervention was implemented by general internists.

Despite the high prevalence of multidrug-resistant organisms in nursing homes, few studies have evaluated the impact of ASPs in this setting. Fleet *et al.* conducted a cluster randomized controlled study in 30 nursing homes in England involving physicians and other nursing home staff and found that an intervention consisting of education and clinical practice guidelines led to a significant 5% reduction in overall antibiotic use, compared to a 5% increase in antibiotic use in the control group [39]. A quasiexperimental trial conducted in 12 nursing homes in North Carolina found that a multicomponent intervention entailing education, guideline dissemination, and prescriber feedback targeting physicians and nurses led to a reduction in antibiotic use of 1.8 prescriptions per 1,000 resident-days [40]. In both studies, nursing home directors volunteered their facilities as study sites for implementing stewardship, and there was involvement of externally contracted physicians, pharmacists, and microbiology laboratories.

As the majority of antibiotic prescribing occurs in the outpatient setting, recent studies have sought to develop methods to extend stewardship principles to this setting. In contrast to interventions performed in the inpatient setting, outpatient interventions typically target provider-level prescribing habits rather than patient-level antibiotic optimization. In one randomized trial in the pediatric outpatient setting, an intervention consisting of education and quarterly audit-and-feedback to providers led to a 6.7% absolute reduction in the prescription of broad-spectrum antibiotics [41]. In another recent randomized controlled trial, a range of behavioral interventions were tested amongst outpatient prescribers in order to assess their effects on antibiotic use for inappropriate indications, such as nonspecific upper respiratory tract infection [42]. Two of the interventions, one which asked providers to justify the indication for antibiotic use and another which compared a provider's prescribing habits to "top-performing" peers, led to significant reductions in antibiotic use.

BUILDING THE STEWARDSHIP TEAM

Given the myriad stewardship strategies that appear to be efficacious in improving antibiotic use, most institutions tailor their ASPs to the needs and infrastructure of the specific institution or patient population. In the acute care setting, a number of key players are involved in the initiation and maintenance of a successful ASP. These key players, along with a brief description of their primary roles, are described below and in Fig. 1.

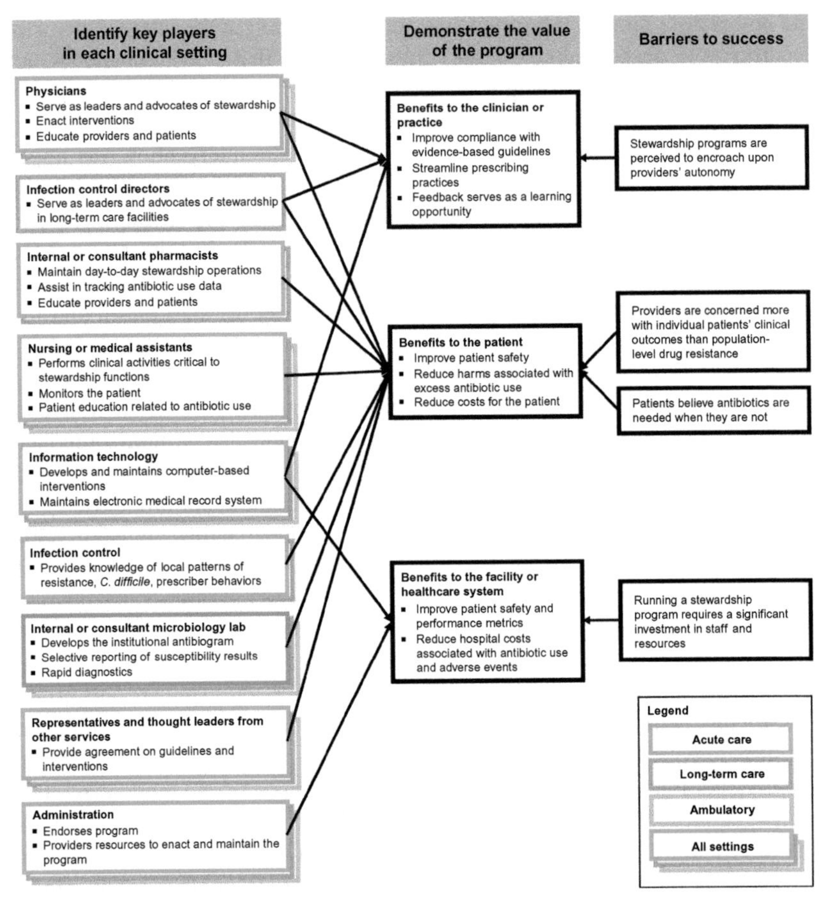

FIG. 1 Summary of the key players involved in the successful implementation of an antimicrobial stewardship program (ASP) in acute care, long-term care, and ambulatory settings (left column). The qualitative benefits of an ASP can be broadly grouped into three categories: benefits to the clinician or practice, benefits to the patient, and benefits to the facility or healthcare system (middle column). Certain benefits may be weighed more heavily than others depending on the key player involved (connections between key players and benefits indicated by arrows). However, familiarity with all three categories of benefits is helpful in order to address common concerns and barriers to success that might otherwise impede the successful implementation of the ASP (right column).

- *Physicians* serve as leaders and advocates of stewardship, back-ups for stewardship pharmacists, and role models for other prescribers and trainees. Physicians also enact interventions and serve as educators to providers and patients alike.
- The *infection control unit* provides data on antibiotic resistance and *C. difficile* infection rates. Infection control programs often have access to information technology that may be useful to the ASP. Furthermore,

infection control, in conjunction with hospital epidemiologists, plays a key role in evaluating the impact of the ASP by monitoring trends in antibiotic prescribing and resistance.

- *Pharmacists* perform many of the same functions as the ASP physician and often perform the majority of day-to-day interventions. They provide a critical link to the pharmacy department and other pharmacists in the institution and often assist in obtaining and tracking antibiotic use data.
- *Nursing* performs many of the clinical activities critical to stewardship functions, including collection of cultures, administration of antibiotics, clinical monitoring of the patient, reporting of adverse events, and patient education [43].
- *Information technology* develops and maintains computer-based interventions and maintains the electronic medical record system that is essential to many stewardship functions. Access to experts in electronic data collection and management are essential to ASPs.
- The *microbiology laboratory* develops the institutional antibiogram and plays a key role in the selective reporting of susceptibility results and in the implementation of rapid diagnostic tests.
- *Representatives and thought leaders from other services* should be actively involved in the development and implementation of the ASP as their agreement on guidelines and interventions will ensure the dissemination of these practices to the wider population of hospital providers.
- *Hospital administration* should be involved from the planning stages through to implementation of the ASP, both to endorse the ASP and its mission and to provide financial and staffing resources.

In contrast to the inpatient setting, long-term care facilities frequently lack the on-site presence of many of the above key players. ASPs in long-term care facilities are often led by the infection control director, although efforts should be made to engage the medical director as well. Engagement of bedside nurses is critical as they make assessments and decisions regarding patient status and whether to send cultures. Involvement of consultant pharmacists and microbiology laboratories can assist the local team with obtaining data about antibiotic use and development of an antibiogram [44]. In the ambulatory setting, patients are usually seen by physicians for short periods of time and may not be followed longitudinally after the first encounter. Thus, time-consuming physician- or pharmacist-led patient-centered interventions may not be feasible. In contrast, point-of-care clinical decision support interventions, such as computer-based interventions, may be particularly efficacious in this setting, necessitating the involvement of informatics experts to develop such a program and maintain an efficient electronic medical record system.

MESSAGING THE IMPORTANCE OF ANTIMICROBIAL STEWARDSHIP

Although we have extensively summarized the benefits of an ASP above, it is important to consider a differential approach to framing these benefits when approaching providers versus administrative representatives. Broadly, the qualitative benefits of an ASP can be grouped into three categories, summarized below and in Fig. 1.

Benefits of the ASP to the patient: It is important to frame stewardship philosophies and practices as both beneficial to patient safety and instrumental in reducing the harms associated with excessive antibiotic use. Providers are invested in the care of their patients and may believe that the "best outcome" for an individual patient requires the use of new, broad-spectrum antibiotics in order to avoid insufficient coverage [45]. Providers should be presented with case studies and scientific data demonstrating the benefits of stewardship practices on patient outcomes (e.g., reductions in drug resistance and *Clostridium difficile* infection rates following a stewardship intervention) as well as emphasizing that stewardship is not associated with patient harms (e.g., shorter courses of antibiotics are not associated with increased rates of recurrent infection). Some providers may worry that patients will interpret the provider's practice of stewardship as a lack of concern for the patient's well-being or an inferior standard of care. To assuage these concerns, providers should reassure patients that the evidence-based practice of stewardship is borne out of concern for the patient's well-being and the desire to avoid unnecessary harms associated with antibiotic overuse.

Benefits of the ASP to the clinician or practice: Some physicians may perceive the ASP as encroaching on their autonomy as an antibiotic prescriber, particularly if the ASP utilizes strict formulary restrictions. In this situation, it is important to highlight the multiple benefits that the ASP would confer on the clinician and the clinician's practice, including streamlining of the clinician's practice such that patients with similar syndromes receive similar treatment and as an opportunity for continued learning if the clinician is amenable to receiving feedback on his or her own prescribing habits.

Benefits of the ASP to the healthcare system: Aside from patient-level benefits, ASPs confer many benefits on the hospital and healthcare system at large. Multiple studies have demonstrated that ASPs reduce drug costs and shorten hospitalizations, leading to significant reductions in antibiotic and hospital costs [11,20]. Compliance with performance metrics related to antibiotic prescribing can improve an institution's reputation and in some instances may be tied to reimbursement. These points may be particularly effective in convincing hospital administration to endorse and provide resources for an ASP.

SUMMARY

The overuse of antibiotics has greatly contributed to the emergence of multidrug-resistant bacteria, which are associated with significant morbidity, mortality, and healthcare costs. ASPs play a vital role in enhancing the appropriate and judicious use of antibiotics, with the goals of curbing the emergence of resistance, improving patient safety and outcomes, and reducing hospital and patient costs. Previous studies, including multiple well-designed randomized trials, have demonstrated the efficacy of a broad range of stewardship strategies in reducing unnecessary antibiotic use, inhibiting the subsequent emergence of resistant organisms and *C. difficile* infection, and reducing antibiotic and hospital costs. Recent studies have further demonstrated that stewardship interventions, previously studied mainly in acute care hospitals, can also be successfully generalized to ambulatory and long-term care settings. Nonetheless, high-quality evidence with regard to optimal stewardship strategies remains sparse, and well-designed clinical trials are needed in this area. Regardless of setting, the successful implementation of an ASP requires the cooperation of a multidisciplinary team.

REFERENCES

[1] Zilberberg MD, Shorr AF, Micek ST, Vazquez-Guillamet C, Kollef MH. Multi-drug resistance, inappropriate initial antibiotic therapy and mortality in Gram-negative severe sepsis and septic shock: a retrospective cohort study. Crit Care 2014;18(6):596.

[2] Bell BG, Schellevis F, Stobberingh E, Goossens H, Pringle M. A systematic review and meta-analysis of the effects of antibiotic consumption on antibiotic resistance. BMC Infect Dis 2014;14:13.

[3] Tacconelli E, De Angelis G, Cataldo MA, Pozzi E, Cauda R. Does antibiotic exposure increase the risk of methicillin-resistant Staphylococcus aureus (MRSA) isolation? A systematic review and meta-analysis. J Antimicrob Chemother 2008;61(1):26–38.

[4] Lawes T, Lopez-Lozano JM, Nebot CA, Macartney G, Subbarao-Sharma R, Dare CR, *et al.* Effects of national antibiotic stewardship and infection control strategies on hospital-associated and community-associated meticillin-resistant Staphylococcus aureus infections across a region of Scotland: a non-linear time-series study. Lancet Infect Dis 2015;15(12):1438–49.

[5] Carmeli Y, Eliopoulos GM, Samore MH. Antecedent treatment with different antibiotic agents as a risk factor for vancomycin-resistant Enterococcus. Emerg Infect Dis 2002;8(8):802–7.

[6] Singh N, Rogers P, Atwood CW, Wagener MM, Yu VL. Short-course empiric antibiotic therapy for patients with pulmonary infiltrates in the intensive care unit. A proposed solution for indiscriminate antibiotic prescription. Am J Respir Crit Care Med 2000;162(2 Pt 1):505–11.

[7] Chastre J, Wolff M, Fagon JY, Chevret S, Thomas F, Wermert D, *et al.* Comparison of 8 vs 15 days of antibiotic therapy for ventilator-associated pneumonia in adults: a randomized trial. JAMA 2003;290(19):2588–98.

[8] Ben-Ami R, Rodriguez-Bano J, Arslan H, Pitout JD, Quentin C, Calbo ES, *et al.* A multinational survey of risk factors for infection with extended-spectrum beta-lactamase-producing enterobacteriaceae in nonhospitalized patients. Clin Infect Dis 2009;49(5):682–90.

[9] Barlam TF, Cosgrove SE, Abbo LM, MacDougall C, Schuetz AN, Septimus EJ, *et al.* Implementing an Antibiotic Stewardship Program: guidelines by the Infectious Diseases Society of America and the Society for Healthcare Epidemiology of America. Clin Infect Dis 2016;62(10):e51–77.

[10] White Jr AC, Atmar RL, Wilson J, Cate TR, Stager CE, Greenberg SB. Effects of requiring prior authorization for selected antimicrobials: expenditures, susceptibilities, and clinical outcomes. Clin Infect Dis 1997;25(2):230–9.

[11] Frank MO, Batteiger BE, Sorensen SJ, Hartstein AI, Carr JA, McComb JS, *et al.* Decrease in expenditures and selected nosocomial infections following implementation of an antimicrobial-prescribing improvement program. Clin Perform Qual Health Care 1997;5(4): 180–8.

[12] Elligsen M, Walker SA, Pinto R, Simor A, Mubareka S, Rachlis A, *et al.* Audit and feedback to reduce broad-spectrum antibiotic use among intensive care unit patients: a controlled interrupted time series analysis. Infect Control Hosp Epidemiol 2012;33(4):354–61.

[13] Yong MK, Buising KL, Cheng AC, Thursky KA. Improved susceptibility of Gram-negative bacteria in an intensive care unit following implementation of a computerized antibiotic decision support system. J Antimicrob Chemother 2010;65(5):1062–9.

[14] Price J, Ekleberry A, Grover A, Melendy S, Baddam K, McMahon J, *et al.* Evaluation of clinical practice guidelines on outcome of infection in patients in the surgical intensive care unit. Crit Care Med 1999;27(10):2118–24.

[15] Coleman RW, Rodondi LC, Kaubisch S, Granzella NB, O'Hanley PD. Cost-effectiveness of prospective and continuous parenteral antibiotic control: experience at the Palo Alto Veterans Affairs Medical Center from 1987 to 1989. Am J Med 1991;90(4):439–44.

[16] Gums JG, Yancey Jr RW, Hamilton CA, Kubilis PS. A randomized, prospective study measuring outcomes after antibiotic therapy intervention by a multidisciplinary consult team. Pharmacotherapy 1999;19(12):1369–77.

[17] Solomon DH, Van Houten L, Glynn RJ, Baden L, Curtis K, Schrager H, *et al.* Academic detailing to improve use of broad-spectrum antibiotics at an academic medical center. Arch Intern Med 2001;161(15):1897–902.

[18] Carling P, Fung T, Killion A, Terrin N, Barza M. Favorable impact of a multidisciplinary antibiotic management program conducted during 7 years. Infect Control Hosp Epidemiol 2003;24(9):699–706.

[19] Evans RS, Pestotnik SL, Classen DC, Clemmer TP, Weaver LK, Orme Jr JF, *et al.* A computer-assisted management program for antibiotics and other antiinfective agents. N Engl J Med 1998;338(4):232–8.

[20] Avdic E, Cushinotto LA, Hughes AH, Hansen AR, Efird LE, Bartlett JG, *et al.* Impact of an antimicrobial stewardship intervention on shortening the duration of therapy for community-acquired pneumonia. Clin Infect Dis 2012;54(11):1581–7.

[21] Li DX, Ferrada MA, Avdic E, Tamma PD, Cosgrove SE. Sustained impact of an antibiotic stewradship intervention for community-acquired pneumonia. Infect Control Hosp Epidemiol 2016;8:1–4 [Epub ahead of print].

[22] Trautner BW, Grigoryan L, Petersen NJ, Hysong S, Cadena J, Patterson JE, *et al.* Effectiveness of an Antimicrobial Stewardship Approach for Urinary Catheter-Associated Asymptomatic Bacteriuria. JAMA Intern Med 2015;175(7):1120–7.

[23] Jenkins TC, Knepper BC, Sabel AL, Sarcone EE, Long JA, Haukoos JS, *et al.* Decreased antibiotic utilization after implementation of a guideline for inpatient cellulitis and cutaneous abscess. Arch Intern Med 2011;171(12):1072–9.

[24] Jenkins TC, Knepper BC, Moore SJ, O'Leary ST, Brooke C, Saveli CC, *et al.* Antibiotic prescribing practices in a multicenter cohort of patients hospitalized for acute bacterial skin and skin structure infection. Infect Control Hosp Epidemiol 2014;35(10):1241–50.

[25] Pogue JM, Mynatt RP, Marchaim D, Zhao JJ, Barr VO, Moshos J, *et al.* Automated alerts coupled with antimicrobial stewardship intervention lead to decreases in length of stay in patients with gram-negative bacteremia. Infect Control Hosp Epidemiol 2014;35(2):132–8.

[26] Landgren FT, Harvey KJ, Mashford ML, Moulds RF, Guthrie B, Hemming M. Changing antibiotic prescribing by educational marketing. Med J Aust 1988;149(11–12):595–9.

[27] Benenson R, Magalski A, Cavanaugh S, Williams E. Effects of a pneumonia clinical pathway on time to antibiotic treatment, length of stay, and mortality. Acad Emerg Med 1999;6(12):1243–8.

[28] Wilde AM, Nailor MD, Nicolau DP, Kuti JL. Inappropriate antibiotic use due to decreased compliance with a ventilator-associated pneumonia computerized clinical pathway: implications for continuing education and prospective feedback. Pharmacotherapy 2012;32(8):755–63.

[29] Feazel LM, Malhotra A, Perencevich EN, Kaboli P, Diekema DJ, Schweizer ML. Effect of antibiotic stewardship programmes on Clostridium difficile incidence: a systematic review and meta-analysis. J Antimicrob Chemother 2014;69(7):1748–54.

[30] Gross R, Morgan AS, Kinky DE, Weiner M, Gibson GA, Fishman NO. Impact of a hospital-based antimicrobial management program on clinical and economic outcomes. Clin Infect Dis 2001;33(3):289–95.

[31] Linkin DR, Fishman NO, Landis JR, Barton TD, Gluckman S, Kostman J, *et al.* Effect of communication errors during calls to an antimicrobial stewardship program. Infect Control Hosp Epidemiol 2007;28(12):1374–81.

[32] Schuts EC, Hulscher ME, Mouton JW, Verduin CM, Stuart JW, Overdiek HW, *et al.* Current evidence on hospital antimicrobial stewardship objectives: a systematic review and meta-analysis. Lancet Infect Dis 2016;16(7):847–56.

[33] Lee TC, Frenette C, Jayaraman D, Green L, Pilote L. Antibiotic self-stewardship: trainee-led structured antibiotic time-outs to improve antimicrobial use. Ann Intern Med 2014; 161(10 Suppl):S53–8.

[34] Guglielmo BJ, Dudas V, Maewal I, Young R, Hilts A, Villmann M, *et al.* Impact of a series of interventions in vancomycin prescribing on use and prevalence of vancomycin-resistant enterococci. Jt Comm J Qual Patient Saf 2005;31(8):469–75.

[35] Drekonja DM, Filice GA, Greer N, Olson A, MacDonald R, Rutks I, *et al.* Antimicrobial stewardship in outpatient settings: a systematic review. Infect Control Hosp Epidemiol 2015;36(2):142–52.

[36] Davey P, Brown E, Fenelon L, Finch R, Gould I, Hartman G, *et al.* Interventions to improve antibiotic prescribing practices for hospital inpatients. Cochrane Database Syst Rev 2005;4: CD003543.

[37] Mehta JM, Haynes K, Wileyto EP, Gerber JS, Timko DR, Morgan SC, *et al.* Comparison of prior authorization and prospective audit with feedback for antimicrobial stewardship. Infect Control Hosp Epidemiol 2014;35(9):1092–9.

[38] Ruttimann S, Keck B, Hartmeier C, Maetzel A, Bucher HC. Long-term antibiotic cost savings from a comprehensive intervention program in a medical department of a university-affiliated teaching hospital. Clin Infect Dis 2004;38(3):348–56.

[39] Fleet E, Gopal Rao G, Patel B, Cookson B, Charlett A, Bowman C, *et al.* Impact of implementation of a novel antimicrobial stewardship tool on antibiotic use in nursing homes: a prospective cluster randomized control pilot study. J Antimicrob Chemother 2014;69(8): 2265–73.

[40] Zimmerman S, Sloane PD, Bertrand R, Olsho LE, Beeber A, Kistler C, *et al.* Successfully reducing antibiotic prescribing in nursing homes. J Am Geriatr Soc 2014;62(5):907–12.
[41] Gerber JS, Prasad PA, Fiks AG, Localio AR, Grundmeier RW, Bell LM, *et al.* Effect of an outpatient antimicrobial stewardship intervention on broad-spectrum antibiotic prescribing by primary care pediatricians: a randomized trial. JAMA 2013;309(22):2345–52.
[42] Meeker D, Linder JA, Fox CR, Friedberg MW, Persell SD, Goldstein NJ, *et al.* Effect of behavioral interventions on inappropriate antibiotic prescribing among primary care practices: a randomized clinical trial. JAMA 2016;315(6):562–70.
[43] Olans RN, Olans RD, DeMaria Jr A. The critical role of the staff nurse in Antimicrobial Stewardship–Unrecognized, but Already There. Clin Infect Dis 2016;62(1):84–9.
[44] Centers for Disease Control and Prevention. The core elements of Antibiotic Stewardship for Nursing Homes: Centers for Disease Control and Prevention; 2016. Available from: http://www.cdc.gov/longtermcare/prevention/antibiotic-stewardship.html.
[45] Metlay JP, Shea JA, Crossette LB, Asch DA. Tensions in antibiotic prescribing: pitting social concerns against the interests of individual patients. J Gen Intern Med 2002;17(2):87–94.

Chapter 3

Quality Indicators and Quantity Metrics of Antibiotic Use

Vera Vlahović-Palčevski* and Inge C. Gyssens,‡**
*University Hospital Rijeka, Rijeka, Croatia
**Radboud University Medical Center, Nijmegen, The Netherlands
‡Hasselt University, Hasselt, Belgium

Antimicrobial stewardship's (AMS) goal by definition refers to coordinated interventions designed to improve and measure the appropriate use of antimicrobials. Thus, an effective AMS program should have developed tools for measuring both the quantity and quality of antimicrobial use. However, the tools have not been standardized, and within various AMS programs across countries, regions, and individual healthcare settings, different measurements have been used.

QUALITY VERSUS PERFORMANCE INDICATORS

A quality indicator is a measurable element of practice performance for which there is evidence or consensus that it can be used to assess the quality, and hence change in the quality, of care provided [1].

Quality indicators infer a judgment about the quality of care provided and should be distinguished from performance indicators, which are statistical devices for monitoring care provided to populations without any necessary inference about quality [2].

MEASURING PERFORMANCE

Antimicrobial drug prescribing is a process in providing healthcare, and its performance measure must be based on a strong foundation of research showing that the process addressed by the measure, when performed correctly, leads to improved clinical outcomes.

Measuring the performance of antimicrobial drug use (i.e., monitoring and surveillance) gives an insight into the patterns, determinants, and outcomes of use. Patterns of use describe the extent and profiles of use and

Antimicrobial Stewardship. http://dx.doi.org/10.1016/B978-0-12-810477-4.00003-9

trends over time that require further qualitative investigation. It enables measuring the effect of stewardship interventions and providing feedback to prescribers and enables regional, national, and international benchmarking. Determinants of use identify reasons that led to prescribing, such as disease prevalence and incidence, socioeconomic factors, drug availability and affordability, prescriber and patient characteristics, etc. Outcomes of antibiotic use besides patient outcomes concern correlations between antibiotic use and resistance, rates of adverse drug reactions, and economic consequences. In addition, antimicrobial pattern of use analysis provides some simple qualitative indices [3].

The Driving re-investment in Research & Development (R&D) and responsible antibiotic use (DRIVE-AB) project is a public–private consortium funded by the EU Innovative Medicines Initiative (IMI). One of the primary objectives was the development of a consensually accepted terminology and a framework to define responsible antibiotic use. Furthermore, the project developed consensually validated quality indicators and quantity metrics for evaluating antibiotic use [4].

QUANTITY METRICS AND QUALITY INDICATORS OF ANTIBIOTIC USE: WHAT IS THE DIFFERENCE?

It is of importance to use a common terminology to understand the difference between quantitative and qualitative measures of antibiotic use. The DRIVE-AB project proposed the following definitions:

A *quantity metric* reflects the volume or the costs of antibiotic use.

Examples are: 'Consumption of antibiotics expressed in DDD per 1000 inhabitants and per day' or 'Consumption of fluoroquinolones expressed as percentage of the total consumption of antibiotics.' The metric only gains value in its comparison (between wards, hospitals, countries).

In contrast, *a quality indicator* reflects the degree in which an antibiotic is correct or appropriate. Examples are: 'Empirical antibiotic therapy should be prescribed according to the local guideline' or 'At least two sets of blood cultures should be taken before starting antibiotic therapy.' The indicator has value of its own.

QUANTITY METRICS

Antimicrobial drug use can be expressed in terms of *cost* (e.g., national currency). This is useful for an overall cost analysis of drug expenditures. However, national and international comparisons based on cost parameters may be misleading and of limited value in the evaluation of the quality of antimicrobial drug use because of price differences among preparations and even same preparations in different settings and countries. Longitudinal studies are disabled due to fluctuations in currency and changes in prices.

Therefore, it is much more appropriate to measure antimicrobial drug use in *volume*. This could be expressed in common physical units, number of packages or tablets, or number of prescriptions. If consumption is presented in terms of grams of active ingredients, drugs with low potency will have a larger proportion of the total than drugs with high potency. Counting numbers of tablets also has disadvantages because strengths of tablets vary, resulting in the low-strength preparations contributing relatively more than high-strength preparations. Packages as the measurement unit are inadequate due to differences in number of units per package and different strengths. Numbers of prescriptions do not give a good expression of total use, unless total amounts of drugs per prescription are also considered. The method may be misleading due to different prescribing regulations.

The prescribed daily dose (PDD) is the average daily amount of a drug that is actually prescribed. It can be determined from prescription studies, medical or pharmacy records, and patient interviews. The main disadvantage is the lack of standardization. Due to huge limitations of these metrics, the WHO recommends using *Defined daily doses (DDDs)* in drug utilization studies. The DDD is the assumed average maintenance dose per day for a drug used for its main indication in adults. It is important to note that DDD is a unit of measurement and does not necessarily reflect the recommended dose or PDD! The advantage of DDD is that it is a stable drug utilization measure that enables comparisons of drug use between countries, regions, and other healthcare settings and examines trends in drug use over time. When there is a substantial discrepancy between the PDD and the DDD, it is important to take this into consideration when evaluating and interpreting drug consumption figures. The main disadvantage is that DDDs have not been assigned for pediatric drug use. Drug consumption data presented in DDDs only give a rough estimate of consumption and not an exact picture of actual use. DDDs provide a fixed unit of measurement independent of price, currencies, package size, and strength, enabling the assessment of trends in drug consumption and the performance of comparisons between population groups [5].

OUTPATIENT AND INPATIENT ANTIBIOTIC USE: HOW TO MEASURE?

International Recommendations

For outpatient antibiotic use, WHO recommends the presentation of the figures as numbers of DDDs per inhabitant per year, which gives an estimate of the number of days for which each inhabitant is, on average, treated annually. For other drugs that are taken for longer periods of time, it is recommended to express drug use as the *number of DDDs per 1000 inhabitants per day (TID)*. Despite this recommendation, a majority of studies found in the literature present antibiotic use as number of DDDs/TID. It provides a rough estimate of the proportion of the population within a defined area treated daily with antibiotics.

For the inpatient antibiotic use, it is suggested to present figures as number of *DDD/100 bed-days*. This metric reflects the percentage of inpatients treated with antibiotics daily in an institution (WHO Collaborating Centre for Drug Statistics Methodology. Guidelines for ATC Classification and DDD Assignment 2016). Expressing antibiotic use by using DDD/100 bed-days allows the benchmarking of antibiotic use regardless of differences in formulary composition, antibiotic potency, or hospital census.

Although the WHO promotes the use of DDDs as metrics of drug use, and this has been widely accepted, many researchers use different metrics, especially in the United States. The recent recommendation by the Infectious diseases Society of America and the Society for Healthcare Epidemiology of America (IDSA/SHEA) for implementing an ASP is to monitor antibiotic use by *days of therapy (DOTs)* in preference to DDDs. The major reason is that DOTs are not impacted by dose adjustments and can be used in both adult and pediatric populations. DOT is the number of days when at least one dose of a medication was administered irrespective of dose or route of administration. Similar to PDDs, expressing drug use in the number of DOTs requires patient-level antibiotic use data, which may not be feasible at every facility [6].

Wide Variations in the Literature

When reporting drug use by the number of defined metric units, the population to which it refers should be clear. In the metric, the numerator measures the amount of antibiotic used (see above), and the denominators control for the size of the population studied. In reports on antimicrobial drug use in the literature, a wide variety of numerators and denominators have been used.

The most commonly used measure of the amount used for *outpatient antibiotic use* is the number of DDDs, treatment courses, or prescriptions per defined population. DDDs and prescriptions are described earlier in the chapter. Treatment courses refer to the number of prescribed courses of unique antibiotics per person. The course implies to consecutive administrations of the same antibiotics with less than a 48-hour window between any two. The method allows comparison of antibiotic prescription patterns. It does not take into account dose, frequency, or route of administration [7].

The same numerators are sometimes used with different denominator, that is, number of treatment courses or prescriptions per physician contact [8].

The most commonly used numerators for expressing *inpatient antibiotic consumption* are DDDs, PDDs, and DOTs. For benchmarking purposes, length of treatment is also used. As with PDDs and DOTs, individual patient data are needed for monitoring antibiotic use in length of treatments. The number of patients exposed to antibiotics is a less commonly used numerator. There are many other rarely used metrics not suitable for benchmarking purposes. A number of different denominators have been used for inpatient antibiotic use. It is important to emphasize that the number of bed-days as the denominator reflecting population size counts the day of admission and the day of discharge as one day. Terms such as patient-days and occupied bed-days are used interchangeably. Another denominator frequently used with DDDs, DOTs, or length of treatment is the number of admissions. It is important not to mix number of bed-days and admissions when assessing the antibiotic pressure. With increasing number of admissions and decreasing length of stay, DDDs per admission may remain stable, while the number of DDDs per 100 bed-days is rising [9,10].

There is obviously a need for standardization of methods for reporting and benchmarking antibiotic use for both inpatient and outpatient settings.

In the DRIVE-AB project, consensus procedures including systematic literature searches and RAND-modified Delphi rounds with experts from several stakeholder groups (medical community, government, and R&D among others) have resulted in a proposed number of metrics for the outpatient and inpatient settings. The data are yet unpublished; however, the list of metrics is already available on the DRIVE-AB website [11].

QUALITY INDICATORS

In the past 45 years, many parameters of importance for optimal quality of antimicrobial therapy have been defined. Ideally, maximal efficacy should be combined with minimal toxicity at the lowest cost. Optimal prescribing quality is dependent on knowledge of many aspects of infectious diseases. Traditionally, quality was measured by an in-depth analysis of medical records, also called audit of practice, which is defined as the analysis of appropriateness of individual prescriptions [12]. Based on the original evaluation criteria of Kunin, Gyssens *et al.* developed and refined an algorithm to facilitate the classification of individual prescriptions in different categories of inappropriate use [13,14]. This algorithm allows an evaluation of each process parameter of importance associated with prescribing antimicrobial drugs. Although this approach is costly in manpower, an audit is certainly the most complete method to judge all aspects of therapy at the bedside. In contrast, a variety of different methods have been used to assess the quality of antibiotic prescriptions at the (hospital) population level. As stated above, quality indicators (QIs) aim to measure the "appropriateness" of a given performance and are based on scientific evidence and/or expert consensus. A classical approach is to develop quality indicators to monitor adherence to evidence-based guidelines. Most work has been done in high-income countries. QIs can be classified as "process," "structure," and "outcome" indicators.

Ideally, QIs have a clear and direct association with patient or economic outcomes. Meaningful patient outcome indicators are: Mortality related to infection (30 days), Length of stay, Prolongation of stay related to infection, Length of stay in intensive care, Readmission or admission of non-outpatients (12 weeks), Nosocomial infection rate, « Cure » (or lack of relapse), and Side effects (allergies, toxicity). Few interventions to evaluate and improve the quality of prescription have analyzed microbiological outcomes. It should be noted that most published studies have been conducted in wards or hospitals with a high antibiotic consumption and that have problems with selected resistant strains. The microbiological outcome is then the elimination of these strains. Risk factors for the development of resistance are still not well known or have not been properly studied, and the impact of infection control measures is often not well documented.

Another returning point of discussion is the selection of experts to appraise the quality indicators derived from the literature. Regardless of the choices made, the background and identity of the experts should be transparent.

OUTPATIENT QUALITY INDICATORS

A systematic review of QIs in primary care has been recently published by Saust *et al.* [15]. A total of 11 studies were found, including 130 quality indicators for diagnosis and antibiotic treatment of infectious diseases in primary care. The authors stated that the majority (72%) of the QIs were focusing on choice of antibiotics, and 22% concerned the decision to prescribe antibiotics. Only few (6%) concerned the diagnostic process, which was considered disappointing. Most QIs were either related to respiratory tract infections or not related to any type of infection.

Recently, the DRIVE-AB project presented a final set of 32 outpatient QIs (OQIs). Several of these outpatient QIs can be adapted to different clinical situations with different numerator and denominator combinations. A total of 20 QIs addressed general practice, 11 OPAT and 1 both settings [11]. The highest appraised outpatient QIs are presented in Table 1.

INPATIENT QIs

QIs have been derived from the Dutch SWAB guidelines on the antibiotic treatment of CAP, sepsis/bacteremia, and complicated urinary tract infections [16,17].

The common methodology for the development of inpatient QIs is a systematic multistep process, which combines evidence and expert opinion. As an example, the Dutch quality indicator development program is described. A RAND-modified Delphi procedure was used to develop a set of QIs to assess the quality of antibiotic use in hospitalized adults treated for a bacterial infection. Potential QIs were retrieved from the literature.

TABLE 1 The highest appraised outpatient quality indicators resulting from the RAND-modified Delphi rounds of the DRIVE-AB project

- *Generic outpatient quality indicators*
  - Outpatients should receive antibiotic therapy compliant with guidelines; this includes, but is not limited to, indication, choice of the antibiotic, duration, dose, and timing
  - Antibiotics in stock should not be beyond the expiry date
  - Antibiotics that are dispensed to outpatients should be adequately labeled (patient name, antibiotic's name, when antibiotics should be taken)
- *OPAT outpatient quality indicators*
  - All OPAT plans should include dose, frequency of administration, and duration of therapy.
  - Administered doses of OPAT intravenous therapy should be documented on a medication card.
  - The OPAT plan should be communicated to the general practitioner at discharge.

OPAT: Outpatient Parenteral Antibiotic Treatment.
http://drive-ab.eu/wp-content/uploads/2014/09/WP1A_Final-QMs-QIs_final.pdf.

In two questionnaire mailings with an in-between face-to-face consensus meeting, an international multidisciplinary expert panel of 17 experts appraised and prioritized these potential QIs. The literature search resulted in a list of 24 potential QIs. Nine QIs describing recommended care at patient level were selected: (1) take 2 blood cultures, (2) take cultures from suspected sites of infection, (3) prescribe empirical antibiotic therapy according to local guideline, (4) change empirical to pathogen-directed therapy, (5) adapt antibiotic dosage to renal function, (6) switch from intravenous to oral, (7) document antibiotic plan, (8) perform therapeutic drug monitoring, and (9) discontinue antibiotic therapy if infection is not confirmed. Two QIs describing recommended care at the hospital level were also selected: (1) a local antibiotic guideline should be present, and (2) these local guidelines should correspond to the national antibiotic guidelines [18]. Subsequently, the researchers subjected these QIs to an applicability test for their clinimetric properties and used these QIs in quality-improvement projects. In a cross-sectional point-prevalence survey, performed in 2011 and 2012, 1890 inpatients from 22 hospitals in the Netherlands treated with antibiotics for a suspected bacterial infection were included, and data were extracted from medical records. In this cohort, the measurability, applicability, reliability, room for improvement, and case mix stability of the previously developed QIs were tested. Low applicability ($\leq$10% of reviewed patients) was found for the QIs 'therapeutic drug monitoring,' 'adapting antibiotics to renal function,' and 'discontinue empirical therapy in case of lack of clinical and/or microbiological evidence of infection'. For the latter, there was a low interobserver agreement (kappa $<$0.4). One QI showed low

TABLE 2 The highest appraised Inpatient Quality Indicators resulting from the RAND-modified Delphi rounds of the DRIVE-AB project

- An antibiotic stewardship program (antibiotic prescribing control program and/or antibiotic prescribing policy) should be in place at the healthcare facility
- An antibiotic plan* should be documented in the medical record at the start of the antibiotic treatment
- The results of bacteriological sensitivities should be documented in the medical records
- The local guidelines should correspond to the national guideline but should be adapted based on local resistance patterns
- Allergy status should be taken into account when antibiotics are prescribed

**Antibiotic plan includes: indication, name, doses, duration, route, and interval of administration.*
http://drive-ab.eu/wp-content/uploads/2014/09/WP1A_Final-QMs-QIs_final.pdf.

improvement potential. The remaining seven QIs had sound clinimetric properties. Case mix correction was necessary for most process QIs. For all QIs, ample room for improvement and large variation between hospitals was found. Establishing the clinimetric properties was essential as 4 of the 11 previously selected QIs showed unsatisfactory properties in the clinical practice test [19]. As shown in this exercise, the testing of the applicability is crucial since the quality of antibiotic use and the process of documenting data is changing over time and may vary according to the setting.

The DRIVE-AB project developed a set of 51 inpatient QIs. The highest appraised inpatient QIs are presented in Table 2. These inpatient QIs are highly standardized and consensually accepted at a global level. They are very generic and, as with the metrics, should be refined and made applicable and measurable according to different settings [11].

CONCLUSION

Standardized quantity metrics and generic, global quality indicators of antibiotic use are available from a variety of sources for the out- and inpatient settings. These metrics and generic indicators should be adapted to the local setting. Testing the clinimetric properties of the QIs is essential to select those indicators that are feasible, valid, and reliable in a specific country or region.

REFERENCES

[1] Lawrence M, Olesen F, *et al.* Indicators of quality health care. Eur J Gen Pract 1997;3:103–8.
[2] Buck D, Godfrey C, Morgan A. Performance indicators and health promotion targets. Discussion paper 150. York: Centre for Health Economics, University of York; 1996.
[3] WHO International Working Group for Drug Statistics Methodology. Introduction to drug utilization research. Oslo: World Health Organization; 2003.

[4] DRIVE-AB (Driving reinvestment in research and development and responsible antibiotic use). Availible at: www.drive-ab.eu [Accessed October 2016].

[5] WHO Collaborating Centre for Drug Statistics Methodology. Guidelines for ATC classification and DDD assignment 2016. Oslo; 2016.

[6] Barlam TF, *et al.* Executive summary: implementing an Antibiotic Stewardship Program: guidelines by the Infectious Diseases Society of America and the Society for Healthcare Epidemiology of America. Clin Infect Dis 2016;62(10):1197–202.

[7] Agency for Healthcare & Research Quality. Possible methods for evaluating antibiotic use. Available at: http://www.ahrq.gov/professionals/quality-patient-safety/patient-safety-resources/resources/cdifftoolkit/cdiffl2tools2c.html [Accessed October 2016]

[8] Bruyndonckx R, *et al.* Exploring the association between resistance and outpatient antibiotic use expressed as DDDs or packages. J Antimicrob Chemother 2015;70(4):1241–4.

[9] Ibrahim OM, Polk RE. Antimicrobial use metrics and benchmarking to improve stewardship outcomes: methodology, opportunities, and challenges. Infect Dis Clin N Am 2014;28(2): 195–214.

[10] Amadeo B, *et al.* Easily available adjustment criteria for the comparison of antibiotic consumption in a hospital setting: experience in France. Clin Microbiol Infect 2010;16(6): 735–41.

[11] DRIVE-AB. http://drive-ab.eu/wp-content/uploads/2014/09/WP1A_Final-QMs-QIs_final.pdf.

[12] Gould IM, Hampson J, Taylor EW, *et al.* Hospital antibiotic control measures in the UK. J Antimicrob Chemother 1994;34:21–42.

[13] Gyssens IC, Van den Broek PJ, Kullberg BJ, Hekster YA, Van der Meer JWM. Optimizing antimicrobial therapy. A method for antimicrobial drug evaluation. J Antimicrob Chemother 1992;30:724–7.

[14] Van der Meer JWM, Gyssens IC. Quality of antimicrobial drug prescription in hospital. Clin Microbiol Infect 2001;7(suppl 6):12–5.

[15] Saust LT, Monrad RN, Hansen MP, Arpi M, Bjerrum L. Quality assessment of diagnosis and antibiotic treatment of infectious diseases in primary care: a systematic review of quality indicators. Scand J Prim Health Care 2016;34(3):258–66.

[16] Schouten JA, Hulscher ME, Wollersheim H, Braspennning J, Kullberg BJ, van der Meer JW, *et al.* Quality of antibiotic use for lower respiratory tract infections at hospitals: (how) can we measure it? Clin Infect Dis 2005;41(4):450–60.

[17] Van den Bosch CM, Hulscher ME, Natsch S, Gyssens IC, Prins JM, Geerlings SE, *et al.* Development of quality indicators for antimicrobial treatment in adults with sepsis. BMC Infect Dis 2014;14:345.

[18] Van den Bosch CM, Geerlings SE, Natsch S, Prins JM, Hulscher ME. Quality indicators to measure appropriate antibiotic use in hospitalized adults. Clin Infect Dis 2015;60(2):281–91.

[19] den Bosch CM Van, Hulscher ME, Natsch S, Wille J, Prins JM, Geerlings SE. Applicability of generic quality indicators for appropriate antibiotic use in daily hospital practice: a cross-sectional point-prevalence multicenter study. Clin Microbiol Infect 2016. pii: S1198-743X(16)30235-X.

Section B

AMS Strategies

Chapter 4

Improving Antimicrobial Prescribing: Input from Behavioral Strategies and Quality Improvement Methods

Marlies E.J.L. Hulscher* and Jeroen Schouten*,**
*Radboud University Medical Center, Nijmegen, The Netherlands
**Canisius Wilhelmina Ziekenhuis, Nijmegen, The Netherlands

STEWARDSHIP INTERVENTIONS

Antimicrobial stewardship is a key approach in the battle against antimicrobial resistance, both in terms of reducing the current burden and the further development and spread of resistance in the future. Through the years, various documents (guidelines, consensus statements, policy statements, etc.) have been published by national and international organizations that provide clear recommendations for developing stewardship programs. These documents often include recommendations on appropriate structural preconditions that should be met when embarking on stewardship, like the establishment of an antimicrobial stewardship team. In addition, they describe various stewardship interventions that can be applied by such teams when aiming at appropriate antimicrobial prescribing practices. Depending on local antimicrobial use, size, staffing, personnel, infrastructures, and available resources, different institutions may need different interventions to combat antimicrobial resistance. Antimicrobial stewardship can be thought of as a menu of interventions that can be designed and adapted to fit the infrastructure of any hospital [1].

Stewardship programs encompass two intrinsically different sets of interventions describing either the "what" or the "how" in stewardship programs. A first set of interventions describes recommended antimicrobial care interventions or antimicrobial prescribing practices that define "appropriate antimicrobial use" in hospital inpatients regarding indication, choice of drug, dose, route, or duration of treatment. Examples of such interventions are

Antimicrobial Stewardship. http://dx.doi.org/10.1016/B978-0-12-810477-4.00004-0

"switch from intravenous to oral antimicrobial therapy" or "streamline therapy" in individual patients when appropriate [2].

A second set of interventions describes interventions to ensure that professionals actually apply these prescribing practices in daily practice. These behavioral change interventions include many different interventions—like the provision of a formulary, prospective, or retrospective audit and feedback; educational meetings; reminders; financial interventions; or the revision of professional roles [3]—that all can be performed to improve appropriate antimicrobial prescribing practices. So, the second set of interventions is applied in professionals to ensure that the first set of interventions is appropriately applied in patients. These behavioral change interventions either directly or indirectly (through interventions targeting the system/organization) target the professional and, overall, restrict or guide toward the more effective professional use of antimicrobials. Table 1 provides more detail on various behavioral change

TABLE 1 Examples of Behavioral Change Interventions to Improve Antimicrobial Prescribing Practices in Daily Practice [3,4]

Persuasive interventions	• Audit and feedback, i.e., a summary of health workers' performance over a specified period of time, given to them in a written, electronic, or verbal format. The summary may include recommendations for clinical action. • Reminders, i.e., manual or computerized interventions that prompt health workers to perform an action during a consultation with a patient, for example, computer decision support systems. • Educational outreach, i.e., personal visits by a trained person to health workers in their own settings to provide information with the aim of changing practice. • Educational meetings and dissemination of educational materials • Formal or informal local consensus processes, for example, agreeing to a clinical protocol to manage a patient group, adapting a guideline for a local health system or promoting the implementation of guidelines.
Restrictive interventions	• Selective reporting of laboratory susceptibilities • Formulary restriction • Requiring prior authorization of prescriptions • Therapeutic substitution • Automatic stop orders • Antimicrobial cycling or rotation
Structural interventions	• Changing from paper to computerized records • Rapid laboratory testing • Computerized decision support systems • Introduction or organization of quality monitoring mechanisms

The definitions for audit and feedback, reminders, educational outreach, educational meetings and consensus processes are drawn verbatim from the following source, with permission: [Taken from Effective Practice and Organization of Care (EPOC). EPOC Taxonomy; 2015. Available at: https://epoc.cochrane.org/epoc-taxonomy and Davey P, Brown E, Charani E, Fenelon L, Gould IM, Holmes A, Ramsay CR, Wiffen PJ, WilcoxM. Interventions to improve antibiotic prescribing practices for hospital inpatients. Cochrane Database of Systematic Reviews 2013, Issue 4. Art. No.: CD003543.]

BOX 1 The Importance of Measurement of Antimicrobial Prescribing in Daily Practice

Depending on the possibilities and resources in a specific hospital or ward, various measurement methods can be used (see Chapter 3). In daily practice, appropriate use is often assessed by performing a point prevalence survey (PPS). During such a cross-sectional measurement of antimicrobial use within a hospital, prescribing practices are measured at one particular point in time within the entire hospital. For each included ward, information is collected for each patient receiving antibiotics on, for example, the indication for antimicrobial prescription and the agent and route prescribed. The advantage of a PPS is that it is a well-accepted, feasible method and that it—by providing a global overview of antimicrobial prescribing—highlights departments or patient groups with potential inappropriate antimicrobial use. The value of its results are, therefore, limited: a PPS merely functions as a quick scan to signal potential problem areas. If such areas are detected, further investigation is always needed. In that case, following the PPS, a measurement needs to be performed in which information is collected on specific prescribing processes (e.g., % of patients in whom the empirical regimen was appropriate) and/or outcomes in a larger group of specified patients over a longer period of time (see also Chapter 3, quality indicators). **NB.** If this specific information on processes or outcomes of appropriate use indicates room for improvement, behavioral strategies need to be selected and applied to improve these aspects of antimicrobial prescribing. The same specific prescribing processes and/or outcomes should be reassessed following the performance of these behavioral strategies (see Chapter 20 for appropriate evaluation approaches).

interventions as described in the literature, in this case the Cochrane review of Davey *et al.* [4].

In this chapter, the application of behavioral change interventions to improve appropriate antimicrobial use is viewed as an example of healthcare quality improvement. In this manner, stewardship refers to "coordinated interventions designed to continuously measure and improve the appropriate use of antimicrobial agents by promoting the selection of the optimal antimicrobial drug regimen including dosing, duration of therapy, and route of administration" [9]. Measurement is critical to identify opportunities for improvement and assess the impact of improvement efforts (see Boxes 1 and 2). For antimicrobial stewardship, measurement may involve the evaluation of both process (Are recommended prescribing practices being followed as expected?) and outcome (Have interventions improved patient or microbiological/ecological outcomes?).

EFFECTIVENESS OF BEHAVIORAL STEWARDSHIP INTERVENTIONS

Behavioral stewardship interventions all aim to optimize professionals' antimicrobial prescribing practices so that patients, throughout their hospital

BOX 2 Example Point Prevalence Study (PPS) and Subsequent Audit

First Phase: Performing a (Two) Yearly PPS Study

A PPS is a cross-sectional measurement of the prescribing behavior within a hospital: prescribing behavior is measured at one point in time within the entire hospital. Professionals assess, for example, how many patients receive antibiotics, which agents, for which indications. In this way, a PPS can identify the departments and/or disease indications that require additional analysis (**bold**):

	General Surgery	Internal Medicine	Urology
Meropenem as empirical prescription in UTI	14%	15%	**48%**
Ceftriaxon as empirical prescription in UTI	65%	65%	34%
Ciprofloxacin as empirical prescription in UTI	17%	10%	16%
Other as empirical prescription in UTI	4%	10%	2%

Second Phase: Additional In-depth Analysis Needed in Specific Departments

In addition to the PPS, one can zoom in on specific departments and/or on specific aspects of prescribing behavior by performing a temporary longitudinal measurement of the quality of prescriptions: an audit. This allows for the evaluation of the quality of prescriptions at a smaller scale and in a more focused and detailed manner in order to gain a clear idea of which improvement goals are a priority (**bold**). During these audits, the focus lies on one or more of the aspects of prescribing behavior, e.g., urine cultures for inpatients with a complicated UTI.

	Urology
Appropriate empirical prescription according to local antibiotic guidelines UTI	**34%**
Appropriate duration of therapy according to local antibiotic guidelines UTI	67%
Appropriate diagnostics (urine cultures) according to local antibiotic guidelines UTI	89%

stay, receive the appropriate regimen for the documented indication at the right time, with the appropriate dose, via the appropriate route, for the appropriate duration. They aim to change the behavior of individual prescribers, that is, the care provided to patients, and, ultimately, to improve patient and microbiological/ecological outcomes. Such behavior change can be reached through professional-oriented, financial, organizational, or regulatory interventions [3].

Many studies have assessed these interventions for improving professionals' prescribing practices, patient outcomes, and microbial outcomes. Various reviews have summarized them [4,10–13,25]. They all conclude that *any* stewardship intervention—whether it is restrictive, persuasive, or structural—can ensure that professionals appropriately use antimicrobials. Although the effects are overall positive, there are large differences in improvement between the various studies that tested similar stewardship interventions. For instance, Davey *et al.* describe how the effect size of interventions that, for example, used dissemination of educational materials as the main intervention, varied between −3.1% and 50.1%. The same phenomenon was seen for the other interventions tested [4].

So, from these systematic reviews, it can be concluded that *any* behavioral stewardship intervention might work to improve professionals' antimicrobial use. How then to select from this menu of effective interventions [1] those interventions that might work best in a specific setting (e.g., hospital or ward)?

THE IMPORTANCE OF UNDERSTANDING THE KEY DRIVERS OF CURRENT PRESCRIBING BEHAVIOR

Numerous change models and theories derived from various disciplines and scientific areas [5,6] describe effective change as the result of a systematic, stepwise approach that needs good preparation and planning (see for an example Box 3 and the paper by Grol [5]).

Looking at these models and theories, a crucial step or principle for successful behavior change recurs through most publications [5,14,15]: the choice of interventions for change should be linked as closely as possible to the results of a problem analysis. By taking the outcomes of this problem analysis into account, a *tailored* mix of interventions can be devised. So, to select from the potentially effective interventions those behavioral stewardship interventions that might work best in a specific hospital/ward, one first has to understand the key drivers of current prescribing behavior. Successful improvement of antimicrobial use requires a problem analysis, that is, a diagnostic analysis to find out the determinants of success or failure to act upon recommended practice. "It is not only microbes that we need to investigate: equally important is a better understanding of our own actions" [16].

Determinants of a prescribing practice are factors that might hinder or help improvements in that practice. The assessment of these determinants, both barriers and facilitators, should inform the choice of behavioral stewardship interventions, for example, education to address a lack of knowledge or reminders if "forgetting to apply the recommended prescribing practice" is the problem. An understanding of the determinants for change is therefore crucial to the selection of effective interventions. In daily practice, however,

BOX 3 A stepwise approach to improvement of patient care: the PLAN-DO-STUDY-ACT cycles for continuous improvement [5,6]

Many models and theories describe a stepwise approach to planning or managing behavioral change, for example, the PRECEDE PROCEED model by Green and Kreuter [26], the stages of change theory by Prochaska and Velicer [27], and the Planned Action model by Graham and Tetroe [28]. A widely accepted 4-step model is the PLAN-DO-STUDY-ACT method that is applied in a cyclical fashion.

PLAN-DO-STUDY-ACT (PDSA) cycles provide a pragmatic structure for testing interventions to iteratively improve patient care. This widely accepted systematic approach creates an ongoing improvement cycle by:

1. identifying improvement aims and generating ideas on how to reach these aims (PLAN);
2. carrying out the improvement plan (DO);
3. analyzing and reflecting on data gathered in a small group of patients (STUDY); and
4. determining whether and what modifications should be made (ACT).

The ACT phase provides input to adapt the previous PLAN phase after which the cycle is repeated. Often, multiple cycles are planned, beginning on a small scale and gradually expanding to the whole patient population or setting.

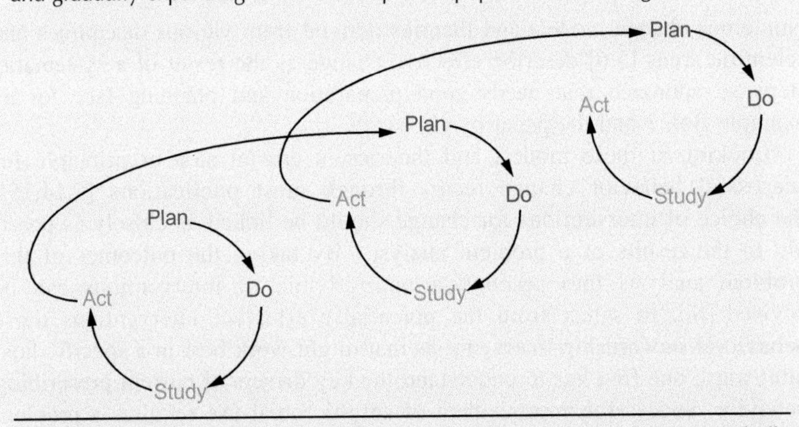

(Source: Reprinted with permission of the Agency for Healthcare Research and Quality; Rockville, Maryland, USA.
"Figure 2. The plan-do-study-act approach to practice improvement." (p6) in: *Eder M, Smith SG, Cappelman J, et al. Improving Your Office Testing Process. A Toolkit for Rapid-Cycle Patient Safety and Quality Improvement. AHRQ Publication No.13-0035. Rockville, MD: Agency for Healthcare Research and Quality; August 2013. https://www.ahrq.gov/sites/default/files/publications/files/officetesting-toolkit.pdf)*

the chosen interventions to improve healthcare are currently mostly based on implicit personal beliefs about human behavior and change [5], in precise opposition to the paradigm of evidence-based medicine [17]. For example, Charani *et al.* concluded [18] in their review on optimizing antimicrobial prescribing in acute care that although qualitative research shows the influence of social norms, attitudes, and beliefs on antimicrobial prescribing behavior [19], these behavioral determinants were not considered while developing improvement interventions.

THE KEY DRIVERS OF CURRENT ANTIMICROBIAL PRESCRIBING BEHAVIOR

Literature shows that antimicrobial prescription is a complex process determined by many factors. The appropriateness of antimicrobial use in hospitals is, for example, influenced by professional knowledge and attitudes, hospital antimicrobial policies, the multiprofessional care delivery system, and differences in sociocultural and socioeconomic factors [7,8] (see Box 4).

For most changes in healthcare, a wide range of determinants influence whether appropriate care is provided or not. Flottorp *et al.* systematically synthesized current frameworks and taxonomies of factors that help or hinder improvements in healthcare [20]. This synthesis includes, for example, the Theoretical Domains Framework, a framework in which 128 explaining constructs from 33 theories of behavior change are grouped into 12 theoretical domains to explain behavior change [21]. Combining all published frameworks, Flottorp *et al.* developed a comprehensive, integrated overview of 57 potential determinants categorized into seven domains. The following categories of determinants are distinguished:

(1) guideline factors (e.g., the clarity of the recommendation, the evidence supporting the recommendation);
(2) individual health professional factors (e.g., awareness and familiarity with the recommendation or the skills needed to adhere);
(3) patient factors (patient preferences or real or perceived needs and demands of the patient);
(4) professional interactions (e.g., opinions and communication among professionals or referral processes);
(5) incentives and resources (e.g., availability of necessary resources or extent to which the information system influences adherence);
(6) capacity for organizational change (e.g., capable leadership or the relative priority given to making necessary changes); and
(7) social, political, and legal factors (e.g., payer or funder policies).

Based on this overview of determinants, they developed a generic checklist that can be used by researchers and others involved in designing behavioral interventions to improve healthcare quality. The checklist guides reflection and data collection on determinants of practice. Thus, stewardship teams could use this checklist when aiming to systematically assess—using semi-structured interviews with individual professionals involved in the prescribing practice, group interviews, questionnaires, and/or observation—the factors that influence a specific prescribing practice in their specific hospital or ward. Users should be aware that the checklist focuses on determinants of one specific recommendation for clinical practice as the relevance

BOX 4 Examples of Potential Determinants Influencing Appropriate Antimicrobial Use in Daily Hospital Practice [7,8]

Sociocultural and socioeconomic factors

- Ideas about health, causes of disease, labeling of illness, attributions, coping strategies, and treatment modalities
- "Uncertainty avoidance" (i.e., unwillingness to accept uncertainty and risks)
- "Power distance" (i.e., willingness to accept that power is unevenly distributed)
- Healthcare funding
- Reimbursement
- Role of pharmaceutical industries
- Availability of antimicrobials (over-the-counter drugs/Internet)

Organizational policies and multiprofessional care-delivery system

- Antibiotic committee
- Infection prevention committee
- Antibiotic booklet
- Antibiotic formulary
- Antibiotic order forms
- Automatic stop orders
- Professionals' collaboration and communication (e.g., courtesy (instead of criticism) to colleagues, ward culture to "never change a winning team")
- Care logistics care and coordination (e.g., use of different guidelines by different wards, antibiotics present on the ward, regular medication rounds and/or supervisor's ward rounds)
- Work intensity
- Senior support

Professional knowledge and attitudes

- Familiarity with or awareness of available evidence or consensus on appropriate antimicrobial use
- Knowledge on infectious diseases
- Knowledge on causative microorganisms and their susceptibility
- Knowledge on effective therapy
- Knowledge on resistance
- Experience/confidence
- Experience/routines
- Diagnostic uncertainty/Fear of "missing things"
- Fear of disciplinary cases
- "Patient's immediate risk outweighs the risk of resistance"
- (Dis)agreement with guidelines

and importance of determinants can vary across different recommendations. In other words, switching and streamlining may be influenced by a different set of factors. Therefore, when the determinants of various prescribing practices need to be assessed, it is necessary to consider each determinant in relationship to each recommendation. For this reason, stewardship teams might consider starting small, that is, to focus improvement efforts on those prescribing practices on those specific wards where there is the greatest potential for benefit. To help prioritize the prescribing practices that warrant stewardship team efforts, teams could, for example, use an additional worksheet developed by Flottorp *et al.*, the worksheet "Prioritization of recommendations" [20].

USE OF THEORIES IN SELECTING AND DEVELOPING BEHAVIORAL STEWARDSHIP INTERVENTIONS

Once the key drivers of a specific prescribing practice are known, behavioral change intervention(s) should be selected by linking potential interventions for change as closely as possible to these results of the problem analysis. Theories on behavioral change play a crucial role in this tailoring principle. Theories relevant to changing healthcare practice enable intervention developers in healthcare to design better interventions to improve patient care [14]. So, if, for example. the systematic assessment of determinants has shown that a lack of knowledge hinders an appropriate switch from intravenous to oral antimicrobial therapy, educational theories would suggest that, for education to be effective, personal targets for improvement and individual learning plans related to the recommended practice should be defined. Thus, theories on behavioral change can be used to generate ideas for planning interventions.

To systematically link behavioral stewardship interventions to the various determinants, a structured approach should be followed. An important example of a theory-based approach is the Intervention Mapping approach [22]. Intervention Mapping is a protocol for the design of interventions that guides developers through a series of steps that assists them in theory-based and evidence-based intervention development. Following a needs assessment and a specification of determinants, theory-based methods that match determinants are selected from the literature, translated into practical interventions, operationalized into plans, implemented, and evaluated. To facilitate the theory-based translation from determinant(s) to intervention(s), Kok *et al.* developed an Intervention Mapping Taxonomy for developing behavioral change interventions [23]. This taxonomy describes behavior change interventions derived from theory to address specific determinants. The authors provide various taxonomy tables that describe interventions to, for example, increase knowledge; to change attitudes, habitual behaviors, skills, or social norms; and to change organizations. This taxonomy might be a helpful stewardship tool: it supports stewardship teams in the theory-based development of a tailored intervention to improve specific prescribing practices in their specific hospital or ward. Two practical examples that might inspire those involved in systematically selecting and developing stewardship interventions can be found in the paper by Grol and colleagues on the use of theory in planning improvement in patient care [14].

For a more pragmatic approach to the systematic selection of tailored behavioral stewardship interventions, stewardship teams could, for example, use the various worksheets and the "definitions questions examples" checklist developed by Flottorp *et al.* [20].

USE OF SYSTEMATIC REVIEWS OF BEHAVIORAL INTERVENTIONS IN SELECTING AND DEVELOPING BEHAVIORAL STEWARDSHIP INTERVENTIONS

Once one or more tailored behavioral stewardship interventions are selected, it is important to find out whether systematic reviews of the effectiveness of these interventions chosen have been published, for example, by checking the Prospero database or the Cochrane Effective Practice and Organisation of Care (EPOC) website (http://epoc.cochrane.org/). EPOC focuses on state-of-the-art reviews of interventions (i.e., various forms of continuing education, financial, organizational, or regulatory interventions) designed to improve professional practice and the delivery of effective health services. Until now, they published over 100 systematic reviews in the Cochrane Library.

So, suppose the stewardship team selected "audit and feedback" as a tailored intervention to address a lack of awareness and to make people conscious of problems in their current performance of the intravenous to oral switch. The next step would be to find out whether systematic reviews of "audit and feedback" have been published and whether any lessons learned could be taken into account. The EPOC database, for example, includes a review by Ivers *et al.* "Audit and feedback: effects on professional practice and healthcare outcomes" [24]. The authors conclude that multivariable metaregression indicated that the effect of using audit and feedback can, among others, be augmented when it is delivered by a supervisor or colleague, when it is delivered in both verbal and written formats, and when it includes both explicit targets and an action plan. To enhance chances of success, the stewardship team could include all these success ingredients in their "audit and feedback" intervention.

CONCLUSION

Antimicrobial prescription is a complex process determined by many factors, which renders changing hospital antimicrobial use into a complex challenge. There is no superior behavioral change intervention, or magic bullet, that works in all circumstances: the challenge lies in systematically building an intervention on the careful assessment of determinants and on a coherent theoretical base while linking determinants to interventions, taking the lessons regarding the effectiveness of various behavioral interventions into account (Box 5).

BOX 5 Example Improving Prescription of Empirical Antibiotic Therapy in CAP in Six Hospitals

STEP 1: How bad is the problem?

Activity: Preintervention measurement of performance on key recommendations for antibiotic prescribing in Community acquired pneumonia (CAP) to choose a recommendation that is most in need of improvement (i.e., low performance and/or large variation between hospitals/wards/professionals)

Outcome: Prescription of empirical therapy for CAP according to guidelines in six Dutch hospitals: 45% (range 5–59)

STEP 2: Which are the determinants/the perceived barriers and facilitators?

Activity 1: Multilevel logistic regression analysis of determinants of appropriate prescribing using preintervention data on empirical therapy and patient, professional, and hospital characteristics

Outcome: Most important determinant in prescribing empirical antibiotic therapy for CAP according to guidelines was recent antibiotic therapy in outpatient setting (<30 days, OR 0.46)

Activity 2: Focus groups and physician interviews to elicit barriers and facilitators

Outcome: "I have been treating patients with this nonguideline-adherent antibiotic since medical school and it is always successful"

Outcome: "Everyone feels safe with a broad- spectrum betalactam antibiotic… colleagues will not quickly criticize you for this choice."

STEP 3: which behavioral intervention to choose?

Activity: Link determinants to potentially effective interventions

Outcome: A local antibiotic guideline is needed to address therapy for patients who were recently treated in the outpatient setting

Outcome: Important peer pressure and "old" habits of prescribing empirical antibiotics prompted to use academic detailing as a part of the improvement strategy

STEP 4: What was the effect of the intervention?

Activity: Evaluate whether the intervention has worked: postintervention measurement + analysis

Outcome: After intervention, a marked improvement in adherence to guidelines was noted in most, but not all, hospitals

Outcome: Finding out which (part of) intervention strategy was most likely to be successful to improve adherence to guidelines (by meticulous process evaluation) is essential to learn for future projects

REFERENCES

[1] Septimus EJ, Owens RC. Need and potential of antimicrobial stewardship in community hospitals. Clin Infect Dis 2011;53(Suppl. 1):S8–S14.

[2] Schuts EC, Hulscher ME, Mouton JW, Verduin CM, Stuart JW, Overdiek HW, *et al.* Current evidence on hospital antimicrobial stewardship objectives: a systematic review and meta-analysis. Lancet Infect Dis 2016;16:847–56.

[3] Effective Practice and Organisation of Care (EPOC). EPOC taxonomy. Available at: https://epoc.cochrane.org/epoc-taxonomy.

[4] Davey P, Brown E, Charani E, Fenelon L, Gould IM, Holmes A, *et al.* Interventions to improve antibiotic prescribing practices for hospital inpatients. Cochrane Database Syst Rev. 4:2013 CD003543.

[5] Grol R. Beliefs and evidence in changing clinical practice. BMJ 1997;315:518–21.

[6] Langley GJ, Moen RD, Nolan KM, Nolan TW, Norman CL, Provost LP. The improvement guide: a practical approach to enhancing organizational performance. second ed. San Francisco, CA: Jossey Bass Publishers; 2009.

[7] Hulscher ME, Grol RP, van der Meer JW. Antibiotic prescribing in hospitals: a social and behavioural scientific approach. Lancet Infect Dis 2010;10:167–75.

[8] Teixeira Rodrigues A, Roque F, Falcão A, Figueiras A, Herdeiro MT. Understanding physician antibiotic prescribing behaviour: a systematic review of qualitative studies. Int J Antimicrob Agents 2013;41:203–12.

[9] Society for Healthcare Epidemiology of America, Infectious Diseases Society of America, Pediatric Infectious Diseases Society. Policy statement on antimicrobial stewardship by the Society for Healthcare Epidemiology of America (SHEA), the Infectious Diseases Society of America (IDSA), and the Pediatric Infectious Diseases Society (PIDS). Infect Control Hosp Epidemiol 2012;33:322–7.

[10] Wagner B, Filice GA, Drekonja D, Greer N, MacDonald R, Rutks I, *et al.* Antimicrobial stewardship programs in inpatient hospital settings: a systematic review. Infect Control Hosp Epidemiol 2014;35(10):1209–28.

[11] Patel D, Lawson W, Guglielmo BJ. Antimicrobial stewardship programs: interventions and associated outcomes. Expert Rev Anti Infect Ther 2008;6:209–22.

[12] Kaki R, Elligsen M, Walker S, Simor A, Palmay L, Danema N. Impact of antimicrobial stewardship in critical care: a systematic Review. J Antimicrob Chemother 2011;66:1223–30.

[13] Patel SJ, Larson EL, Kubin CJ, Saiman L. A review of antimicrobial control strategies in hospitalized and ambulatory pediatric populations. Pediatr Infect Dis J 2007;26:531–7.

[14] Grol RPTM, Bosch MC, Hulscher MEJL, Eccles MP, Sensing M. Planning and studying improvement in patient care. The use of theoretical perspectives. Milbank Q 2007;85:93–138.

[15] Grimshaw JM, Eccles MP, Lavis JN, Hill SJ, Squires JE. Knowledge translation of research findings. Implement Sci 2012;7:50.

[16] Infectious diseases and the future: policies for Europe A non-technical summary of an EASAC report, European Public Health and Innovation Policy for Infectious Disease: The View from EASAC, http://www.easac.eu/fileadmin/Reports/Infectious_Diseases/Easac_11_IDF.pdf [Accessed 15 July 2016].

[17] Shojania KG, Grimshaw JM. Evidence-based quality improvement: the state of the science. Health Aff (Millwood) 2005;24:138–50.

[18] Charani E, Edwards R, Sevdalis N, Alexandrou B, Sibley E, Mullett D, *et al.* Behavior change strategies to influence antimicrobial prescribing in acute care: a systematic review. Clin Infect Dis 2011;53:651–62.

[19] Broom A, Broom J, Kirby E. Cultures of resistance? A Bourdieusian analysis of doctors' antibiotic prescribing. Soc Sci Med 2014;110:81–8.

[20] Flottorp SA, Oxman AD, Krause J, Musila NR, Wensing M, Godycki-Cwirko M, *et al.* A checklist for identifying determinants of practice: a systematic review and synthesis of

frameworks and taxonomies of factors that prevent or enable improvements in healthcare professional practice. Implement Sci 2013;8:35.

[21] Michie S, Johnston M, Abraham C, Lawton R, Parker D, Walker A. Making psychological theory useful for implementing evidence based practice: a consensus approach. Qual Saf Health Care 2005;14:26–33.

[22] Bartholomew LK, Parcel GS, Kok G. Intervention mapping: a process for developing theory- and evidence-based health education programs. Health Educ Behav 1998;25:545–63.

[23] Kok G, Gottlieb NH, Peters GJ, Mullen PD, Parcel GS, Ruiter RA, *et al.* A taxonomy of behavior change methods: an intervention mapping approach. Health Psychol Rev 2016;10:297–312.

[24] Ivers N, Jamtvedt G, Flottorp S, Young JM, Odgaard-Jensen J, French SD, *et al.* Audit and feedback: effects on professional practice and healthcare outcomes. Cochrane Database Syst Rev 2012;6. CD000259.

[25] Davey P, Marwick CA, Scott CL, Charani E, McNeil K, Brown E, *et al.* Interventions to improve antibiotic prescribing practices for hospital inpatients. Cochrane Database Syst Rev 2017(2). CD003543.

[26] Green LW, Kreuter MW. Health promotion planning: an educational and environmental approach. Palo Alto, CA: Mayfield Publishing Company, 1991.

[27] Prochaska JO, Velicer WF. The transtheoretical model of health behavior change. Am J Health Promot 1997;12:38–48.

[28] Graham ID, Tetroe J. Getting evidence into policy and practice: perspective of a health research funder. J Can Acad Child Adolesc Psychiatry 2009;18:46–50.

Chapter 5

Education of Healthcare Professionals on Responsible Antimicrobial Prescribing

Oliver J. Dyar* and Bojana Beović,‡**
**Karolinska Institutet, Stockholm, Sweden*
***University Medical Centre Ljubljana, Ljubljana, Slovenia*
‡*Faculty of Medicine, University of Ljubljana, Ljubljana, Slovenia*

INTRODUCTION

Antimicrobials are among the most commonly used type of drugs within healthcare, prescribed daily by healthcare workers across a wide range of clinical specialties, both in inpatient and outpatient settings. Antimicrobials have traditionally been prescribed by physicians, but a growing range of professionals are involved in prescribing decisions today, including pharmacists, midwives, and nurses. Education lies at the heart of efforts to improve prescribing; education can improve knowledge and attitudes among prescribers and help make them aware of the broader contextual and cultural factors that affect prescribing behaviors.

In this chapter, we will discuss the structure and content of education for antimicrobial prescribers throughout their careers and suggest strategies for how educational efforts can be improved in local settings. Table 1 presents the target prescriber groups and phases of education on antimicrobial prescribing and stewardship included in this chapter.

UNDERGRADUATE EDUCATION OF PRESCRIBERS

Many doctors are expected to be able to prescribe antimicrobials immediately after completing their undergraduate studies and will often do so on a daily basis without direct supervision; this is true for pharmacists in some countries too. This highlights the importance of ensuring that undergraduate students receive education on both general antibiotic use and prudent antibiotic use [1]. Several recent studies have revealed that students do not feel ready for

Antimicrobial Stewardship. http://dx.doi.org/10.1016/B978-0-12-810477-4.00005-2

TABLE 1 Healthcare Prescribers Who Should be Included in Education on Responsible Antimicrobial Prescribing

Phase of Education	Physicians	Pharmacists	Nurse Prescribers, Midwives
Undergraduate studies	+	+	+
Internship/foundation	+	N/A	N/A
Specialty training	+	+[a]	+[a]
CME/CPD	+	+	+

[a] *If existing.*

TABLE 2 Perceived Need for More Education Among Undergraduate Medical Students

Study	Region	Percentage Who Want or Need More Education on Antibiotic Use
Minen *et al.* [12]	USA	78%
Abbo *et al.* [13]	USA	90%
Huang *et al.* [14]	China	89%
Dyar *et al.* [15]	Europe	74%
Student-PREPARE 2015[a]	Europe	67%

[a] *Currently unpublished.*

their prescribing duties and still feel they need more education on these topics (Table 2).

There are several advantages to improving education at the undergraduate level:

- Students are engaged in a learning mindset
- Students are encouraged to integrate different parts of their curricula
- Students have compulsory assessments, defined curricula, and national licensing requirements which can act as complementary reinforcing mechanisms

Finally, it is usually easier to *shape* behaviors early than it is to *change* them at a later stage.

WHAT CONTENT SHOULD BE INCLUDED FOR UNDERGRADUATES?

Students should graduate with an understanding that antibiotics are a unique class of drugs with negative externalities that affect individual patients and the population health. They must become aware that prescribers have dual responsibilities: optimizing therapy for an individual patient and preserving antimicrobial efficacy and minimizing resistance for the same and future patients [1]. Undergraduate education should provide students with the principles that they can use to address both of these responsibilities in practice.

Table 3 lists topic areas, concepts, and principles that should be included in undergraduate education for antimicrobial prescribers. This table highlights the fields and disciplines in which these principles might be taught in a typical course for medical students; antibiotic use crosses many disciplines, so coordination in curriculum planning is essential. A recent study of curricula across European medical schools showed that many principles of responsible antibiotic use are poorly covered at present [2]. There were wide variations within countries, in part due to a lack of national frameworks. Similarly, surveys of undergraduate students consistently identify certain topics as insufficiently covered, such as surgical antibiotic prophylaxis, planning the duration of antibiotic treatments, and combination therapy [Student-PREPARE 2015 (unpublished)].

Another challenge for course organizers to consider is the hidden curriculum that students will inevitably be exposed to [3]. Students in a typical medical school may have up to 10 h of scheduled educational sessions related to responsible antimicrobial use; in reality, students will have far more exposure during their courses to the practices of antimicrobial prescribers who have not been directly tasked with educating them on responsible antimicrobial use. Students may receive conflicting messages when the explicit curriculum and this hidden curriculum are not harmonized.

HOW SHOULD CONTENT BE STRUCTURED AND DELIVERED IN AN UNDERGRADUATE PROGRAM?

The structure of undergraduate programs is heterogeneous: integration of education on responsible antibiotic use must be tailored to the individual setting. Many medical schools follow a pattern of teaching a foundation of preclinical sciences for the first 2–3 years, followed by a clinical focus in the remaining years; in other schools, patient-centered problem-based learning may be the main approach used from the start, and this is more easily suited to including responsible prescribing in the earliest phases of medical education. Some of the topics in Table 3 may be more appropriate earlier in an undergraduate program (mechanisms of actions of antibiotics, selection of resistant bacteria), whereas others may be better suited to later parts of a program after students

TABLE 3 Elements of Education on Prudent Antibiotic Prescribing

Topic	Concept, Understanding	Field, Discipline	Principles, Learning Outcomes, Competencies
Bacterial resistance	Selection, mutation	(Micro) biology, genetics	• Extent, causes of bacterial resistance in pathogens (low antibiotic concentration, longtime exposure of microorganisms to antibiotics is driving resistance) • Extent, causes of bacterial resistance in commensals and the phenomenon of overgrowth (e.g., *Clostridium difficile* infection, yeast infection)
		Epidemiology	• Epidemiology of resistance, accounting for local variations and importance of surveillance (differences between wards, countries...)
	Hygiene	Infection control—mostly microbiology	• Spread of resistant organisms
Antibiotics	Mechanisms of action of antibiotics/resistance Toxicity	Pharmacology	• Broad- vs. narrow-spectrum antibiotics, preferred choice of narrow-spectrum drugs • Combination therapy (synergy, limiting emergence of resistance, broaden the spectrum)
	Costs	Ethics, public health, pharmacology	• Collateral damage of antibiotic use (toxicity, cost) • Consequences of bacterial resistance • Lack of development of new antibiotics (limited arsenal)
Diagnosis of infection	Infection/inflammation	Physiology, microbiology, immunology, infectious diseases	• Interpretation of clinical and laboratory biological markers • Fever and C-Reactive Protein (CRP) elevation are also a sign of inflammation, not per se of an infection

	Isolation, identification of bacteria, viruses, and fungi	(Micro) biology	• Practical use of point-of-care tests (e.g., urine dipstick, streptococcal rapid antigen diagnostic test in tonsillitis…) • Importance of taking microbiological samples for culture before starting antibiotic therapy
	Susceptibility to antibiotics	Microbiology, infectious diseases	• Interpretation of basic microbiological investigations (Gram stain, culture, PCR, serology…)
Treatment of infection	Indication for antimicrobials	Clinical microbiology, infectious diseases organ specialty	• Definitions and indications of empiric/directed therapy vs. prophylaxis • Clinical situations when not to prescribe an antibiotic: ○ Colonization vs. infection (e.g., asymptomatic bacteriuria) ○ Viral infections (e.g., acute bronchitis) ○ Inflammation vs. infection (e.g., fever without a definite diagnosis in a patient with no severity criteria)
Prevention of infection		Pharmacotherapy, surgery, anesthesiology, clinical microbiology, infectious diseases	• Surgical antibiotic prophylaxis: indication, choice, duration (short), timing
Medical record keeping	Choice, Duration, Timing	Clinical medicine	• Documentation of antimicrobial indication in clinical notes • Recording (planned) duration or stop date
Prescribing antibiotics: initially	Empiric therapy (local guide, antibiotic booklet…) Diagnostic uncertainty	Clinical microbiology, infectious diseases, organ specialists Clinical pharmacology	• Best bacteriological guess for empiric therapy • Choice in case of prior use of antibiotics when selecting an antibiotic for empiric therapy • Choosing the dose and interval of administration (basic principles of PK/PD) • Estimating the shortest possible adequate duration

Continued

TABLE 3 Elements of Education on Prudent Antibiotic Prescribing—Cont'd

Topic	Concept, Understanding	Field, Discipline	Principles, Learning Outcomes, Competencies
Prescribing antibiotics: targeted therapy	Communication with the microbiology laboratory Value of specialist consultation in infectious diseases or microbiology	Clinical microbiology, infectious diseases, organ specialists Hospital pharmacy	• Reassessment of the antibiotic prescription around day 3 • Streamlining/de-escalation once microbiological results are known • IV-oral switch (bioavailability of antibiotics) • Therapeutic drug monitoring to ensure adequate drug levels (e.g., vancomycin)
Prescribing antibiotics: standard of care	The importance of guidelines in clinical practice	Clinical medicine, organ specialists	• Prescribing antibiotic therapy according to national/local practice guidelines
	Quality indicators of antibiotic use	Quality institute	• Audit and feedback assessing prescribing practice using quality indicators
Communication skills	Discussion techniques	Psychology, clinical medicine	• Explaining to the patient the absence of an antibiotic prescription • Education of patients regarding prudent antibiotic use (comply with the doctors' prescription, no self-medication...)

From Pulcini C, Gyssens IC. How to educate prescribers in antimicrobial stewardship practices. Virulence 2013;4:192–202.

have gained sufficient clinical exposure (targeted therapy, standards of care). Sessions on responsible antibiotic use in primary care should be included since in some countries up to half of all medical students will choose to specialize in primary care.

Many formats can be used for educational sessions for both undergraduate and postgraduate training. Table 4 presents a brief description of these types of activities and their efficacy (in the context of continuing medical education).

Lecture sessions are often an important component early in undergraduate courses. During the later years, small group case-based teaching is particularly

TABLE 4 Types of Educational Interventions and Their Efficacy

Type of Education	Effectiveness	Comment
Educational lectures	modest	Passive education of a larger group of attendees, easy to organize, usually not expensive, more effective when repetitive
Interactive educational lectures	moderate	Educational lectures including participation of the attendees
Interactive small group sessions	moderate	Seminar sessions with up to 15 attendees
Printed material	modest	Leaflets, banners, brochures with information and motivating content
Reminders	moderate	Alerts prompting health professional to perform or avoid some action, includes computer-supported decisions
Guidelines, clinical pathways	modest to moderate	Comprehensive recommendation based on best evidence and their local implications (pathways)
Audit and feedback	moderate	Review of current practice followed by advice to the clinician
Educational outreach visits	moderate to high	Personal educational visits of professionals to a physician or a group of physicians
Interactive web-based educational programs	moderate	Efficacy depends on the type of e-learning: passive or active

Modified from Cisneros JM, Cobo J, San Juan R, Montejo M, Farinase MC. Education on antibiotic use. Education systems and activities that work. Enferm Infecc Microbiol Clin 2013;31(Suppl. 4): 31–37; Ohl CA, Luther VP. Health-care provider education as a tool to enhance antibiotic stewardship practices. Infect Dis Clin North Am 2014;28:177–93.

valuable and can be strengthened by including student reflection and feedback. Such sessions can promote in-depth discussions around gray cases, in which it is possible to highlight the role of responsible prescribing in the context of clinical uncertainty. Encouraging students to maintain a portfolio of antibiotic prescribing interactions may be a valuable supplement for these discussions. Local and national antibiotic guidelines can also be useful tools for students to make active comparisons with observed practices. Elective modules in clinical microbiology and infectious diseases are available for students in some medical schools. These students could be encouraged to become peer teachers, organizing student-led teaching sessions for their fellow students on responsible antibiotic use.

Toward the end of many undergraduate medical school programs, there are now short summary courses with the explicit aim of preparing students to transition to their future practice as a junior doctor. These courses are an excellent opportunity to include dedicated sessions on responsible antibiotic use, synthesizing material and experiences from previous diverse components in the undergraduate program.

Tips for undergraduate course organizers:

1. Try mapping the suggested curriculum topics in Table 3 to the existing curriculum at your medical school. Is anything missing?
2. Find out what organ specialists are teaching and practicing to understand the hidden curriculum that students are exposed to
3. Make incremental changes: start with organizing seminars for final-year students on stewardship principles and listen to their concerns
4. Encourage use of an antibiotic prescribing portfolio, particularly in primary care, coupled with meetings for reflection

POSTGRADUATE EDUCATION OF PRESCRIBERS

The structure and curricula of postgraduate education vary between countries but in most cases starts with a foundation period or internship, followed by specialist training in a clinical discipline. Throughout these periods of internship and specialization, young doctors shape their medical behavior based on knowledge as well as the cultural environment at their workplaces. After completing specialty training, medical doctors in many countries are included in a system of continuing medical education (CME) and continuing professional development (CPD).

MEDICAL DOCTORS IN TRAINING (INTERNSHIP, SPECIALIZATION)

The start of a young doctor's professional career is extremely challenging. They are facing new responsibilities in an environment full of uncertainty and emotions. Drug prescribing is one of the tasks accompanied by the highest levels of uncertainty [4]. The curricula leading to certification within a specialty differ between countries, but with the partial exception of general/family medicine, they all lead to more specialized knowledge, which means leaving behind to some extent the more generic parts of medicine such as clinical microbiology and infection. Young doctors use a variety of sources to inform their antimicrobial prescribing, including guidelines, ID physicians, the Sanford guide, and other colleagues. Modern smartphone and internet based tools are also frequently used [5]. The pharmaceutical industry is not perceived as a major source of information [6,7].

Several studies have investigated the knowledge, attitudes, beliefs, and practices of young doctors in the field of antimicrobial prescribing and resistance. The basic determinant of young doctors' antibiotic prescribing is knowledge, which has frequently been reported as poor. In particular, young doctors have little knowledge of local resistance rates in common bacteria and underestimate the rate of antibiotic misuse [8]. Longer time spent in training and greater amounts of clinical experience do not seem to be associated with improvements in knowledge; this may be explained by the paucity of specific education during training. Some studies have even shown that knowledge among residents is higher than among senior staff members [7]. Culture is another important factor that influences the prescribing practices of young doctors. The domination of cultural aspects over rational decisions has been shown in studies from many different parts of the world and is not limited only to young doctors. A qualitative study identified a set of cultural rules that determine antibiotic prescribing in hospitals in the United Kingdom and form a so-called "prescribing etiquette." This prescribing etiquette consists of the identification of young doctors with the clinical team they work with, the hierarchical organization of clinical teams, the autonomous position of senior doctors who rely more on their experience than policies and guidelines, and a culture of noninterference when an antimicrobial has been prescribed by a peer [9].

Most young doctors are aware that they need more education in antimicrobial prescribing and stewardship. They believe that antimicrobial resistance is more of a problem on a global and national level and less in the place in which they work. At the same time, they think that inappropriate prescribing is more of a problem among other doctors than themselves. Aside from infectious diseases, there are currently very few specializations that explicitly incorporate education on antimicrobial resistance and stewardship into their training curricula or requirements. In Scotland, the Doctors Online Training System is used as a web-based educational resource for all young doctors, and it includes a module on responsible prescribing. This module has also been expanded to offer education to pharmacists and other nonphysician providers.

CONTINUING MEDICAL EDUCATION AND CONTINUING PROFESSIONAL DEVELOPMENT

CME-CPD educational systems exist for doctors in many countries. The content and structure of CME-CPD is far less defined than that of specialty training. In most cases, CME consists of various courses and conferences tightly linked to core specialty knowledge. A more recent concept is CPD, which incorporates the concept of CME and goes beyond it to support the improvement of all aspects of a medical practitioner's performance. For example, it could require not only knowledge in infectious diseases, clinical microbiology, and pharmacotherapy, but also changes in attitudes and behaviors toward antimicrobial prescribing. Several studies have shown that the cultural aspects influencing antibiotic prescribing play an important role throughout the medical careers of physicians [1].

TYPES OF EDUCATIONAL ACTIVITIES FOR HEALTHCARE PROFESSIONALS

A variety of educational activities in CME have been described and evaluated (Table 4). Studies evaluating the efficacy of CME interventions have shown that including multiple media and multiple techniques is more effective than single CME interventions; similarly, repetition leads to more sustained improvements in knowledge and practices. Many studies have focused on specific interventions to improve prescribing, but overall, there is still no consensus on the skills needed for appropriate prescribing. In a recent systematic review of antibiotic stewardship interventions in outpatients, most of the educational activities studied were multifaceted and yielded mixed results. Implementation of guidelines and teaching communication skills was more effective in decreasing the use of antimicrobials and the choice of antibiotic [10].

The use of e-learning is increasing in popularity in medical education. The best known examples of e-learning platforms come from the United Kingdom, United States, and Australia. The University of Dundee and the British Society for Antimicrobial Chemotherapy launched a massive open online course in 2015. In the United States, the Get Smart about Antibiotics initiative offers various tools for the education of professionals and patients. Online educational tools are also available through the ECDC Antibiotic Awareness Day initiative. The advantages and disadvantages of e-learning are summarized in Table 5.

EDUCATION OF OTHER HEALTHCARE PROFESSIONALS IN ANTIMICROBIAL STEWARDSHIP

Pharmacists and nurses are frequently involved in antimicrobial stewardship activities that are not limited to prescribing. As examples, pharmacists may participate in preauthorization and audit/feedback strategies, whilst nurse

TABLE 5 **Advantages and Disadvantages of e-Learning**

Advantages	Disadvantages
Wide availability	Laborious preparation
Low cost	Time consuming for students
Flexible schedule for students	Less personal contact
Access to the content at various location	Lack of technical infrastructure
Potential access to high-level experts	Language and cultural context problems
Background for further discussion	
Opportunity to establish personal learning network	
Simulation of workplace	

Modified from Rocha-Pereira N, Lafferty N, Nathwani D. Educating healthcare professionals in antimicrobial stewardship: can online-learning solutions help? J Antimicrob Chemother 2015;70:3175–77.

responsibilities include timely administration of antimicrobials and close monitoring of adverse events. Educational efforts to support the wide variety of nonprescribing activities are described in the chapters on the role of nurses (Chapter 11) and the role of pharmacists (Chapter 10).

COMPETENCIES IN ANTIMICROBIAL STEWARDSHIP

Competencies are sets of knowledge and skills needed to deliver desired results and have been proposed as a method to enhance different healthcare professionals' roles in antimicrobial stewardship activities. Antimicrobial prescribing and stewardship competencies have been developed in the United Kingdom by the Advisory Committee on Antimicrobial Resistance and Healthcare Associated Infection (ARHAI) of the Department of Health, with the wide support of all interested professional societies. The UK competencies consist of five dimensions: infection control and prevention, antimicrobial resistance and antimicrobials, antimicrobial prescribing, antimicrobial stewardship, and monitoring and education. The competencies may be used by individuals or institutions to help improve antimicrobial prescribing. As generic competencies, they are supposed to serve as a template for the development of the competencies for various professional groups and to be included in the educational curricula of all healthcare professionals [11]. Similar competencies are currently being developed by the European Society of Clinical Microbiology and Infectious Diseases.

KEY ISSUES

- The most important determinants of antimicrobial prescribing in young doctors in training are knowledge and "prescribing etiquette"
- The majority of antimicrobial prescribers do not receive any education in antimicrobial prescribing or stewardship during their specialty training
- Many types of education intervention show some efficacy, but there is no consensus on the most effective intervention to improve antibiotic prescribing
- Education in antimicrobials stewardship for pharmacists and nurses is very limited

STEPS FORWARD

- Inclusion of teaching on antimicrobial stewardship and prescribing knowledge and skills in internship/foundation period and specialty curricula: discussion with competent authorities for specialty training on the national and local level, enhancement of the interdisciplinary approach
- Inclusion of antimicrobial stewardship and prescribing topics in CPD: discussion with professional organizations involved in CME/CPD and healthcare authorities, emphasis on the quality of care and patient safety
- Organization of education in antimicrobial stewardship for pharmacists and nurses: discussion of the topic with the providers of undergraduate and postgraduate education
- Education in antimicrobial prescribing and stewardship should draw on the principles of behavior shaping and changing

Resources for undergraduate and postgraduate training in antimicrobial prescribing and stewardship:

ESGAP OVLC collection of educational resources: http://esgap.escmid.org/?page_id=330

An Antibiotic Stewardship Curriculum for Medical Students: http://www.wakehealth.edu/AS-Curriculum/

PAUSE learning platform: http://www.pause-online.org.uk/

Stanford MOOC: *Antimicrobial Stewardship optimization of Antibiotic Practices:* https://www.coursera.org/course/antimicrobial

University of Dundee and BSAC MOOC: https://www.futurelearn.com/courses/antimicrobial-stewardship

Centers for Disease Control and Prevention: Get Smart programmes and observances: http://www.cdc.gov/getsmart/

European Centre for Disease Prevention and Control: European Antibiotic Awareness Day: www.ecdc.europa.eu/en/eaad/Pages/Home.aspx

NHS Education for Scotland: Doctors Online Training System: http://www.nes.scot.nhs.uk/education-and-training.aspx

REFERENCES

[1] Pulcini C, Gyssens IC. How to educate prescribers in antimicrobial stewardship practices. Virulence 2013;4:192–202.

[2] Pulcini C, Wencker F, Frimodt-Møller N, Kern WV, Nathwani D, Rodríguez-Baño J, *et al.* European survey on principles of prudent antibiotic prescribing teaching in undergraduate students. Clin Microbiol Infect 2015;21:354–61.

[3] Marinker M. Myth, paradox and the hidden curriculum. Med Educ 1997;31:293–8.

[4] Mattick K, Kelly N, Rees C. A window into the lives of junior doctors: narrative interviews exploring antimicrobial prescribing experiences. J Antimicrob Chemother 2014;69: 2274–83.

[5] May L, Gudger G, Armstrong P, Brooks G, Hinds P, Rahul Bhat R, *et al.* Multisite exploration of clinical decision making for antibiotic use by emergency medicine providers using quantitative and qualitative methods. Infect Control Hosp Epidemiol 2014;35: 1114–25.

[6] Srinivasan A, Song X, Richards A, Sinkowitz-Cochran R, Cardo D, Rand C. A survey of knowledge, attitudes, and beliefs of house staff physicians from various specialties concerning antimicrobial use and resistance. Arch Intern Med 2004;164:1451–6.

[7] Abbo L, Sinkowitz-Cochran R, Smith L, Ariza-Heredia E, Gomez-Marın O, Srinivasan A, *et al.* Faculty and resident physicians' attitudes, perceptions, and knowledge about antimicrobial use and resistance. Infect Control Hosp Epidemiol 2011;32:714–8.

[8] Chaves NJ, Cheng AC, Runnegar N, Kirschner J, Lee T, Buising K. Analysis of knowledge and attitude surveys to identify barriers and enablers of appropriate antimicrobial prescribing in three Australian tertiary hospitals. Int Med J 2014;44:568–74.

[9] Charani E, Castro-Sanchez E, Sevdalis N, Kyratsis Y, Drumright L, Shah N, *et al.* Understanding the determinants of antimicrobial prescribing within hospitals: the role of "Prescribing Etiquette." Clin Infect Dis 2013;57:188–96.

[10] Drekonja DM, Filice GA, Greer N, Olson A, MacDonald R, Rutks I, *et al.* Antimicrobial stewardship in outpatient settings: a systematic review. Infect Control Hosp Epidemiol 2014;36:142–52.

[11] Oredope D, Cookson B, Fry C. on behalf of the Advisory Committee on Antimicrobial Resistance and Healthcare Associated Infection Professional Education Subgroup. Developing the first national antimicrobial prescribing and stewardship competences. J Antimicrob Chemother 2014;69:2886–8.

[12] Minen MT, Duquaine D, Marx MA, Weiss D. A survey of knowledge, attitudes, and beliefs of medical students concerning antimicrobial use and resistance. Microb Drug Resist 2010;16:285–9.

[13] Abbo LM, Cosgrove SE, Pottinger PS, Pereyra M, Sinkowitz-Cochran R, Srinivasan A, *et al.* Medical students' perceptions and knowledge about antimicrobial stewardship: how are we educating our future prescribers? Clin Infect Dis 2013;57:631–8.

[14] Huang Y, Gu J, Zhang M, Ren Z, Yang W, Chen Y, *et al.* Knowledge, attitude and practice of antibiotics: a questionnaire study among 2500 Chinese students. BMC Med Educ 2013;13:163. http://dx.doi.org/10.1186/1472-6920-13-163.

[15] Dyar OJ, Pulcini C, Howard P, Nathwani D. ESGAP (ESCMID Study Group for Antibiotic Policies). European medical students: a first multicentre study of knowledge, attitudes and perceptions of antibiotic prescribing and antibiotic resistance. J Antimicrob Chemother 2014;69:842–6.

FURTHER READING

[1] Cisneros JM, Cobo J, San Juan R, Montejo M, Farinase MC. Education on antibiotic use. Education systems and activities that work. Enferm Infecc Microbiol Clin 2013;31(Suppl. 4):31–7.

[2] Ohl CA, Luther VP. Health-care provider education as a tool to enhance antibiotic stewardship practices. Infect Dis Clin North Am 2014;28:177–93.

[3] Rocha-Pereira N, Lafferty N, Nathwani D. Educating healthcare professionals in antimicrobial stewardship: can online-learning solutions help? J Antimicrob Chemother 2015;70: 3175–7.

Chapter 6

Rapid Diagnostics and Biomarkers for Antimicrobial Stewardship

José R. Paño Pardo
Hospital Clínico Universitario "Lozano Blesa", Instituto de Investigación Sanitaria Aragón, Zaragoza, Spain

INTRODUCTION

Two of the main barriers to optimal antimicrobial use are the uncertainty prescribers face when making prescribing decisions and how they tolerate it. Infections commonly have nonspecific presentations, resembling those of multiple noninfectious diseases. Frequently, neither the causative agent nor its susceptibility profile can be anticipated based on the infection's clinical presentation. This leads to a substantial number of unnecessary antibiotic treatments, to the delay of appropriate antimicrobial therapy, and to the avoidable use of broad-spectrum antibiotics that foster antimicrobial resistance.

Rapid diagnostics and biomarkers are tests that aim to minimize some of the uncertainties (which patients benefit from antibiotics, which are the causative microorganisms and their susceptibility patterns) during the antimicrobial prescribing process. Rapid diagnostic tests (RDTs) are those tests that aim to diagnose the cause of a given infection with turnaround times significantly shorter than those provided by conventional ones. A biomarker is a biological measurable parameter that can be used as surrogate of a pathological process (i.e., bacterial infection) or its course [1]. Thus, from an antimicrobial stewardship standpoint, RDT and biomarkers are tools to assist clinicians to make better antimicrobial prescribing decisions. To assist prescribers, they should be reliable and timely available during the prescribing process and easy to interpret and to use.

Antimicrobial Stewardship. http://dx.doi.org/10.1016/B978-0-12-810477-4.00006-4

RAPID DIAGNOSTIC TESTS

Why Do We Need Rapid Diagnostic Tests?

One of the main principles of antimicrobial stewardship is to appropriately "target the pathogen," with the aim of reducing unnecessary antibiotic pressure and, thus, minimizing the ecological impact of antimicrobial therapy. However, targeted antimicrobial therapy requires timely identification of the causative pathogen(s).

For over a century, microbiologists have relied on culture-based methods to identify and characterize microorganisms in the laboratory. Conventional culture-based methods are composed of several sequential steps, which, broadly, include (a) sampling, (b) inoculation and incubation in appropriate media, (c) identification, and (d) susceptibility testing. As a result, final turnaround time of conventional microbiological tests usually ranges between 2 and 7 days, depending on several biological and logistic factors. These turnaround times are too long to guide timely antimicrobial prescribing, especially in the case of severe infections like sepsis.

Novel Technologies to Speed up Microbial Identification

Recent progress in diagnostic technologies has enabled a significant reduction in time needed to identify and characterize microorganisms from several days to hours or even minutes [2]. Along this chapter, we will focus on those approaches that can already be incorporated into clinical practice, namely, nucleic acid amplification tests (NAAT), MALDI-TOF, and antigen detection.

NAAT, mostly based on polymerase chain reaction (PCR), are molecular techniques that amplify either DNA or RNA (reverse transcriptase PCR or RT-PCR) of targeted pathogens when present in biological samples. These methods are most frequently used to identify microorganisms present in biological samples, but they can also amplify antibiotic resistance encoding genes, providing valuable input on their antimicrobial susceptibility. They can be applied directly to biological samples (blood, CSF, and sputum) or on samples in which microorganisms have already been detected after incubation (e.g., positive blood culture bottles). NAAT may provide qualitative or quantitative results (e.g., real-time PCR), and they may target both single and multiple pathogens (e.g., multiplex PCR). Interestingly, they may be processed in platforms that minimize biohazard and the technical training required to operate them to the extent that in some instances, they can be run at the point of care (POC).

Ascertaining the clinical value of positive NAAT results can be challenging especially in samples from nonsterile fluids given the low level of detection of these tests. In addition, contamination from positive control material, from the normal flora of the test operators, and from the environment (i.e., nonclosed systems) can occur, and it is difficult to detect. NAAT costs vary among the different available platforms, depending on several factors (e.g., if it is a single sample test or if it is batched), but they are usually higher than those associated with conventional methods. Integration with laboratory workflow is another relevant issue that has to be addressed locally.

Today, matrix-assisted laser desorption/ionization time of flight (MALDI-TOF) has become the standard method for bacterial and fungal identification in many clinical microbiology laboratories in the developed world. Specific software rapidly (few minutes), cheaply, and accurately provides a presumptive microbial identification on the basis of the mass spectra generated from the sample obtained of a colony. Although (per test) costs of microbial identification are low, on-site access to MALDI-TOF equipment is something that may not be feasible in low-resource settings.

Microbial antigen detection in biological samples such as patients' blood or urine, mainly through immunochromatography, is another approach to rapid microbiological diagnosis. These assays are usually displayed in easy-to-use kits (dipsticks) targeting a specific pathogen that can be applied at the point of care, providing results in minutes.

Application of Rapid Diagnostic Tests in Different Clinical Scenarios

RDTs can be applied to different samples in several clinical scenarios, such as blood in the case of sepsis, respiratory specimens for community-acquired pneumonia (CAP), and cerebrospinal fluid (CSF) for central nervous system (CNS) infections. Their performance and limitations in these settings will be approached next.

Blood cultures and conventional microbiological characterization is the most common approach to etiological diagnosis in sepsis, providing turnaround times ranging from 2 to 7 days. NAAT can be directly applied to blood at the time infection is suspected (Table 1) or to broth when bacterial growth has been detected after incubation. In both instances, a multiple-pathogen testing strategy is needed. Direct NAAT or NAAT/MALDI-TOF applied on positive blood culture bottles has repeatedly proved efficacious to significantly decrease the time to etiologic diagnosis in sepsis [3–5]. Nevertheless, in the absence of a coordinated effort with the local antimicrobial stewardship program (e.g., bacteremia team), the information provided by these tests does not shorten the time to appropriate therapy, as observed in the randomized clinical trial by Banerjee *et al.* [3]. For a comprehensive summary of the studies that have assessed RDT to optimize blood cultures, see the recently published review by Banerjee *et al.* (Table 2) [25].

Pneumococcal urinary antigen test is a rapid immunochromatographic assay that is more sensitive than sputum and blood cultures (60% vs. 8.7% and 18.2%, respectively) for the diagnosis of pneumococcal CAP providing results in minutes [26]. Nevertheless, despite its high specificity, availability

TABLE 1 Available NAAT Platforms That Can Be Applied Directly on Blood Sampled in Patients With Suspected Sepsis

Assay	Technology	Volume Input (mL)	Number of Pathogens Detected (Library)	Limit of Detection (CFU)	Turnaround Time (h)
SeptiFast	Real-time PCR	1.5	19 bac/6 f	3–30	6
Iridica	PCR +electrospray ionization mass spectrometry	5	>750 bac >200 f >130 v		6
SepsiTest	16S rDNA PCR +sequencing	1–5	>300	2–460	8
Looxster/ Vyoo	PCR +electrophoresis/ microarray	5	34 b 7 f	5–100	7
Magicplex	Nested real-time PCR	1	90		6
T2Candida T2Bacteria	PCR+NMR	2	5f ?	1–3	3
Polaris/Idylla	Real-time PCR	5–10	10 b 6 f	5–10	2

PCR, polymerase chain reaction; *bac*, bacteria; *f*, fungi; *v*, virus; *CFU*, colony-forming units.

of pneumococcal urinary antigen test is not necessarily associated with optimal antimicrobial therapy, as observed in a multicenter study in France in which less than half of the patients with CAP and a positive pneumococcal urinary antigen test received targeted therapy, existing wide variations among institutions [27]. Legionella urinary antigen test also has high sensitivity and specificity for *Legionella pneumophila* serotype 1, responsible for over 80% of community-acquired infections caused by Legionella species [28,29].

As CAP can be caused by a wide array of pathogens, a multiple-pathogen NAAT (mainly multiplex PCR) significantly increases the likelihood of reaching an etiologic diagnosis as recently proved by Gadsby *et al.* [30]. Nevertheless, multiplex molecular-based assays for CAP have several limitations. First, respiratory samples are not available in all patients, and molecular tests frequently do not provide input on the antimicrobial susceptibility pattern. Second, detection of a pathogen in a respiratory sample by molecular methods can be difficult to interpret since it can be either colonizing or causing infection; although, quantitative molecular methods (real-time PCR) can help to

TABLE 2 Studies Evaluating Rapid Blood Culture Diagnostic Tests, Implementation Methods, and Clinical Outcomes

Technology	Test/Targets	Study Design	Outcomes of Rapid Test Compared With Standard Methods	Antimicrobial Stewardship Intervention	Reference
Pathogen-specific PCR	*mecA* PCR	Single center, prospective pre-/postintervention	Decreased time to optimal therapy, other outcomes not assessed	Audit and feedback by ID pharmacist 6 days/week	[6]
	Cepheid GeneXpert MRSA	Single center, pre-/postintervention with retrospective evaluation of preintervention group	Decreased time to optimal therapy, LOS, mortality, cost	Audit and feedback by ID pharmacist from Monday to Friday during the day	[7]
	Cepheid Xpert MRSA/SA	Single center, retrospective pre-/postintervention	Decreased time to optimal therapy, other outcomes not assessed	None	[8]
	mecA PCR	Single center, retrospective, pre-/postintervention	Decreased time to optimal therapy and LOS	None	[9]
	BD GeneOhm StaphSR	Multicenter, retrospective, pre-/postintervention	No impact on time to optimal therapy	None	[10]

Continued

TABLE 2 Studies Evaluating Rapid Blood Culture Diagnostic Tests, Implementation Methods, and Clinical Outcomes—Cont'd

Technology	Test/Targets	Study Design	Outcomes of Rapid Test Compared With Standard Methods	Antimicrobial Stewardship Intervention	Reference
Panel-based PCR	BioFire blood culture identification test	Single center, prospective, randomized, controlled trial	Decreased time to optimal therapy. No differences in LOS, mortality, adverse events, cost	Audit and feedback by ID pharmacist or physician 24/7; treatment guidance comments included in microbiology result report	[3]
	BioFire blood culture identification test; *Enterococcus* and *Staphylococcus* species, excluding coagulase-negative staphylococci considered true bacteremias	Single center, pre-/postintervention with historical preintervention group	Decreased time to optimal therapy, mortality, cost, LOS	Rapid testing and audit and feedback by ID pharmacists both performed once daily from Monday to Friday daytime	[11]
Panel-based nucleic acid microarray	Verigene BC-GP	Single center, retrospective, pre-/postintervention	Decreased time to optimal therapy, LOS, cost	Audit and feedback by ID pharmacist from Monday to Friday daytime	[12]
	Verigene BC-GP	Single center, pre-/postintervention with retrospective evaluation of preintervention group	Decreased time to optimal therapy, mortality, cost	ID physicians involved in all patients with positive BCs on weekdays	[13]

	Verigene BC-GP	Multicenter, pre/postintervention, with retrospective evaluation of preintervention group	Decreased time to optimal therapy, LOS, and cost	Audit and feedback by ID pharmacist during daytime	[14]
	Verigene BC-GP	Single center, retrospective, pre-/postintervention	Decreased time to optimal therapy. No differences in LOS, mortality	Treatment guidance comments included in microbiology result report	[15]
	Verigene BC-GN	Single center, retrospective, pre-/postintervention	Decreased time to optimal therapy for ESBLs, LOS, mortality, cost	Audit and feedback by ID pharmacist daily	[16]
PNA-FISH	*Candida albicans*	Single center, retrospective, pre-/postintervention	Decreased time to optimal therapy, cost	Audit and feedback by stewardship team once daily when rapid test results available	[17]
	S. aureus	Single center, retrospective, pre-/postintervention	Decreased LOS, cost	Audit and feedback by stewardship team once daily when rapid test results available	[17]
	Enterococcus species	Single center, pre/postintervention with retrospective evaluation of preintervention group	Decreased time to optimal therapy, mortality	Audit and feedback by stewardship team seven days/week	[18]

Continued

TABLE 2 Studies Evaluating Rapid Blood Culture Diagnostic Tests, Implementation Methods, and Clinical Outcomes—Cont'd

Technology	Test/Targets	Study Design	Outcomes of Rapid Test Compared With Standard Methods	Antimicrobial Stewardship Intervention	Reference
	S. aureus	Single center, retrospective, pre-/postintervention	No difference in time to optimal therapy or LOS	None	[19]
	Yeast Traffic Light (five *Candida* species)	Single center, pre/postintervention with retrospective evaluation of preintervention group	Decreased time to optimal therapy, cost, faster rate of culture clearance	Audit and feedback by ID pharmacist seven days/week from 7 AM to 9:30 PM	[20]
MALDI-TOF mass spectrometry	Gram-negative bacilli	Single center, pre-/postintervention with retrospective evaluation of preintervention group	Decreased time to optimal therapy, LOS, mortality	Audit and feedback by ID pharmacist 24/7	[21]
	Resistant Gram-negative bacilli	Single center, pre-/postintervention with retrospective evaluation of preintervention group	Decreased time to optimal therapy, LOS, mortality	Audit and feedback by ID pharmacist 24/7	[22]

		Single center, pre-/postintervention with historical preintervention group	Decreased time to optimal therapy, LOS, mortality, recurrent bacteremia	Real-time audit and feedback by stewardship team. Stewardship team notifications through electronic pages from 6 AM to 11:30 PM and email 24 h/day	[4]
	Gram-negative bacilli	Single center, prospective observational, no control group	Decreased time to optimal therapy, other outcomes not assessed	All included subjects had infectious diseases consultation	[23]
	S. aureus (MALDI-ToF mass spectrometry vs. MALDI-TOF mass spectrometry plus Gene Xpert MRSA)	Single center, prospective, randomized, not blinded	Decreased time to optimal therapy. Other outcomes not assessed	None	[23]
	Bacteria and yeast	Single center, pre-/postintervention with retrospective evaluation of preintervention group	Decreased time to optimal therapy. No differences in LOS, mortality, readmission	Audit and feedback by ID pharmacist from Monday to Friday, from 8 AM to 4 PM	[24]

LOS, length of stay; *PNA-FISH,* peptide nucleic acid fluorescence in situ hybridization; *ESBL,* extended-spectrum β-lactamase; *ID,* infectious diseases.

address this issue. Finally, the impact of new diagnostics for CAP on clinical decisions and patient outcomes has not been evaluated yet.

Meningitis and encephalitis can also be caused by a wide array of pathogens, and especially, these types of patients frequently receive prolonged antibiotic combinations. Therefore, rapid etiologic diagnosis for these syndromes could really help to reduce inappropriate antibiotic treatment and improve patient outcomes. Multiple-pathogen detection in cerebrospinal fluid (CSF) is feasible with molecular-based methods (multiplex PCR) [31]. Unfortunately, as yet, these panels do not include all the potential causative pathogens, and there is not enough information about its performance. As a result, antimicrobial therapy should be continued in patients with high suspicion of bacterial or fungal meningitis despite a negative test result, until results of conventional tests are available.

Selection and Implementation of Rapid Diagnostic Tests at a Local Level

The availability of rapid diagnostics does not necessarily lead to improvements in antimicrobial prescribing since knowledge (microbiological input) does not equal behavior (prescribing). To translate rapid diagnostics into better prescribing, it is necessary to take into account several issues at a local level.

As the number of available rapid tests is growing rapidly, selecting which rapid test to implement locally is critical. Local priorities in regards to antimicrobial prescribing (e.g., severe sepsis, septic shock, pneumonia, and meningitis) and the prevalence of pathogens and their associated mechanisms of resistance have to be considered. How the RDT will integrate with laboratory workflow (e.g., whether the test needs validation by a microbiologist, whether it is done 24/7 or it is batched…) is also of paramount relevance when selecting an RDT.

RDTs by themselves have a variable impact on antimicrobial prescribing. This is why the involvement of the antimicrobial management team (AMT) is critical. AMT may contribute to educate medical staff on the indications and performance of the new test and in choosing the appropriate antimicrobial therapy according to RDT results. Communication of novel diagnostic tests results to prescribing physicians, and also AMT is essential. Assessing the number of orders of the new RDT and its level of appropriate therapy, the time to effective therapy, and clinical outcomes (e.g., mortality and time to discharge) is essential, and AMT may be of great help in this task.

BIOMARKERS

The Ideal Biomarker for Clinical Use in Infectious Diseases

In the field of infectious diseases, biomarkers are biological surrogates of infection or predictors of certain outcomes. Although the term biomarker most frequently refers to biochemical indicators, they can also be anatomical (e.g., size of a vegetation), physiological (e.g., persistent fever), and genetic. Indeed, the identification of genetic biomarkers is promising as shown by Sweeney *et al.*, who assessed the expression of a set of 11 genes using gene expression arrays to differentiate bacterial from viral infections with very high performance [32].

The utility of biomarkers with the purpose of antimicrobial stewardship is to assist clinicians in prescribing decisions in a context of clinical uncertainty. Some relevant prescribing decisions in which biomarkers can assist are whether to initiate empirical antimicrobial therapy (helping to differentiate infectious from noninfectious presentations or bacterial from viral infections) or when to safely stop antimicrobial therapy. In order to be useful for this purpose, biomarkers must have a good diagnostic or prognostic performance. Biomarker diagnostic and prognostic performance obviously depend on their sensitivity and specificity. It is very importantly also on the pretest probability of disease (diagnosis) or outcome (prognosis), which necessarily has to be taken into account by clinicians on a patient-by-patient basis. In other words, biomarker performance greatly depends in the context in which they are used. In addition to their diagnostic and prognostic performance, other features that increase biomarker value are acceptability to patients, availability to clinicians, ease of interpretation, short turnaround time, and acceptable cost.

There are dozens of potentially useful biomarkers (Table 3), but only very few of them have been tested systematically and/or are available in clinical practice. In this chapter, we will focus on the two most broadly studied biomarkers, C-reactive protein (CRP) and procalcitonin (PCT).

C-Reactive Protein

CRP is an acute-phase protein of hepatic origin whose levels rise in response to acute tissue injury, regardless of the etiology (infection, trauma, and inflammation). Nevertheless, in the appropriate clinical setting (i.e., low risk

TABLE 3 Potentially Useful Biomarkers for Infectious Diseases

Biomarker
• White blood cell count • Eosinophil count (inverse) • Fibrinogen • Erythrocyte sedimentation rate (ESR) • C-reactive protein (CRP) • Procalcitonin (PCT) • Interleukin-6 • sTREM-1 • Soluble urokinase-type plasminogen activator receptor (suPAR) • Proadrenomedullin (pro-ADM) • Presepsin

of severe bacterial infection and antibiotic overuse), CRP can be used as a biomarker of infection.

The setting in which the role of CRP has been more extensively explored are acute respiratory infections, for which there is significant antibiotic overuse. CRP is the only one biomarker in infectious diseases that is available as a POC test, and the available evidence for its role guiding the initiation of antimicrobial therapy for acute respiratory tract infections (ARTI) in primary care was recently assessed in a Cochrane review [33]. Six studies, three randomized clinical trials and three cluster randomized clinical trials, were included. Three of them included upper (URTI) and lower respiratory tract infections (LRTI) and three only included LRTI. A pooled analysis (3284 participants) showed a reduction in the use of antibiotics among patients assigned to CRP (631/1685; 37.4%) as compared with those assigned to standard care (785/1599; 49%) with a pooled RR of 0.78 (IC95 0.66–0.92). Nevertheless, these results need to be interpreted with caution since a high level of heterogeneity was observed, indicating that the pooled studies included different populations and distinct effects were evaluated. No deaths were observed in any of the trials and no significant differences in the patient outcomes assessed (number of patients needing a reconsultation in the following 28 days, days to symptom resolution and patient satisfaction).

Procalcitonin

Procalcitonin (PCT), the prohormone of calcitonin, is a peptide that is released into the bloodstream in response to bacterial toxins, leading to rapidly elevated serum levels in patients with bacterial infections, but it has a higher cost and is not available as a POC test.

A systematic review on the available evidence (14 trials with 4221 patients) about the use of procalcitonin to decide whether to initiate antimicrobial therapy in patients with ARTI in different clinical settings showed a significantly reduced duration of antibiotic exposure (median from 8 to 4 days) across all the different clinical settings and diagnoses without significant differences in patient outcomes (death or treatment failure) [34]. In many of these studies, the observed reduced antimicrobial use occurred despite the fact that the suggested algorithm was frequently overruled, mainly in regards to initiation of antimicrobial therapy in the intensive care unit (ICU), underscoring the role of PCT as a guide rather than as a rule.

A recently published large randomized clinical trial in ICUs of 15 Dutch hospitals, including 1546 patients in which antibiotics have already been started on the basis of a suspected infection, explored the role of a PCT-based algorithm to guide duration of antimicrobial therapy. Median consumption of antibiotics (7.5 DDD vs. 9.3 DDD), median duration of treatment (5 days vs. 7 days), and 28-day mortality rate in the intention to treat (20% vs. 25%) and in the per protocol (20% vs. 27%) analysis were lower in the procalcitonin group than in the control group with only conventional diagnostics.

Compliance with the procalcitonin protocol, in the first 48 h after the stop threshold was reached, was 97% [35].

The role of CRP to identify patients that would benefit from antibiotics in the emergency department (ED) or in the inpatient setting has been less extensively studied than that of procalcitonin. Nevertheless, in the only randomized trial comparing a CRP and a procalcitonin-guided strategy in patients with severe sepsis or septic shock, no significant differences were found in antibiotic exposure or patient outcomes [36].

Implementation of a Biomarker-Based Strategy to Guide Antimicrobial Therapy

CRP and PCT can be used as an aid to identify patients requiring antibiotics and a monitor to determine the need for continuation. Nevertheless, both are far from being perfect biomarkers and have to be used to minimize clinical uncertainty in a holistic way, in combination with all other relevant available input. Thus, it is essential to consider the pretest probability of infection and patients' individual circumstances, such as the existence of any form of immune compromise and the source of infection. In order to improve the impact of biomarkers, mainly CRP and PCT, on antimicrobial prescribing, practicing clinicians must have an understanding of their basic biology, clinical utility, and limitations and learn that as of yet, no biomarker can substitute judicious clinical assessment. Educating clinicians on whom, when, and how to use biomarkers clearly falls within the scope of the activities and interventions of antimicrobial stewardship teams [6].

CONCLUSIONS

RDT and biomarkers (mainly CRP and PCT) are tools that can assist clinicians to better assess patients with suspected or confirmed infections and can be of help to improve antimicrobial prescribing (increasing the opportunities and decreasing time to targeted therapy, as well as helping to guide initiation and cessation of therapy) both in the inpatient and in the outpatient setting. In order to be useful, they have to be contextualized and interpreted carefully, considering all other relevant input. To augment their impact on antimicrobial prescribing to reduce inappropriate antibiotic therapy, their implementation has to be adapted to local circumstances, and AMT should be involved.

REFERENCES

[1] Dupuy A-M, Philippart F, Péan Y, Lasocki S, Charles PE, Chalumeau M, *et al.* Role of biomarkers in the management of antibiotic therapy: an expert panel review: I—currently available biomarkers for clinical use in acute infections. Ann Intensive Care 2013;3(1):22.

[2] Bauer KA, Perez KK, Forrest GN, Goff DA. Review of rapid diagnostic tests used by antimicrobial stewardship programs. Clin Infect Dis 2014;59(Suppl 3):S134–45.

[3] Banerjee R, Teng CB, Cunningham SA, Ihde SM, Steckelberg JM, Moriarty JP, *et al.* Randomized trial of rapid multiplex polymerase chain reaction-based blood culture identification and susceptibility testing. Clin Infect Dis 2015;61:1071–80.
[4] Huang AM, Newton D, Kunapuli A, Gandhi TN, Washer LL, Isip J, *et al.* Impact of rapid organism identification via matrix-assisted laser desorption/ionization time-of-flight combined with antimicrobial stewardship team intervention in adult patients with bacteremia and candidemia. Clin Infect Dis 2013;57(9):1237–45.
[5] Vincent JL, Brealey D, Libert N, Abidi NE, O'Dwyer M, Zacharowski K, *et al.* Rapid diagnosis of infection in the critically ill, a multicenter study of molecular detection in bloodstream infections, pneumonia, and sterile site infections. Crit Care Med 2015;43(11):2283–91.
[6] Carver PL, Lin SW, DePestel DD, Newton DW. Impact of *mecA* gene testing and intervention by infectious disease clinical pharmacists on time to optimal antimicrobial therapy for *Staphylococcus aureus* bacteremia at a University Hospital. J Clin Microbiol 2008;46(7):2381–3.
[7] Bauer KA, West JE, Balada-Llasat JM, Pancholi P, Stevenson KB, Goff DA. An antimicrobial stewardship program's impact with rapid polymerase chain reaction methicillin-resistant *Staphylococcus aureus/S. aureus* blood culture test in patients with *S. aureus* bacteremia. Clin Infect Dis 2010;51(9):1074–80.
[8] Parta M, Goebel M, Thomas J, Matloobi M, Stager C, Musher DM. Impact of an assay that enables rapid determination of *Staphylococcus* species and their drug susceptibility on the treatment of patients with positive blood culture results. Infect Control Hosp Epidemiol 2010;31(10):1043–8.
[9] Nguyen DT, Yeh E, Perry S, *et al.* Real-time PCR testing for *mecA* reduces vancomycin usage and length of hospitalization for patients infected with methicillin-sensitive staphylococci. J Clin Microbiol 2010;48(3):785–90.
[10] Frye AM, Baker CA, Rustvold DL, *et al.* Clinical impact of a real-time PCR assay for rapid identification of staphylococcal bacteremia. J Clin Microbiol 2012;50(1):127–33.
[11] Pardo J, Klinker KP, Borgert SJ, Butler BM, Giglio PG, Rand KH. Clinical and economic impact of antimicrobial stewardship interventions with the FilmArray blood culture identification panel. Diagn Microbiol Infect Dis 2016;84(2):159–64.
[12] Sango A, McCarter YS, Johnson D, Ferreira J, Guzman N, Jankowski CA. Stewardship approach for optimizing antimicrobial therapy through use of a rapid microarray assay on blood cultures positive for Enterococcus species. J Clin Microbiol 2013;51(12):4008–11.
[13] Suzuki H, Hitomi S, Yaguchi Y, *et al.* Prospective intervention study with a microarray-based, multiplexed, automated molecular diagnosis instrument (Verigene system) for the rapid diagnosis of bloodstream infections, and its impact on the clinical outcomes. J Infect Chemother 2015;21(12):849–56.
[14] Box MJ, Sullivan EL, Ortwine KN, *et al.* Outcomes of rapid identification for gram-positive bacteremia in combination with antibiotic stewardship at a community-based hospital system. Pharmacotherapy 2015;35(3):269–76.
[15] Beal SG, Thomas C, Dhiman N, *et al.* Antibiotic utilization improvement with the Nanosphere Verigene Gram-Positive Blood Culture assay. Proc (Bayl Univ Med Cent) 2015;28(2):139–43.
[16] Walker T, Dumadag S, Lee CJ, *et al.* Clinical impact of laboratory implementation of Verigene BC-GN microarray-based assay for detection of Gram-negative bacteria in positive blood cultures. J Clin Microbiol 2016;54(7):1789–96.

[17] Forrest GN, Mankes K, Jabra-Rizk MA, *et al.* Peptide nucleic acid fluorescence in situ hybridization-based identification of Candida albicans and its impact on mortality and antifungal therapy costs. J Clin Microbiol 2006;44(9):3381–3.

[18] Forrest GN, Roghmann MC, Toombs LS, *et al.* Peptide nucleic acid fluorescent in situ hybridization for hospital-acquired enterococcal bacteremia: delivering earlier effective antimicrobial therapy. Antimicrob Agents Chemother 2008;52(10):3558–63.

[19] Holtzman C, Whitney D, Barlam T, Miller NS. Assessment of impact of peptide nucleic acid fluorescence in situ hybridization for rapid identification of coagulase-negative staphylococci in the absence of antimicrobial stewardship intervention. J Clin Microbiol 2011;49(4):1581–2.

[20] Heil EL, Daniels LM, Long DM, Rodino KG, Weber DJ, Miller MB. Impact of a rapid peptide nucleic acid fluorescence in situ hybridization assay on treatment of *Candida* infections. Am J Health Syst Pharm 2012;69(21):1910–4.

[21] Perez KK, Olsen RJ, Musick WL, *et al.* Integrating rapid pathogen identification and antimicrobial stewardship significantly decreases hospital costs. Arch Pathol Lab Med 2013;137(9):1247–54.

[22] Perez KK, Olsen RJ, Musick WL, *et al.* Integrating rapid diagnostics and antimicrobial stewardship improves outcomes in patients with antibiotic-resistant Gram-negative bacteremia. J Infect 2014;69(3):216–25.

[23] Clerc O, Prod'hom G, Vogne C, Bizzini A, Calandra T, Greub G. Impact of matrix-assisted laser desorption ionization time-of-flight mass spectrometry on the clinical management of patients with Gram-negative bacteremia: a prospective observational study. Clin Infect Dis 2013;56(8):1101–7.

[24] Malcolmson C, Ng K, Hughes S, *et al.* Impact of matrix-assisted laser desorption and ionization time-of-flight and antimicrobial stewardship intervention on treatment of bloodstream infections in hospitalized children. J Pediatric Infect Dis Soc 2016.

[25] Banerjee R, Özenci V, Patel R. Individualized approaches are needed for optimized blood cultures. Clin Infect Dis 2016;63:1332–9.

[26] Molinos L, Zalacain R, Menendez R, Reyes S, Capelastegui A, Cillóniz C, *et al.* Sensitivity, specificity, and positivity predictors of the pneumococcal urinary antigen test in community-acquired pneumonia. Ann Am Thorac Soc 2015;12(10):1482–9.

[27] Blanc V, Mothes A, Smetz A, Timontin I, Guardia MD, Billiemaz A, *et al.* Severe community-acquired pneumonia and positive urinary antigen test for S. pneumoniae: amoxicillin is associated with a favourable outcome. Eur J Clin Microbiol Infect Dis 2015;34:2455–61.

[28] Helbig JH, Uldum SA, Bernander S, Luck PC, Wewalka G, Abraham B, *et al.* Clinical utility of urinary antigen detection for diagnosis of community-acquired, travel-associated, and nosocomial Legionnaires' disease. J Clin Microbiol 2003;41(2):838–40.

[29] Yu VL, Plouffe JF, Pastoris MC, Stout JE, Schousboe M, Widmer A, *et al.* Distribution of Legionella species and serogroups isolated by culture in patients with sporadic community-acquired legionellosis: an international collaborative survey. J Infect Dis 2002;186(1):127–8.

[30] Gadsby NJ, Russell CD, McHugh MP, Mark H, Conway Morris A, Laurenson IF, *et al.* Comprehensive molecular testing for respiratory pathogens in community-acquired pneumonia. Clin Infect Dis 2016;62(7):817–23.

[31] Leber AL, Everhart K, Balada-Llasat J-M, Cullison J, Daly J, Holt S, *et al.* Multicenter evaluation of the BioFire FilmArray meningitis encephalitis panel for the detection of bacteria, viruses and yeast in cerebrospinal fluid specimens. J Clin Microbiol 2016;54:2251–61.

[32] Sweeney TE, Wong HR, Khatri P. Robust classification of bacterial and viral infections via integrated host gene expression diagnostics. Sci Transl Med 2016;8(346):346ra91.

[33] Aabenhus R, Jensen J-US, Jørgensen KJ, Hróbjartsson A, Bjerrum L. Biomarkers as point-of-care tests to guide prescription of antibiotics in patients with acute respiratory infections in primary care. Cochrane Database Syst Rev 2014;11:CD010130.

[34] Schuetz P, Müller B, Christ-Crain M, Stolz D, Tamm M, Bouadma L, *et al.* Procalcitonin to initiate or discontinue antibiotics in acute respiratory tract infections. Cochrane Database Syst Rev 2012;9:CD007498.

[35] de Jong E, van Oers JA, Beishuizen A, Vos P, Vermeijden WJ, Haas LE, *et al.* Efficacy and safety of procalcitonin guidance in reducing the duration of antibiotic treatment in critically ill patients: a randomised, controlled, open-label trial. Lancet Infect Dis 2016;16:819–27.

[36] Oliveira CF, Botoni FA, Oliveira CRA, Silva CB, Pereira HA, Serufo JC, *et al.* Procalcitonin versus C-reactive protein for guiding antibiotic therapy in sepsis: a randomized trial. Crit Care Med 2013;41(10):2336–43.

Chapter 7

Pharmacokinetic and Pharmacodynamic Tools to Increase Efficacy

Mahipal G. Sinnollareddy*,, Menino O. Cotta‡,§ and Jason A. Roberts‡,§**

*University of South Australia, Adelaide, SA, Australia

**The Canberra Hospital, Canberra, ACT, Australia

‡The University of Queensland, Brisbane, QLD, Australia

§Royal Brisbane and Women's Hospital, Brisbane, QLD, Australia

INTRODUCTION

Antimicrobial resistance is a significant health issue worldwide. Globally, there is an increase in microbial resistance to existing antimicrobial agents, and at the same time, there are dwindling numbers of novel antimicrobial agents. This has led to the development of strategies by various government agencies across the world in order to improve use of current antimicrobials, reduce the emergence of resistance, and encourage the development of new agents. Antimicrobial stewardship (AMS) is one of the strategies widely adapted to optimize antimicrobial use. AMS is a systematic approach that includes not only limiting inappropriate use but also promotion of optimized antimicrobial selection, dosing, route, and duration of therapy [1,2]. In regards to antimicrobial dosing, exposure-response and exposure-resistance relationships play an important role in order to maximize clinical/microbiological cure and minimize resistance.

Over the last three decades, significant progress has been made in elucidating exposure-response relationships of antimicrobials [3,4]. Typically, the magnitude of antimicrobial exposure (concentrations) required for clinical effectiveness in humans is similar to that described in animal infection models [4]. However, dose-response relationships cannot be extrapolated to humans from these models primarily due to differences in pharmacokinetics (PK). These differences demand a thorough understanding of PK in humans in order to choose appropriate doses that achieve optimal exposures. Furthermore, it is

Antimicrobial Stewardship. http://dx.doi.org/10.1016/B978-0-12-810477-4.00007-6

essential to understand variations in PK in certain patient groups (e.g., neonates/pediatrics, neutropenic patients, burn patients, trauma patients, and the critically ill) and infections in sanctuary sites (e.g., lung infections, central nervous system infections, orthopedic infections, and endocarditis) owing to the physiological changes that could influence PK and subsequently dose-response relationships [5–7].

Another important factor in determining target exposures is pharmacodynamics (PD). The minimum inhibitory concentration (MIC) of an antimicrobial for a given pathogen is an important component for quantification of antimicrobial PD. For pathogens with decreased susceptibility (i.e., increased MICs), higher exposures are most likely required to achieve optimal outcomes. This chapter provides an overview of the PK/PD principles, strategies for dose optimization in difficult-to-treat patient groups, use of therapeutic drug monitoring (TDM), and approaches to optimize dosing strategies using PK/PD tools.

PK/PD OF ANTIMICROBIALS: EFFICACY AND RESISTANCE

PK defines the relationship between the dose of an antimicrobial administered and the observed concentration over a period of time in blood and other body fluids or tissues of interest. Maximum concentration (C_{max}), minimum concentration (C_{min}), and overall exposure, that is, area under the concentration-time curve over 24 h (AUC_{0-24}), are widely used in defining the exposure of antimicrobials. PD describes the relationship between the observed concentration and the consequent effect, that is, microbial killing or growth inhibition for antimicrobials. It follows that concentration is the key variable that links PK and PD for a given antimicrobial agent being used to treat an infection caused by a particular pathogen. Antimicrobial efficacy is broadly categorized according to three different indices:

1. Time-dependent antimicrobials—efficacy is best described as the time for which the free or unbound concentration of the drug is maintained above the MIC for a pathogen ($fT_{>MIC}$).
2. Concentration-dependent antimicrobials—ratio of C_{max} to MIC is the predictive index for efficacy (C_{max}/MIC).
3. Concentration-dependent with time-dependent antimicrobials—for these antimicrobials, ratio of the exposure (AUC_{0-24}) of free drug to the MIC is the best predictive index associated with efficacy ($fAUC_{0-24}$/MIC).

Table 1 describes the PK/PD indices, expected PK changes in critically ill, and dosing strategies for commonly used antibiotics and antibiotic classes. The denominator in all PK/PD indices is MIC and hence defines exposure required to achieve numerator (PK) to attain target exposure. Clinicians will most likely face challenges to obtain the PK targets in the presence of pathogens with reduced susceptibility. There is little a clinician can do to

TABLE 1 PK/PD Indices, Expected PK Changes in Critically Ill, and Dosing Strategies for Commonly Used Antibiotic Classes

	PK/PD Target[a]	PK Changes in Critically Ill		Loading Dose Required	Dose-Optimization Strategies
		Increased Vd	*Augmented renal clearance*		
Beta-lactam antibiotics	$fT_{>MIC} = 100\%$	Yes	Yes[b]	No	Adjust based on renal function Critically ill—prolonged infusions, TDM-based approach
Vancomycin (teicoplanin)	Total $AUC_{0-24}/MIC > 400$ $C_{min} > 20$ mg/L	Yes	Yes[b]	Yes	Adjust based on renal function using TDM approach Critically ill—higher loading doses required
Aminoglycosides	C_{max}/MIC 8–10	Yes	Yes[c]	Yes	Adjust based on renal function using TDM approach Critically ill—higher doses required
Fluoroquinolones	Gram-negative bacteria total $AUC_{0-24}/MIC > 125$ Gram-positive bacteria $fAUC_{0-24}/MIC > 30$	No	Yes[d]	No	Dose reduction may not be necessary in renal impairment
Oxazolidinones (linezolid)	Total AUC_{0-24}/MIC of 80–120 and 100% $T_{>MIC}$	No	Yes[d]	No	Dose reduction is not necessary in renal impairment Critically ill—TDS or continuous infusion may be required TDM could be useful

Continued

TABLE 1 PK/PD Indices, Expected PK Changes in Critically Ill, and Dosing Strategies for Commonly Used Antibiotic Classes—Cont'd

	PK/PD Target	PK Changes in Critically Ill		Loading Dose Required	Dose-Optimization Strategies
		Increased Vd	*Augmented renal clearance*		
Lipopeptides (daptomycin)	Total AUC_{0-24}/MIC 788–1460	Yes	Yes[b]	No	Adjust based on renal function and toxic effects Critically ill—may require loading doses and higher maintenance doses
Polymyxins (colistin)	$f\,AUC_{0-24}$/MIC 37–46[e]	Yes	Unclear[d]	Yes	Adjust based on renal function and nephrotoxicity
Glycylcyclines (tigecycline)	AUC/MIC >6[f]	No	No	Yes	Higher loading and maintenance doses

[a] *It is important to highlight that the magnitude of PK/PD targets could be different based on type of pathogen for a particular antimicrobial agent, for example, fluoroquinolones.*
[b] *Demonstrated.*
[c] *Likely.*
[d] *Poorly understood.*
[e] *No clinical data.*
[f] *For intra-abdominal infections.*
It should also be noted that the magnitude of targets required for efficacy and suppression of resistance are different for antimicrobials.
TDS, three times daily administration; *TDM*, therapeutic drug monitoring.

modify PD (i.e., MIC) in the clinical setting; however, dosing can be altered or adjusted to account for altered PK to achieve the required PK/PD target index.

DOSE OPTIMISATION STRATEGIES

Dosing regimens extrapolated from dose-finding studies in noncritically ill patients do not account for the altered PK behavior observed in some patients such as the critically ill, burn patients, and neutropenic adult/pediatric patients. These patients tend to have physiological changes that alter PK behavior including increases in drug volume of distribution and clearance (high clearance associated with augmented renal clearance and low clearance associated with acute kidney injury). Consequently, dosing practices that do not account for these altered PK may reduce the likelihood of attaining PK/PD targets and, possibly, clinical outcomes. To improve the probability of target attainment, two main approaches could be utilized in the clinical setting—alternative dosing strategies and/or a TDM-based approach to dose adaptation.

Alternative Dosing Strategies

Based on the various antimicrobial PK/PD characteristics, alternative dosing strategies have been proposed in numerous PK studies and associated dosing simulations to improve the PK/PD target attainment.

As an example, high-dose extended-interval (once daily dosing in normal renal function) dosing strategy has been investigated extensively to maximize the concentration-dependent killing and the postantibiotic effect of aminoglycosides. Numerous meta-analyses have been conducted with mixed results. Two meta-analyses [8,9] have shown statistically significant difference in clinical efficacy with high-dose extended-interval compared with traditional dosing, whereas another analysis in adults [10] and one in children have shown no difference in clinical efficacy [11]. One meta-analysis in adults has suggested that high-dose extended-interval strategy might reduce nephrotoxicity [10], whereas other meta-analyses have shown nonsignificant trends or no difference in nephrotoxicity outcomes. None of the meta-analyses have shown any significant difference in ototoxicity between the two modalities. The reasons for such disparities in the analyses could be due to significant heterogeneity among the trials included, administration of additional antibiotics reflecting contemporary clinical practice, and problems associated with extracting and combining data during the analyses. Overall, high-dose extended-interval aminoglycoside dosing appears to provide improved efficacy and safety compared with traditional dosing. Moreover, this approach simplifies administration and may potentially minimize cost. As such, high-dose extended-interval dosing of aminoglycosides is considered standard approach in the clinical setting and particularly useful in patients with severe infections and changes in fluid balance [12]. Where there are changes in renal function that is associated with altered aminoglycoside clearance, high-dose extended-interval is still appropriate, but the timing of the redosing changes depending on the renal function (e.g., 36-, 48- or 72-h dosing may be required).

For the glycopeptide vancomycin, the use of a 24-h continuous infusion has been proposed as a strategy to increase the frequency of PK/PD target attainment because of its predominantly time-dependent kill characteristics. Continuous infusion has not been shown to be clinically superior to intermittent bolus dosing but has been associated with lower risk of nephrotoxicity based on a meta-analysis [13]. These findings were confirmed in a recent meta-analysis, which also included a recent randomized controlled trial (RCT) [14,15]. A vancomycin steady-state concentration of 20–25 mg/L for continuous infusion appears to result in a similar AUC_{0-24}/MIC as the intermittent infusion trough concentrations of 15–20 mg/L and are the commonly employed therapeutic targets in the clinical setting [13]. It should be noted that a loading dose is strongly recommended to rapidly achieve target concentrations. A loading dose of 30–35 mg/kg and a maintenance dose of at least 35 mg/kg/day have been suggested in critically ill patients with creatinine clearances of 100 mL/min/1.73 m^2 to attain a target trough concentration of 20 mg/L [16]. Teicoplanin has some similarities to vancomycin, although the clinical PK/PD target has been defined as a trough concentration >20 mg/L assuming that infections in critically ill patients are severe infections [17]. Loading doses are also highly important for optimal dosing of teicoplanin [17,18].

In accordance with the PK/PD index associated with efficacy, using continuous infusion when dosing linezolid has been proposed to optimize target attainment in critically ill patients with sepsis and ventilator-associated pneumonia [19–21]. Use of a continuous infusion has consistently been associated with attainment of significantly higher AUC_{0-24}/MIC and steady-state concentrations. However, there is a lack of clinical outcome information with continuous infusion of linezolid in comparison with intermittent infusion. Additionally, in order to minimize delay in reaching therapeutic concentrations when dosing linezolid as a continuous infusion, an initial loading dose is most likely required [19–21]. There is no data to usurp intermittent dosing of linezolid, typically 600 mg 12-hourly assuming normal renal function, although data highlighting the role of TDM for maximizing achievement of concentrations associated with efficacy [22] suggest that TDM may become more widespread in the future.

Beta-lactams exhibit wide variations in PK among critically ill patients [23,24]. Prolonged infusions (PI) such as by continuous infusion or extended infusion (EI—infusing the dose over 3–4 h rather than a traditional 30-min infusion) have been proposed as a way to maximize PK/PD target attainment while minimizing the PK variations [25]. Earlier meta-analyses have not shown any difference in clinical cure or mortality between PI and intermittent dosing [26–28]. Several issues need to be highlighted with the RCTs included in those meta-analyses. Disease severity was generally low with mortality rates far lower than those reported in the critical care literature [26]; most of the bacterial isolates were highly susceptible, and the use of higher doses in the bolus dosing group [29–32] may have masked any potential benefits of continuous infusion.

In the last few years, more robust RCTs have been conducted to elucidate the clinical benefit of administering beta-lactam antibiotics (predominantly meropenem and piperacillin-tazobactam) as a continuous infusion in the management of severe sepsis [33–35]. Infection with highly susceptible pathogens, inclusion of patients with noninfectious diagnoses, and inclusion of patients receiving renal replacement therapies in these RCTs might have reduced the potential benefits of continuous infusion. A recent open-label RCT that did not include patients receiving renal replacement therapy and including a third of patients with isolated pathogens with high MICs (i.e., *P. aeruginosa* or *A. baumannii*) demonstrated a higher clinical cure rate and reduced number of ventilator-free days [35]. These findings were echoed in a more recent meta-analysis that included the three RCTs involving patients with severe sepsis and demonstrated, for the first time, survival benefit as well [36]. Collectively, these results support the conduct of a definitive multicenter phase 3 RCT involving specific subsets of critically ill patients such as those infected with pathogens with higher MICs, particular infection sites (e.g., pneumonia), and normal renal function.

Standard dosing of daptomycin (i.e., 6 mg/kg 24-hourly) has resulted in less than optimal exposures in a cohort of critically ill patients owing to increased volume of distribution and augmented clearance [37]. Higher doses (i.e., 8–12 mg/kg 24-hourly) or equivalent fixed doses of 750 mg or 1000 mg have a higher probability of achieving PK/PD targets among sepsis patients with MRSA bacteremia [37]. It is likely that critically ill patients would require doses $\geq$8 mg/kg/day (or $\geq$750 mg/day) to achieve effective exposures [37]. However, the clinical benefit of such dosing strategies needs to be tested via RCTs.

For tigecycline, there is increasing plausibility that a higher $fAUC_{0\text{–}24}$/MIC may be necessary to optimize efficacy in patients with hospital-acquired pneumonia [38]. In an observational study, there was higher mortality in patients treated with standard dosing of tigecycline (i.e., 100 mg loading dose followed by 50 mg twice daily) compared with high dose (i.e., 200 g loading dose followed by 100 mg twice daily). Coupled with this, there is emerging evidence that the high-dose regimen is associated with improved clinical cure when used in combination therapy with colistin for multidrug-resistant Gram-negative bacterial infections in critically ill patients [38,39].

Colistin, a polymyxin antibiotic, is increasingly being used as the last-line treatment against infections caused by multidrug-resistant Gram-negative bacteria. It is administered parenterally and by inhalation as its inactive prodrug colistin methane sulfonate. Loading-dose strategies (6–9 million IU of colistin methane sulfonate) have been employed because of delay (day 2 or 3 of treatment) in achieving target colistin exposures and were generally below 1 mg/L after the first dose in early PK studies [40–42]. Various dosing recommendations/guidelines have been developed and validated in critically ill patients [42,43]. It should be noted that combination therapy would be considered beneficial for highly resistant pathogens (colistin MIC $\geq$1 mg/L) in patients with normal renal function.

The experience with the above antimicrobials highlights that using alternative dosing strategies—change in dose (loading and maintenance dose), frequency, and/or method of administration—could be a useful tool to optimize empirical dosing in a subset of patients likely to have altered PK and/or infections caused by less susceptible pathogens.

Therapeutic Drug Monitoring and Approaches for Dose Modification

Use of TDM in individual patients is a mode of personalized dosing for increasing the probability of therapeutic success while minimizing the risk of drug toxicity and emergence of resistance. Among antimicrobials, vancomycin and gentamicin have been historically subjected to TDM because of toxicity (mainly nephrotoxicity). However, with increasing understanding about the exposure-response relationships and the emergence of resistance secondary to suboptimal exposures [44], the focus of TDM has also included improving the probability of target PK/PD attainment. For antimicrobials not usually associated with TDM such as the beta-lactams, linezolid, and daptomycin, TDM may be most beneficial in cases of extreme PK variability as seen in critically ill patients.

TDM involves measuring antimicrobial concentrations, interpreting the measured concentration in the context of PK/PD, and then dose-adjusting as necessary. In the last few years, emphasis has been placed on improving and adapting these techniques in clinical setting to increase the likelihood of obtaining therapeutic concentrations. Broadly, these approaches can be placed into three categories: 1) linear/nonlinear regression, 2) population PK-based nomograms, and 3) Bayesian dose adaptation.

In linear/nonlinear regression approach, PK parameters are calculated based on the concentrations measured at different time points in a dosing interval using mathematical regression. These parameters are then compared with the established PK/PD targets, and then, doses are adjusted typically proportionally as needed. A lack of population PK data and the necessity for measuring at least two concentration points are considered limitations with this approach.

Dosing nomograms for antimicrobials, particularly for vancomycin and gentamicin, are widely used in the clinical setting. Nomograms are developed using population PK models and display a range of concentrations at different time points within the sampling time. The measured concentration of the prescribed antimicrobials is compared with the nomogram at the time point for the measured concentration. Dose adaptation is then performed to ensure further concentrations are within the therapeutic range without taking individual PK parameters into consideration and thus called an *a priori* dosing approach. Therefore, it is important to ensure an individual patient is represented by the population on which a nomogram is developed. It becomes even more difficult in critically ill patients where there are dynamic physiological changes.

A Bayesian approach incorporates both population PK parameters (*a priori*, similar to nomogram method) and patient-specific PK (*a posteriori*, similar to regression approach) to more accurately predict doses required for the attainment of therapeutic concentrations. It is important to ensure an individual patient is represented by the population PK included in the model. This method offers additional advantages such as use of only one measured concentration to predict dose, predicting an appropriate initial dose based on patient information, increasing the possibility of target attainment prior to steady state, and use of patient prior information to determine a future dose that meets the target concentrations. Many programs are available that employ different techniques to individualize antimicrobial dosing in clinical setting [7]. Fig. 1 displays different dosing strategies for patients with different illness severity and approaches to TDM.

The majority of data on the impact of TDM on efficacy and safety has focused on the aminoglycosides. In general, PK monitoring and dose adjustments compared with no intervention are associated with increased time in the therapeutic range. TDM using Bayesian software (i.e., with adaptive feedback) resulted in reduced length of hospital stay (6 days) and nephrotoxicity (a 10% reduction) in one randomized trial [45]. Another RCT, using a

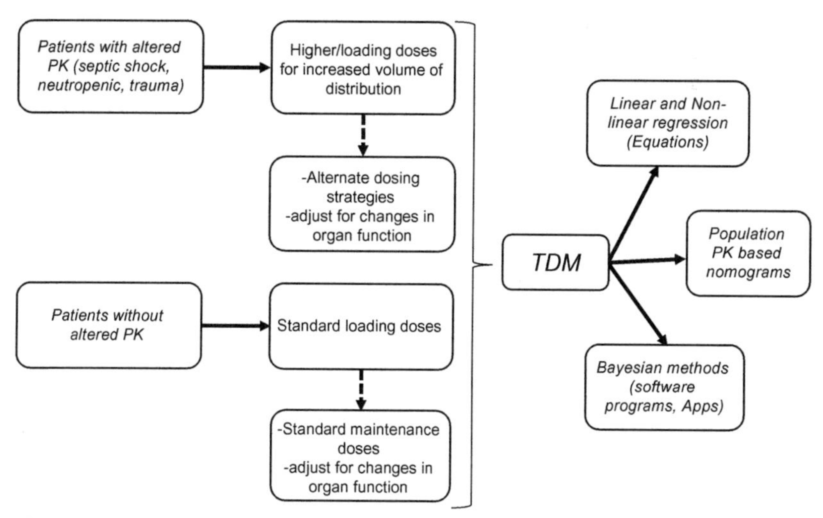

FIG. 1 Schematic flowchart illustrating dosing options in patients with different illness severity and TDM.

nomogram approach, reported a survival benefit (12% vs. 60% mortality with and without PK intervention, respectively) in severely ill hospitalized patients using pharmacist-supported PK intervention [46]. Other studies have reported clinical success [47] and reduced nephrotoxicity [48] using PK monitoring-based dose adjustment. In comparison, Bayesian approaches have been shown to outperform the other approaches described above by increasing the percentage (from 68% to 77% with Bayesian approach) of patients attaining the target concentrations for aminoglycosides [49].

Only one RCT has described the impact of PK monitoring-based dose-adjustment approach for vancomycin [50]. No difference in clinical efficacy was demonstrated in this cohort of patients with hematologic malignancies, but the incidence of nephrotoxicity was reduced (9% in control group vs 0% in TDM group) in the TDM group. More recently, a meta-analysis including one RCT and five cohort studies reported a significantly higher clinical cure and reduced nephrotoxicity in TDM groups [51]. However, no significant difference in duration of vancomycin therapy, length of hospital stay, or mortality was reported between TDM and non-TDM groups. For vancomycin, currently, trough concentrations are measured and used as a surrogate for AUC_{0-24}. The underlying assumption is that a trough between 15 and 20 mg/L will produce the target AUC_{0-24}. The discordance between trough and AUC has been acknowledged with potentially unnecessary increases in doses based on trough concentrations [52]. It is recommended to measure the AUC using either Bayesian programs or analytic equations using measured peak and trough concentrations to avoid underestimating AUC (if trough concentrations are used as a surrogate) leading to unnecessary toxicity [53].

The utility of beta-lactam TDM has been recently reviewed [24]. In general, the available evidence, although limited, suggests that TDM-based approaches to perform dose adjustment increase PK/PD target attainment compared with conventional dosing approaches. The benefits may be most evident in patients with severe sepsis and those infected with high MIC pathogens. In a prospective study, Scaglione *et al.* [54] employed feedback dose alteration method in patients with hospital-acquired pneumonia and reported a significant reduction in clinical failure, length of hospital stay, duration on mechanical ventilation, and mortality in the TDM group. The study antibiotics were amikacin, ceftazidime, cefotaxime, ciprofloxacin, and levofloxacin, with cephalosporins, ceftazidime and cefotaxime used in almost 50% of patients in the study. However, until the effect of beta-lactam TDM on clinical outcomes and resistance can be evaluated robustly via RCT, its role remains equivocal.

In the last few years, linezolid [22,55,56] and daptomycin [57,58] have been subject to TDM due to large interpatient and intrapatient variability in AUC_{0-24} and trough concentrations, respectively. Some authors have also purported TDM for colistin, considering the 20-fold variation in concentration with a 5-fold range in the daily dose [59].

Unfortunately, the wider application of TDM is impeded by the lack of availability of assays to expedite analysis, rapid microbiological diagnostic methods and susceptibility quantification, and technical know-how on the interpretation and implementation of the data. Linear regression and Bayesian methods are recommended for more accurate dosing regimens; however, they require additional resources such as information technology and health-care personnel with background training in PK.

CONCLUSION

Improving dosing techniques based on applying antimicrobial PK/PD knowledge and using TDM-based dose adaptation may have significant value for optimizing antimicrobial therapy. Indeed dose optimization should be considered a vital tool for ensuring good AMS practice. Further research, most likely in the way of RCTs, is required so as to quantify improvements in PK/PD target attainment and clinical outcomes and evaluating the impact on antimicrobial resistance.

REFERENCES

[1] Barlam TF, Cosgrove SE, Abbo LM, MacDougall C, Schuetz AN, Septimus EJ, *et al.* Implementing an Antibiotic Stewardship Program: Guidelines by the Infectious Diseases Society of America and the Society for Healthcare Epidemiology of America. Clin Infect Dis 2016;62:e51–77.

[2] Duguid M, Cruickshank M (Eds). Antimicrobial stewardship in Australian hospitals. Sydney (AU): ACSQHC; 2011. Australian Commission on Safety and Quality in Health Care, Sydney, 2010. http://www.safetyandquality.gov.au/wp-content/uploads/2011/01/Antimicrobial-stewardship-in-Australian-Hospitals-2011.pdf [Cited 15 August 2016].

[3] Craig WA. Pharmacokinetic/pharmacodynamic parameters: rationale for antibacterial dosing of mice and men. Clin Infect Dis 1998;26:1–10.

[4] Ambrose PG, Bhavnani SM, Rubino CM, Louie A, Gumbo T, Forrest A, *et al.* Pharmacokinetics-pharmacodynamics of antimicrobial therapy: it's not just for mice anymore. Clin Infect Dis 2007;44:79–86.

[5] Roberts JA, Lipman J. Pharmacokinetic issues for antibiotics in the critically ill patient. Crit Care Med 2009;37:840–51.

[6] Roberts JA, Roberts MS, Robertson TA, Dalley AJ, Lipman J. Piperacillin penetration into tissue of critically ill patients with sepsis-bolus versus continuous administration? Crit Care Med 2009;37:926–33.

[7] Roberts JA, Abdul-Aziz MH, Lipman J, Mouton JW, Vinks AA, Felton TW, *et al.* Individualised antibiotic dosing for patients who are critically ill: challenges and potential solutions. Lancet Infect Dis 2014;14:498–509.

[8] Bailey TC, Little JR, Littenberg B, Reichley RM, Dunagan WC. A meta-analysis of extended-interval dosing versus multiple daily dosing of aminoglycosides. Clin Infect Dis 1997;24:786–95.

[9] Munckhof WJ, Grayson ML, Turnidge JD. A meta-analysis of studies on the safety and efficacy of aminoglycosides given either once daily or as divided doses. J Antimicrob Chemother 1996;37:645–63.

[10] Barza M, Ioannidis JP, Cappelleri JC, Lau J. Single or multiple daily doses of aminoglycosides: a meta-analysis. BMJ 1996;312:338–45.

[11] Contopoulos-Ioannidis DG, Giotis ND, Baliatsa DV, Ioannidis JP. Extended-interval aminoglycoside administration for children: a meta-analysis. Pediatrics 2004;114:e111–8.

[12] de Montmollin E, Bouadma L, Gault N, Mourvillier B, Mariotte E, Chemam S, *et al.* Predictors of insufficient amikacin peak concentration in critically ill patients receiving a 25 mg/kg total body weight regimen. Intensive Care Med 2014;40:998–1005.

[13] Cataldo MA, Tacconelli E, Grilli E, Pea F, Petrosillo N. Continuous versus intermittent infusion of vancomycin for the treatment of Gram-positive infections: systematic review and meta-analysis. J Antimicrob Chemother 2012;67:17–24.

[14] Hao JJ, Chen H, Zhou JX. Continuous versus intermittent infusion of vancomycin in adult patients: a systematic review and meta-analysis. Int J Antimicrob Agents 2016;47:28–35.

[15] Schmelzer TM, Christmas AB, Norton HJ, Heniford BT, Sing RF. Vancomycin intermittent dosing versus continuous infusion for treatment of ventilator-associated pneumonia in trauma patients. Am Surg 2013;79:1185–90.

[16] Roberts JA, Taccone FS, Udy AA, Vincent JL, Jacobs F, Lipman J. Vancomycin dosing in critically ill patients: robust methods for improved continuous-infusion regimens. Antimicrob Agents Chemother 2011;55:2704–9.

[17] Nakamura A, Takasu O, Sakai Y, Sakamoto T, Yamashita N, Mori S, *et al.* Development of a teicoplanin loading regimen that rapidly achieves target serum concentrations in critically ill patients with severe infections. J Infect Chemother 2015;21:449–55.

[18] Pea F, Brollo L, Viale P, Pavan F, Furlanut M. Teicoplanin therapeutic drug monitoring in critically ill patients: a retrospective study emphasizing the importance of a loading dose. J Antimicrob Chemother 2003;51:971–5.

[19] Adembri C, Fallani S, Cassetta MI, Arrigucci S, Ottaviano A, Pecile P, *et al.* Linezolid pharmacokinetic/pharmacodynamic profile in critically ill septic patients: intermittent versus continuous infusion. Int J Antimicrob Agents 2008;31:122–9.

[20] Boselli E, Breilh D, Caillault-Sergent A, Djabarouti S, Guillaume C, Xuereb F, *et al.* Alveolar diffusion and pharmacokinetics of linezolid administered in continuous infusion to critically ill patients with ventilator-associated pneumonia. J Antimicrob Chemother 2012;67:1207–10.

[21] Boselli E, Breilh D, Rimmele T, Djabarouti S, Toutain J, Chassard D, *et al.* Pharmacokinetics and intrapulmonary concentrations of linezolid administered to critically ill patients with ventilator-associated pneumonia. Crit Care Med 2005;33:1529–33.

[22] Pea F, Furlanut M, Cojutti P, Cristini F, Zamparini E, Franceschi L, *et al.* Therapeutic drug monitoring of linezolid: a retrospective monocentric analysis. Antimicrob Agents Chemother 2010;54:4605–10.

[23] Goncalves-Pereira J, Povoa P. Antibiotics in critically ill patients: a systematic review of the pharmacokinetics of beta-lactams. Crit Care 2011;15:R206.

[24] Sime FB, Roberts MS, Peake SL, Lipman J, Roberts JA. Does beta-lactam pharmacokinetic variability in critically ill patients justify therapeutic drug monitoring? A systematic review. Ann Intensive Care 2012;2:35.

[25] Roberts JA, Paul SK, Akova M, Bassetti M, De Waele JJ, Dimopoulos G, *et al.* DALI: defining antibiotic levels in Intensive care unit patients: are current beta-lactam antibiotic doses sufficient for critically ill patients? Clin Infect Dis 2014;58:1072–83.

[26] Roberts JA, Webb S, Paterson D, Ho KM, Lipman J. A systematic review on clinical benefits of continuous administration of beta-lactam antibiotics. Crit Care Med 2009; 37:2071–8.

[27] Tamma PD, Putcha N, Suh YD, Van Arendonk KJ, Rinke ML. Does prolonged β-lactam infusions improve clinical outcomes compared to intermittent infusions? A meta-analysis and systematic review of randomized, controlled trials. BMC Infect Dis 2011;11:181.

[28] Kasiakou SK, Sermaides GJ, Michalopoulos A, Soteriades ES, Falagas ME. Continuous versus intermittent intravenous administration of antibiotics: a meta-analysis of randomised controlled trials. Lancet Infect Dis 2005;5:581–9.

[29] Angus BJ, Smith MD, Suputtamongkol Y, Mattie H, Walsh AL, Wuthiekanun V, *et al.* Pharmacokinetic-pharmacodynamic evaluation of ceftazidime continuous infusion vs intermittent bolus injection in septicaemic melioidosis. Br J Clin Pharmacol 2000;50:184–91.

[30] Hanes SD, Wood GC, Herring V, Croce MA, Fabian TC, Pritchard E, *et al.* Intermittent and continuous ceftazidime infusion for critically ill trauma patients. Am J Surg 2000;179:436–40.

[31] Nicolau DP, McNabb J, Lacy MK, Quintiliani R, Nightingale CH. Continuous versus intermittent administration of ceftazidime in intensive care unit patients with nosocomial pneumonia. Int J Antimicrob Agents 2001;17:497–504.

[32] Rafati MR, Rouini MR, Mojtahedzadeh M, Najafi A, Tavakoli H, Gholami K, *et al.* Clinical efficacy of continuous infusion of piperacillin compared with intermittent dosing in septic critically ill patients. Int J Antimicrob Agents 2006;28:122–7.

[33] Dulhunty JM, Roberts JA, Davis JS, Webb SA, Bellomo R, Gomersall C, *et al.* Continuous infusion of beta-lactam antibiotics in severe sepsis: a multicenter double-blind, randomized controlled trial. Clin Infect Dis 2013;56:236–44.

[34] Dulhunty JM, Roberts JA. A multicenter randomized trial of continuous versus intermittent beta-lactam infusion in severe sepsis. Am J Respir Crit Care Med 2015;192:1298–305.

[35] Abdul-Aziz MH, Sulaiman H, Mat-Nor MB, Rai V, Wong KK, Hasan MS, *et al.* Beta-Lactam Infusion in Severe Sepsis (BLISS): a prospective, two-centre, open-labelled randomised controlled trial of continuous versus intermittent beta-lactam infusion in critically ill patients with severe sepsis. Intensive Care Med 2016;42:1535–45.

[36] Roberts JA, Abdul-Aziz MH, Davis JS, Dulhunty JM, Cotta MO, Myburgh J, *et al.* Continuous versus intermittent beta-lactam infusion in severe sepsis: a meta-analysis of individual patient data from randomized trials. Am J Respir Crit Care Med 2016;194:681–91.

[37] Falcone M, Russo A, Venditti M, Novelli A, Pai MP. Considerations for higher doses of daptomycin in critically ill patients with methicillin-resistant Staphylococcus aureus bacteremia. Clin Infect Dis 2013;57:1568–76.

[38] Ramirez J, Dartois N, Gandjini H, Yan JL, Korth-Bradley J, McGovern PC. Randomized phase 2 trial to evaluate the clinical efficacy of two high-dosage tigecycline regimens versus imipenem-cilastatin for treatment of hospital-acquired pneumonia. Antimicrob Agents Chemother 2013;57:1756–62.

[39] Vicari G, Bauer SR, Neuner EA, Lam SW. Association between colistin dose and microbiologic outcomes in patients with multidrug-resistant gram-negative bacteremia. Clin Infect Dis 2013;56:398–404.

[40] Garonzik SM, Li J, Thamlikitkul V, Paterson DL, Shoham S, Jacob J, *et al.* Population pharmacokinetics of colistin methanesulfonate and formed colistin in critically ill patients from a multicenter study provide dosing suggestions for various categories of patients. Antimicrob Agents Chemother 2011;55:3284–94.

[41] Mohamed AF, Karaiskos I, Plachouras D, Karvanen M, Pontikis K, Jansson B, *et al.* Application of a loading dose of colistin methanesulfonate in critically ill patients: population pharmacokinetics, protein binding, and prediction of bacterial kill. Antimicrob Agents Chemother 2012;56:4241–9.

[42] Plachouras D, Karvanen M, Friberg LE, Papadomichelakis E, Antoniadou A, Tsangaris I, *et al.* Population pharmacokinetic analysis of colistin methanesulfonate and colistin after intravenous administration in critically ill patients with infections caused by gram-negative bacteria. Antimicrob Agents Chemother 2009;53:3430–6.

[43] Dalfino L, Puntillo F, Mosca A, Monno R, Spada ML, Coppolecchia S, *et al.* High-dose, extended-interval colistin administration in critically ill patients: is this the right dosing strategy? A preliminary study. Clin Infect Dis 2012;54:1720–6.

[44] Roberts JA, Kruger P, Paterson DL, Lipman J. Antibiotic resistance—what's dosing got to do with it? Crit Care Med 2008;36:2433–40.

[45] van Lent-Evers NA, Mathot RA, Geus WP, van Hout BA, Vinks AA. Impact of goal-oriented and model-based clinical pharmacokinetic dosing of aminoglycosides on clinical outcome: a cost-effectiveness analysis. Ther Drug Monit 1999;21:63–73.

[46] Whipple JK, Ausman RK, Franson T, Quebbeman EJ. Effect of individualized pharmacokinetic dosing on patient outcome. Crit Care Med 1991;19:1480–5.

[47] Streetman DS, Nafziger AN, Destache CJ, Bertino Jr AS. Individualized pharmacokinetic monitoring results in less aminoglycoside-associated nephrotoxicity and fewer associated costs. Pharmacotherapy 2001;21:443–51.

[48] Bartal C, Danon A, Schlaeffer F, Reisenberg K, Alkan M, Smoliakov R, *et al.* Pharmacokinetic dosing of aminoglycosides: a controlled trial. Am J Med 2003;114:194–8.

[49] Avent ML, Teoh J, Lees J, Eckert KA, Kirkpatrick CM. Comparing 3 methods of monitoring gentamicin concentrations in patients with febrile neutropenia. Ther Drug Monit 2011;33:592–601.

[50] Fernandez de Gatta MD, Calvo MV, Hernandez JM, Caballero D, San Miguel JF, Dominguez-Gil A. Cost-effectiveness analysis of serum vancomycin concentration monitoring in patients with hematologic malignancies. Clin Pharmacol Ther 1996;60:332–40.

[51] Ye ZK, Tang HL, Zhai SD. Benefits of therapeutic drug monitoring of vancomycin: a systematic review and meta-analysis. PLoS One 2013;8:e77169.

[52] Neely MN, Youn G, Jones B, Jelliffe RW, Drusano GL, Rodvold KA, *et al.* Are vancomycin trough concentrations adequate for optimal dosing? Antimicrob Agents Chemother 2014;58:309–16.

[53] Pai MP, Neely M, Rodvold KA, Lodise TP. Innovative approaches to optimizing the delivery of vancomycin in individual patients. Adv Drug Deliv Rev 2014;77:50–7.

[54] Scaglione F, Esposito S, Leone S, Lucini V, Pannacci M, Ma L, *et al.* Feedback dose alteration significantly affects probability of pathogen eradication in nosocomial pneumonia. Eur Respir J 2009;34:394–400.

[55] Pea F, Viale P, Cojutti P, Del Pin B, Zamparini E, Furlanut M. Therapeutic drug monitoring may improve safety outcomes of long-term treatment with linezolid in adult patients. J Antimicrob Chemother 2012;67:2034–42.

[56] Zoller M, Maier B, Hornuss C, Neugebauer C, Dobbeler G, Nagel D, *et al.* Variability of linezolid concentrations after standard dosing in critically ill patients: a prospective observational study. Crit Care 2014;18:R148.

[57] Reiber C, Senn O, Muller D, Kullak-Ublick GA, Corti N. Therapeutic drug monitoring of daptomycin: a retrospective monocentric analysis. Ther Drug Monit 2015;37:634–40.

[58] Pai MP, Russo A, Novelli A, Venditti M, Falcone M. Simplified equations using two concentrations to calculate area under the curve for antimicrobials with concentration-dependent pharmacodynamics: daptomycin as a motivating example. Antimicrob Agents Chemother 2014;58:3162–7.

[59] Landersdorfer CB, Nation RL. Colistin: how should it be dosed for the critically ill? Semin Respir Crit Care Med 2015;36:126–35.

Chapter 8

The Use of Computerized Decision Support Systems to Support Antimicrobial Stewardship Programs

Karin Thursky
University of Melbourne, Melbourne, VIC, Australia
Peter MacCallum Cancer Centre, Melbourne, VIC, Australia

INTRODUCTION

Computerized decision support may be defined as access to knowledge stored electronically to aid patients, carers, and service providers in making decisions on health-care. Key infectious diseases bodies support the use of computerized decision support systems (CDSS) as potentially useful tools in antibiotic stewardship programs [1].

At a patient level, antibiotic prescribing requires a complex sequence of decisions based on uncertain and poorly structured information often from a variety of sources [2,3]. The ability to reduce the complexity of decisions around antibiotic prescribing is observed in senior clinicians [3], and patient outcomes are improved with infectious disease consultation [4–6]. CDSS can improve the quality of decision-making in less experienced doctors, primarily through reducing task complexity [3]. At a hospital AMS program level, the stewardship workforce's primary aim is to identify those patients that need early review so that optimization of care can be performed and to undertake surveillance and reporting.

Decision support systems range from mobile applications to approval systems, electronic medical records (EMRs), computerized physician order entry (CPOE), and advanced decision support [7,8]. Although CDSS appear beneficial for improving the quality of prescribing [9–12], the impact on patient outcome and antimicrobial resistance is much less certain [13]. While they have been shown to be effective in reducing the costs of antibiotic prescribing

Antimicrobial Stewardship. http://dx.doi.org/10.1016/B978-0-12-810477-4.00008-8

(per patient or institution) [14–18] and hospitalization [14,19–21], there are few rigorous cost-effectiveness studies.

Moreover, there are many barriers (and facilitators) for implementation and uptake of these systems [22,23]. These are not well understood, and the literature is lacking about implementation or measures to monitor the unintended consequences. Finally, as AMS programs are multidisciplinary and need to be effective across the health-care system, no single CDSS is likely to be a solution.

PASSIVE DECISION SUPPORT AND MOBILE APPLICATIONS

Probably the most CDSS are electronic guidelines that may or may not be integrated with other systems and that do not present patient specific information to the clinician. The Internet or intranet is frequently used as a convenient and effective way to control and distribute clinical practice guidelines. In the United States, UpToDate was identified as the most commonly used resource for learning about antimicrobial prescribing in a survey of medical students [24].

The recent upturn in smartphone use in the general population has been matched by increased development of smartphone apps geared toward use in health-care, including in the field of antimicrobial stewardship [25] (see Table 1). They are likely to increasingly influence antimicrobial prescribing practices due to the availability at the bedside [26]. Clinical systems are increasingly becoming mobile device compatible so that access to data to support decision-making is possible.

Several leading institutions have developed mobile phone apps to support AMS programs, and these are preferred to traditional intranet guidelines by medical staff [27]. While third-party apps are readily available, the knowledge bases may not support local practices or clinical guidelines, although some applications (see Table 1) support local customization. A disadvantage of mobile apps is that the user must initiate updates on their own device, which may lead to potential issues with version control. Their impact on prescribing appropriateness is less certain as prescribing decisions are often made by senior doctors and while on ward rounds [28].

BACK-END COMPUTERIZED DECISION SUPPORT AND SURVEILLANCE SYSTEMS

Antimicrobial CDSS may be asynchronous (do not provide decision support at the time of prescribing), utilizing knowledge-based expert systems that monitor pathology, pharmacy, or microbiology results. These systems may monitor (1) antibiotic choice (based on microbiology results), (2) antibiotic dosing and monitoring (based on pathology results), and (3) antibiotic resistance and simultaneous microbiology surveillance [29]. These systems

TABLE 1 Example of Mobile Applications Supporting AMS or Antimicrobial Prescribing

Imperial Antibiotic Prescribing Policy (IAPP) App	Imperial College London
Sanford Guide Online	Sanford Guide
John Hopkins Antibiotic Guide	Johns Hopkins Medicine
Microguide	Horizon Strategic Partners, the United Kingdom
2015 EMRA Antibiotic Guide	Emergency Medicines Residents Association, the United States
UptoDate Online	UptoDate, Wolters Kluwer Health, the United States
Antibiotic pocket	Borm Bruckmeier Publishing, LLC Medica
Spectrum MD—Localized Antimicrobial Stewardship	Spectrum Mobile Health Inc.
Impact	Infection Control Branch, Centre for Health Protection, Department of Health, Hong Kong
CBR-based Imperial Antibiotic Prescribing Policy (ENIAPP) smartphone app	Imperial College London

can generate reports of prescriptions where there is potential therapeutic mismatch, overlapping coverage, or inappropriate route or administration [16,21,30,31].

Infection prevention systems (both homegrown [32,33] and commercial) integrate the electronic patient record, microbiology, pathology, and sometimes radiology results (see Table 2). These systems have been developed to assist in the identification of patients at high risk of nosocomial infection, to monitor antimicrobial resistance, and to assist with routine surveillance activities. EMRs can send information to these third-party software vendors for data collation and reporting. The licensing costs of these systems can be significant and a barrier for administrations already dealing with the costs of EMR/CPOE.

While electronic surveillance systems are able to mine large amounts of data and are capable of providing real-time alerts for infection prevention, they do not necessarily save time [34]. In fact, there are several studies that have shown that the increased information flow requires interpretation and

TABLE 2 Commercial Infection Prevention Surveillance Systems

Safety Surveillor	Premier HealthCare, Charlotte, North Carolina
TheraDoc	Premier HealthCare, Charlotte, North Carolina
MedMined	CareFusion, BD, New Jersey
Sentri7	Wolters Kluwer
ICNET	ICNET Systems, Illinois
RL Solutions	RL, Ontario, Canada
QC Pathfinder	Vecna Technologies, Cambridge, Massachusetts

triaging, leading to substantial number of nonactionable alerts [16]. In an evaluation of a system in a 652-bed hospital, 8571 alerts were generated in 791 patients over 5 months. Only 30% of alerts were actionable and required 2–3 h per day reviewing and 1–2 h intervention and documentation [35].

ANTIMICROBIAL RESTRICTION/APPROVAL BASED SYSTEMS

Computerized antimicrobial authorization or approval systems for antimicrobials are an important strategy for AMS and have been shown to be very effective in reducing consumption of targeted antimicrobials and reducing drug costs [36–40]. Computerized antimicrobial authorization or approval systems generate alerts or reminders leading to an action by the AMS program such as postprescription review. Computerized antimicrobial authorization or approval systems do not function in isolation and require close collaboration with pharmacy, clinical microbiology, and ID staff to be successful. Importantly, electronic approvals allow for easy data extraction and auditing of antimicrobial use, thereby facilitating feedback to individual prescribers, units, and hospital committees.

Australia is unique in its widespread uptake of computerized antimicrobial authorization or approval systems. These systems have been successfully implemented in sites without electronic health record (EHR) or electronic prescribing system (EPS) and have streamlined the workflow for AMS programs [36,37]. The Guidance program (Royal Melbourne Hospital) has been adopted in over 60 sites including public, private, and regional hospitals. The program supports a bundle of interventions including formulary support, restricted indications for target antimicrobials, access to national guidelines, and administration alerts by pharmacists if drugs are given without an approval, targeted postprescription review, feedback, and reporting. The system has been associated with reduced drug consumption and use of broad-spectrum antimicrobials [36,41], improved resistance patterns in gram-negative isolates in the ICU [42], improved prescribing for community-acquired pneumonia in ED [43], and achieved acceptable usability for clinicians [44]. The success of an approval system to support AMS programs has led to a broad recommendation for their adoption that is supported by dedicated national accreditation standards for AMS.

ELECTRONIC PRESCRIBING SYSTEMS (WITH OR WITHOUT CDSS)

EMRs and CPOE systems can effectively be harnessed to support AMS. There is reasonable evidence that CPOE is very effective in reducing procedural error rates (~60%) (e.g., illegible orders) but less effective in reducing clinical error (wrong dose and wrong drugs) rates particularly in the absence of CDSS [45].

Cost-effectiveness studies have demonstrated that CPOE particularly with CDSS are likely to lead to long-term savings due to reduction in adverse drug events, readmissions, and reduced health-care costs [46–50]. Even so, the high cost of CPOE has led to a slow uptake in many countries, compared with the United States where government financial incentives for "meaningful use" led to widespread deployment leading to an increase from <10% to >70% of prescriptions being written electronically in a few years [51]. In 2014, ~35% of English hospitals had begun the implementation of CDSS functionality in at least one ward or hospital department [22]. Many hospitals lack the foundations required for successful implementation and are still in a state of transition between paper-based medical records and EMRs.

Almost all commercial CPOE systems are associated with front-end decision support that can support AMS such as default values, routes of administration, dose and frequencies, drug-allergy alerts, and drug-drug interaction alerts. CDSS content including AMS modules provided by commercial vendors may not always meet the requirements of an institution, and significant resources are required to modify these systems and then implement them effectively [52–55] (Table 2). EMR/CPOE can support a bundled approach to AMS with antimicrobial restriction, dosing recommendations, automated stop orders, rule based alerts, and order sets for disease conditions. One study has demonstrated reduction in mortality, LOS, and readmission for patients admitted with community acquired pneumonia (CAP) using an evidenced-based order set [48].

However, poorly implemented systems and systems without associated CDSS may be associated with patient harm [56,57]. The potential safety risks of 10 CPOE systems (four inpatient, six outpatient, homegrown, and commercial systems) were examined by the Brigham and Women's Hospital Center for Patient Safety Research and Practice [45,58]. Several important errors that could impact on antimicrobial prescribing are the following:

1. Large numbers of drugs and dosing combinations and dangerous autocomplete directions that displaced or contradicted original intended orders.
2. Failure to transmit CPOE medication discontinuation orders to outpatient pharmacies.

3. Inconsistent clinical decision support design implementation and firing. There were very high rates of override (>90%) for many alerts due to poor design.
4. Off-the-shelf commercial drug databases were poorly designed to meet the needs of sites, leading to local, extensive customization that was difficult to maintain with software releases.

Most systems do not fully support the task complexity of AMS; the ability to collate multiple sources of patient data in a useful format for postprescription review or reporting has led software vendors to develop a bespoke solution or to interface with a third-party tool (see Table 2).

ADVANCED DECISION SUPPORT

There are few publications of advanced CDSS that support antimicrobial prescribing and that have been successfully implemented outside of the originating institution.

A landmark publication in the New England Journal of Medicine by the LDS hospital in Utah demonstrated the efficacy of a computerized antibiotic assistant in the ICU [14]. The fully computerized hospital enabled predictive models to be developed from stepwise logistic regression models of the patient database. These models provide population-based probabilities of infections in relation to variables from clinical, radiology, and pathology data. Local epidemiological data and variables from matched patients from the previous 5-year period were used if clinical data were incomplete or unavailable. One of the striking findings of this study was that only 46% of antibiotic recommendations were followed, compared with 94% of antibiotic dosing suggestions. They published extensively on a wide range of CDSS tools for the management of infections, infection control surveillance, surgical prophylaxis, and adverse drug events [59,60].

Intelligent presentation of information to prescribers or to the stewardship service (such as dashboards) [61] can impact on prescribing, even in the absence of complex CDSS. Sintchenko *et al.* evaluated the impact of a handheld device on antibiotic prescribing in a before-and-after study in a single ICU over a 12-month period [20]. When the same information was provided as a handheld tool, the most frequent reasons for using the system were the microbiology reports (53%), followed by antibiotic guidelines (22%), antibiogram (16%), and ventilator associated pneumonia (VAP) risk calculator (9%). Despite the infrequent use of the CDSS, there was a reduction in both total and broad-spectrum antibiotics. Another Australian study of a homegrown CDSS in ICU also demonstrated that access to intelligently presented microbiology results was a major driver of use and acceptability to clinicians [62].

TREAT (Treat Systems, Denmark) uses data available within the first few hours of infection presentation in a causal probabilistic network to predict sites of infection and specific pathogens. The CDSS for empirical antimicrobial therapy was implemented and studied in a cluster randomized trial across three wards across three countries (Israel, Denmark, and Germany). The program was shown to improve appropriateness of empirical antimicrobial therapy and mortality at 6 months [63,64].

Machine learning poses a promising technology to support AMS, as it can utilize free text in EMRs, pathology, radiology reports, and prescriptions. These systems use supervised learning to establish the knowledge base of classification rules. A Canadian CDSS was augmented with machine learning capabilities to identify inappropriate prescriptions such as dose and dosing frequency adjustments, discontinuation of therapy, early switch from intravenous to oral therapy, and redundant antimicrobial spectrum [65]. A text-mining tool for predicting pulmonary invasive fungal infection from CT chest reports was more effective that traditional manual methods and led to earlier detection [66].

Detection and management of sepsis should be a key focus for CDSS. A recent systematic review of eight studies found that automated sepsis alerts derived from electronic health data may improve care processes but tend to have poor positive predictive value (20.5%–53.8%) and negative predictive value (76.5%–99.7%) and do not improve mortality or length of stay [67]. Nevertheless, an automated, real-time surveillance algorithm that aggregated, normalized, and analyzed patient data from disparate clinical systems and delivered early sepsis alerts to nurses and treatment advice to clinicians via mobile devices and portals was associated with a significant reduction in mortality [68].

IMPLEMENTATION BARRIERS

Information flows for antimicrobial decision-making and administration are complex [69] and highlights the critical importance of human factors design to ensure that systems meet the needs of all health-care workers (nurses, pharmacists, and doctors) interacting with these systems. Similarly, the literature is sparse on organizational context of implementation of health IT systems. Many systems developed over years by technology champions, coevolved in institutions, and hence adapted to the environment and culture of the institution. Boonstra and Broekhuis categorize the barriers influencing the adoption of an EMR. The change process—local champions (project leaders) and an organization that supports innovation, incentives, and participation—will determine the final uptake [70]. Sites with successful advanced CDSS reported a common set of factors—very strong leadership with a clear long-term commitment, a commitment to improving clinical processes by enlisting clinician support and involving the clinicians in all stages of the development process [71].

It is essential that AMS team members are formally engaged during the scoping, functional specification, and implementation of an EMR or CPOE system. The eHealth Research group in the University of Edinburgh has published several useful publications around implementation of large-scale IT systems. A key point relevant to AMS is that while multimodular systems offer better usability, stand-alone systems provide greater flexibility and opportunity for innovation, particularly in relation to interoperability with external systems and in relation to customizability to the needs of different user groups [22,23,72].

Implementation of an IT solution requires technical readiness (integration requirements and IT infrastructure), financial and human resources (project officer and AMS members), skills (training needs) process readiness (project planning, system implementation, and evaluation planning), and administrative readiness (executive support and clinical champions).

The major barrier for effective reporting and surveillance of AMS interventions is the current lack of systems interoperability. Standardization of clinical data systems, semantic interoperability, use of standard terminologies (e.g., SNOMED CT) and messaging standards (e.g., HL7), or the use of unified patient record numbers are ultimately required to support national approaches for AMS. Data security is of paramount importance with the move toward wireless, mobile technologies, and cloud computing.

CONCLUSIONS

There is a move toward electronic medication management in the acute health-care setting in Europe, the United States, and Australia, with government initiatives to modernize the health-care information technology infrastructure. Commercial EMR and CPOEs will be implemented across many institutions within the next 5 years requiring substantial organizational change and funding. Commercial CPOE/EMR systems currently have variable decision support capability to support AMS with most attributes relating to direct patient care. Other types of CDSS such as those provided by dedicated third-party systems (that can be integrated into the EMR), or those that are stand-alone in hospitals with/without EMR are likely to remain a cost-effective alternative. The complex requirements of AMS mean that no single solution is likely to be sufficient (Table 3).

TABLE 3 Opportunities for AMS Interventions Using CDSS

CDSS	Type	Intervention Opportunities	Advantages and Disadvantages
Smartphone applications	Passive	Dissemination of guidelines • Disease-based • Drug-based Calculators (dosing) Antibiograms	Rapid dissemination Useful for hospitals with poor IT infrastructure Not usually integrated to hospital systems Uncontrolled use including version control May not influence prescribing of senior clinicians
Infection prevention surveillance systems	Back-end (asynchronous)	Pharmacy $\pm$ laboratory integration Rule based alerts are • Drug-bug mismatches • Double coverage Monitor restricted drug usage Surveillance—real-time alerts	Support an organizational approach Can be integrated with an EMR Requires substantial resources to review reports and to determine clinically relevant alerts that need action Requires dedicated EFT to support Commercial systems can be expensive
Approval systems (stand-alone)	Front-end	Enforce formulary May be preprescription or postprescription (after dispensing) Enforce-approved indications by drug Educational opportunity for the prescriber Can include clinical decision support Reports and feedback	Can work well in the absence of EHR or EPS Support an organizational approach to AMS Best combined with an antimicrobial team to review patients at a time period postapproval 24–48 h

Continued

TABLE 3 Opportunities for AMS Interventions Using CDSS—Cont'd

CDSS	Type	Intervention Opportunities	Advantages and Disadvantages
CPOE	Front-end	Alerts • Drug-drug interactions • Dosing • Drug-allergy Restriction prompts Automated stop orders (e.g., surgical prophylaxis) Order sets (CAP and sepsis)	Will reduce transcription errors but incorrect choice or indication may not impacted Best combined with CDSS Patient-centered rather than organizational Requires additional resources to develop customized AMS reports Alert fatigue is very common
EMR	Front-end	Error alerts – allergy, dosing, medicine–medicine interactions Chart abstraction tools to screen and identify patients at risk for sepsis, or collate information for AMS (medicines, results) Pre-prescription restriction rules Record AMS recommendations and interventions Support order sets for syndromes (e.g community-acquired pneumonia, sepsis) Alerts and triggers to identify patients suitable for intravenous-to-oral switch, or AMS review Care protocols (templates or phased order sets)	Eliminates the cost of external vendor Real-time interventions and alerts possible Allows for retrieval of data for research But Requires substantial institutional investment up front Significant time required by hospital IT to create the tools Templates must be incorporated into EMR at each site Less responsive to change

Advanced CDSS		Causal probabilistic network of pathogens, by specimen type or underlying conditions of patient Case-based probability Machine learning algorithms Pathogen prediction	Complex, usually "homegrown" systems Currently in early phase of adoption

Abstracted from published reviews of systems used to support antimicrobial prescribing [7,8,23].

REFERENCES

[1] Kullar R, Goff DA. Transformation of antimicrobial stewardship programs through technology and informatics. Infect Dis Clin North Am 2014;28(2):291–300.

[2] Kushniruk A, Patel V, Fleiszer D. Analysis of medical decision making: a cognitive perspective on medical informatics, In: Proceedings—the annual symposium on computer applications in medical care; 1995. p. 193–7.

[3] Sintchenko V, Coiera EW. Which clinical decisions benefit from automation? A task complexity approach. Int J Med Inform 2003;70:309–16.

[4] Butt AA, Al Kaabi N, Saifuddin M, Krishnanreddy KM, Khan M, Jasim WH, *et al.* Impact of infectious diseases team consultation on antimicrobial use, length of stay and mortality. Am J Med Sci 2015;350(3):191–4.

[5] Tissot F, Calandra T, Prod'hom G, Taffe P, Zanetti G, Greub G, *et al.* Mandatory infectious diseases consultation for MRSA bacteremia is associated with reduced mortality. J Infect 2014;69(3):226–34.

[6] Takakura S, Fujihara N, Saito T, Kimoto T, Ito Y, Iinuma Y, *et al.* Improved clinical outcome of patients with Candida bloodstream infections through direct consultation by infectious diseases physicians in a Japanese university hospital. Infect Control Hosp Epidemiol 2006;27(9):964–8.

[7] Thursky K. Use of computerized decision support systems to improve antibiotic prescribing. Expert Rev Anti Infect Ther 2006;4(3):491–507.

[8] Sintchenko V, Coiera E, Gilbert GL. Decision support systems for antibiotic prescribing. Curr Opin Infect Dis 2008;21(6):573–9.

[9] Hunt DL, Haynes RB, Hanna SE, Smith K. Effects of computer-based clinical decision support systems on physician performance and patient outcomes: a systematic review. JAMA 1998;280(15):1339–46.

[10] Kaushal R, Shojania KG, Bates DW. Effects of computerized physician order entry and clinical decision support systems on medication safety: a systematic review. Arch Intern Med 2003;163(12):1409–16.

[11] Shiffman RN, Liaw Y, Brandt CA, Corb GJ. Computer-based guideline implementation systems: a systematic review of functionality and effectiveness. J Am Med Inform Assoc 1999;6(2):104–14.

[12] Walton R, Dovey S, Harvey E, Freemantle N. Computer support for determining drug dose: systematic review and meta-analysis. BMJ 1999;318(7189):984–90.

[13] Baysari MT, Lehnbom EC, Li L, Hargreaves A, Day RO, Westbrook JI. The effectiveness of information technology to improve antimicrobial prescribing in hospitals: a systematic review and meta-analysis. Int J Med Inform 2016;92:15–34.

[14] Evans RS, Pestotnik SL, Classen DC, Clemmer TP, Weaver LK, Orme JF, *et al.* A computer-assisted management program for antibiotics and other antiinfective agents. N Engl J Med 1998;338:232–8.

[15] Pestotnik SL, Classen DC, Evans RS, Burke JP. Implementing antibiotic practice guidelines through computer-assisted decision support: clinical and financial outcomes. Ann Intern Med 1996;124(10):884–90.

[16] Jozefiak ET, Lewicki JE, Kozinn WP. Computer-assisted antimicrobial surveillance in a community teaching hospital. Am J Health Syst Pharm 1995;52(14):1536–40.

[17] Shojania KG, Yokoe D, Platt R, Fiskio J, Ma'luf N, Bates DW. Reducing vancomycin use utilizing a computer guideline: results of a randomized controlled trial. J Am Med Inform Assoc 1998;5(6):554–62.

[18] Schentag JJ, Ballow CH, Fritz AL, Paladino JA, Williams JD, Cumbo TJ, *et al.* Changes in antimicrobial agent usage resulting from interactions among clinical pharmacy, the infectious disease division, and the microbiology laboratory. Diagn Microbiol Infect Dis 1993;16(3):255–64.

[19] Burton ME, Ash CL, Hill Jr DP, Handy T, Shepherd MD, Vasko MR. A controlled trial of the cost benefit of computerized bayesian aminoglycoside administration. Clin Pharmacol Ther 1991;49(6):685–94.

[20] Sintchenko V, Iredell JR, Gilbert GL, Coiera E. Handheld computer-based decision support reduces patient length of stay and antibiotic prescribing in critical care. J Am Med Inform Assoc 2005;12(4):398–402.

[21] Barenfanger J, Short MA, Groesch AA. Improved antimicrobial interventions have benefits. J Clin Microbiol 2001;39(8):2823–8.

[22] Cresswell K. Evaluation of implementation of health IT. Stud Health Technol Inform 2016;222:206–19.

[23] Cresswell K, Mozaffar H, Shah S, Sheikh A. A systematic assessment of review to promoting the appropriate use of antibiotics through hospital electronic prescribing systems. Int J Pharm Pract 2017;25(1):5–17.

[24] Abbo LM, Cosgrove SE, Pottinger PS, Pereyra M, Sinkowitz-Cochran R, Srinivasan A, *et al.* Medical students' perceptions and knowledge about antimicrobial stewardship: how are we educating our future prescribers? Clin Infect Dis 2013;57(5):631–8.

[25] Goff DA. iPhones, iPads, and medical applications for antimicrobial stewardship. Pharmacotherapy 2012;32(7):657–61.

[26] Panesar P, Jones A, Aldous A, Kranzer K, Halpin E, Fifer H, *et al.* Attitudes and behaviours to antimicrobial prescribing following introduction of a smartphone app. PLoS One 2016;11(4):e0154202.

[27] Charani E, Kyratsis Y, Lawson W, Wickens H, Brannigan ET, Moore LS, *et al.* An analysis of the development and implementation of a smartphone application for the delivery of antimicrobial prescribing policy: lessons learnt. J Antimicrob Chemother 2013;68(4): 960–7.

[28] Bartlett JG, editor. E-prescribing of antibiotics: using the Hopkins Antibiotic Guide. Washington, DC: ICAAC; 2004.

[29] Schiff GD, Klass D, Peterson J, Shah G, Bates DW. Linking laboratory and pharmacy: opportunities for reducing errors and improving care. Arch Intern Med 2003;163(8): 893–900.

[30] Grau S, Monterde J, Carmona A, Drobnic L, Salas E, Marin M, *et al.* Monitoring of antimicrobial therapy by an integrated computer program. Pharm World Sci 1999;21(4):152–7.

[31] Martinez MJ, Freire A, Castro I, Inaraja MT, Ortega A, Del Campo V, *et al.* Clinical and economic impact of a pharmacist-intervention to promote sequential intravenous to oral clindamycin conversion. Pharm World Sci 2000;22(2):53–8.

[32] Evans RS, Burke JP, Classen DC, Gardner RM, Menlove RL, Goodrich KM, *et al.* Computerized identification of patients at high risk for hospital-acquired infection. Am J Infect Control 1992;20(1):4–10.

[33] Heininger A, Niemetz AH, Keim M, Fretschner R, Doring G, Unertl K. Implementation of an interactive computer-assisted infection monitoring program at the bedside. Infect Control Hosp Epidemiol 1999;20(6):444–7.

[34] Grota PG, Stone PW, Jordan S, Pogorzelska M, Larson E. Electronic surveillance systems in infection prevention: organizational support, program characteristics, and user satisfaction. Am J Infect Control 2010;38(7):509–14.

[35] Hermsen ED, VanSchooneveld TC, Sayles H, Rupp ME. Implementation of a clinical decision support system for antimicrobial stewardship. Infect Control Hosp Epidemiol 2012;33(4):412–5.

[36] Buising KL, Thursky KA, Robertson MB, Black JF, Street AC, Richards MJ, *et al.* Electronic antibiotic stewardship—reduced consumption of broad-spectrum antibiotics using a computerized antimicrobial approval system in a hospital setting. J Antimicrob Chemother 2008;62(3):608–16.

[37] Grayson ML, Melvani S, Kirsa SW, Cheung S, Korman AM, Garrett MK, *et al.* Impact of an electronic antibiotic advice and approval system on antibiotic prescribing in an Australian teaching hospital. Med J Aust 2004;180(9):455–8.

[38] Chan YY, Lin TY, Huang CT, Deng ST, Wu TL, Leu HS, *et al.* Implementation and outcomes of a hospital-wide computerised antimicrobial stewardship programme in a large medical centre in Taiwan. Int J Antimicrob Agents 2011;38(6):486–92.

[39] Potasman I, Naftali G, Grupper M. Impact of a computerized integrated antibiotic authorization system. Isr Med Assoc J 2012;14(7):415–9.

[40] Agwu AL, Lee CK, Jain SK, Murray KL, Topolski J, Miller RE, *et al.* A World Wide Web-based antimicrobial stewardship program improves efficiency, communication, and user satisfaction and reduces cost in a tertiary care pediatric medical center. Clin Infect Dis 2008;47(6):747–53.

[41] Cairns KA, Jenney AW, Abbott IJ, Skinner MJ, Doyle JS, Dooley M, *et al.* Prescribing trends before and after implementation of an antimicrobial stewardship program. Med J Aust 2013;198(5):262–6.

[42] Yong MK, Buising KL, Cheng AC, Thursky KA. Improved susceptibility of gram negative bacteria in an intensive care unit following implementation of a computerised antibiotic decision support system; 2009 [unpublished].

[43] Buising KL, Thursky KA, Black JF, MacGregor L, Street AC, Kennedy MP, *et al.* Improving antibiotic prescribing for adults with community acquired pneumonia: does a computerised decision support system achieve more than academic detailing alone?—a time series analysis. BMC Med Inform Decis Mak 2008;8:35.

[44] Zaidi ST, Marriott JL, Nation RL. The role of perceptions of clinicians in their adoption of a web-based antibiotic approval system: do perceptions translate into actions? Int J Med Inform 2008;77(1):33–40.

[45] Schiff GD, Amato MG, Eguale T, Boehne JJ, Wright A, Koppel R, *et al.* Computerised physician order entry-related medication errors: analysis of reported errors and vulnerability testing of current systems. BMJ Qual Saf 2015;24(4):264–71.

[46] Westbrook JI, Gospodarevskaya E, Li L, Richardson KL, Roffe D, Heywood M, *et al.* Cost-effectiveness analysis of a hospital electronic medication management system. J Am Med Inform Assoc 2015;22(4):784–93.

[47] Nuckols TK, Asch SM, Patel V, Keeler E, Anderson L, Buntin MB, *et al.* Implementing computerized provider order entry in acute care hospitals in the United States could generate substantial savings to society. Jt Comm J Qual Patient Saf 2015;41(8): 341–50.

[48] Krive J, Shoolin JS, Zink SD. Effectiveness of evidence-based pneumonia CPOE order sets measured by health outcomes. Online J Public Health Inform 2015;7(2):e211.

[49] Vermeulen KM, van Doormaal JE, Zaal RJ, Mol PG, Lenderink AW, Haaijer-Ruskamp FM, *et al.* Cost-effectiveness of an electronic medication ordering system (CPOE/CDSS) in hospitalized patients. Int J Med Inform 2014;83(8):572–80.

[50] Forrester SH, Hepp Z, Roth JA, Wirtz HS, Devine EB. Cost-effectiveness of a computerized provider order entry system in improving medication safety ambulatory care. Value Health 2014;17(4):340–9.

[51] Schiff GD, Hickman TT, Volk LA, Bates DW, Wright A. Computerised prescribing for safer medication ordering: still a work in progress. BMJ Qual Saf 2016;25(5):315–9.

[52] Pogue JM, Potoski BA, Postelnick M, Mynatt RP, Trupiano DP, Eschenauer GA, *et al.* Bringing the "power" to Cerner's PowerChart for antimicrobial stewardship. Clin Infect Dis 2014;59(3):416–24.

[53] Kullar R, Goff DA, Schulz LT, Fox BC, Rose WE. The "epic" challenge of optimizing antimicrobial stewardship: the role of electronic medical records and technology. Clin Infect Dis 2013;57(7):1005–13.

[54] Cook PP, Rizzo S, Gooch M, Jordan M, Fang X, Hudson S. Sustained reduction in antimicrobial use and decrease in methicillin-resistant *Staphylococcus aureus* and *Clostridium difficile* infections following implementation of an electronic medical record at a tertiary-care teaching hospital. J Antimicrob Chemother 2011;66(1):205–9.

[55] Phansalkar S, Wright A, Kuperman GJ, Vaida AJ, Bobb AM, Jenders RA, *et al.* Towards meaningful medication-related clinical decision support: recommendations for an initial implementation. Appl Clin Inform 2011;2(1):50–62.

[56] Ash JS, Berg M, Coiera E. Some unintended consequences of information technology in health care: the nature of patient care information system-related errors. J Am Med Inform Assoc 2004;11(2):104–12.

[57] Magrabi F, Ong MS, Runciman W, Coiera E. Patient safety problems associated with healthcare information technology: an analysis of adverse events reported to the US Food and Drug Administration. AMIA Ann Symp Proc 2011;2011:853–7.

[58] Administration USFaD. Computerised prescriber order entry medication safety (CPOEMS): Uncovering and learning from issues and errors; 2015.

[59] Larsen RA, Evans RS, Burke JP, Pestotnik SL, Gardner RM, Classen DC. Improved perioperative antibiotic use and reduced surgical wound infections through use of computer decision analysis. Infect Control Hosp Epidemiol 1989;10(7):316–20.

[60] Pestotnik SL, Evans RS, Burke JP, Gardner RM, Classen DC. Therapeutic antibiotic monitoring: surveillance using a computerized expert system. Am J Med 1990;88(1):43–8.

[61] Waitman LR, Phillips IE, McCoy AB, Danciu I, Halpenny RM, Nelsen CL, *et al.* Adopting real-time surveillance dashboards as a component of an enterprisewide medication safety strategy. Jt Comm J Qual Patient Saf 2011;37(7):326–32.

[62] Thursky KA, Buising KL, Bak N, MacGregor L, Street AC, MacIntyre CR, *et al.*, The impact of a real-time microbiology browser with computer-assisted decision support for antibiotic prescription in an intensive care unit. Royal Australian College of Physicians annual scientific meeting, Canberra; 2004.

[63] Paul M, Andreassen S, Tacconelli E, Nielsen AD, Almanasreh N, Frank U, *et al.* Improving empirical antibiotic treatment using TREAT, a computerized decision support system: cluster randomized trial. J Antimicrob Chemother 2006;58(6):1238–45.

[64] Paul M, Nielsen AD, Goldberg E, Andreassen S, Tacconelli E, Almanasreh N, *et al.* Prediction of specific pathogens in patients with sepsis: evaluation of TREAT, a computerized decision support system. J Antimicrob Chemother 2007;59(6):1204–7.

[65] Beaudoin M, Kabanza F, Nault V, Valiquette L. Evaluation of a machine learning capability for a clinical decision support system to enhance antimicrobial stewardship programs. Artif Intell Med 2016;68:29–36.

[66] Ananda-Rajah MR, Martinez D, Slavin MA, Cavedon L, Dooley M, Cheng A, *et al.* Facilitating surveillance of pulmonary invasive mold diseases in patients with haematological malignancies by screening computed tomography reports using natural language processing. PLoS One 2014;9(9):e107797.

[67] Makam AN, Nguyen OK, Auerbach AD. Diagnostic accuracy and effectiveness of automated electronic sepsis alert systems: a systematic review. J Hosp Med 2015;10(6):396–402.

[68] Manaktala S, Claypool SR. Evaluating the impact of a computerized surveillance algorithm and decision support system on sepsis mortality. J Am Med Inform Assoc 2017;24(1):88–95.

[69] Thursky KA, Mahemoff M. User-centered design techniques for a computerised antibiotic decision support system in an intensive care unit. Int J Med Inform 2007;76(10):760–8.

[70] Boonstra A, Broekhuis M. Barriers to the acceptance of electronic medical records by physicians from systematic review to taxonomy and interventions. BMC Health Serv Res 2010;10:231.

[71] Kawamoto K, Houlihan CA, Balas EA, Lobach DF. Improving clinical practice using clinical decision support systems: a systematic review of trials to identify features critical to success. BMJ 2005;330(7494):765.

[72] Cresswell KM, Bates DW, Sheikh A. Ten key considerations for the successful optimization of large-scale health information technology. J Am Med Inform Assoc 2017;24(1): 182–7.

Chapter 9

The Role of Microbiology Laboratory in Promoting Antimicrobial Stewardship

Füsun Can* and Onur Karatuna**
**Koç University School of Medicine, Istanbul, Turkey*
***Acibadem University School of Medicine, Istanbul, Turkey*

INTRODUCTION

Microbiology laboratories play an important role in guidance to clinicians for appropriate prescription of antimicrobial therapy. Inappropriate antibiotic prescribing results in poor patient outcomes, increased treatment costs, and increased risk of superinfections by multiresistant organisms with the selection of resistant clones. Physicians usually select the most appropriate therapy of choice based on information from clinical microbiology laboratories. However, information generated by laboratories is not always sufficient because of prolonged time for identification and antibiotic susceptibility testing or failure in detection of causative agent. The goal of antimicrobial stewardship (AMS) is to improve patient outcomes while encouraging rational use of antibiotics. Clinical microbiology laboratories support clinical and economic impact of AMS with early and accurate reporting of the causative agent and antibiograms. Also, a successful combination of antimicrobial stewardship and infection-control stewardship programs limits the emergence of antimicrobial-resistant bacteria.

RAPID DIAGNOSTIC TESTS (RDTs)

The introduction of rapid diagnostic tests provided new opportunities for the diagnosis of infections over the past decade. These methods significantly reduce the time for pathogen identification and antimicrobial susceptibility from days to hours compared with conventional techniques. In most hospital microbiology laboratories, the average time for identification and antimicrobial susceptibility test results is approximately 40 h by traditional

Antimicrobial Stewardship. http://dx.doi.org/10.1016/B978-0-12-810477-4.00009-X

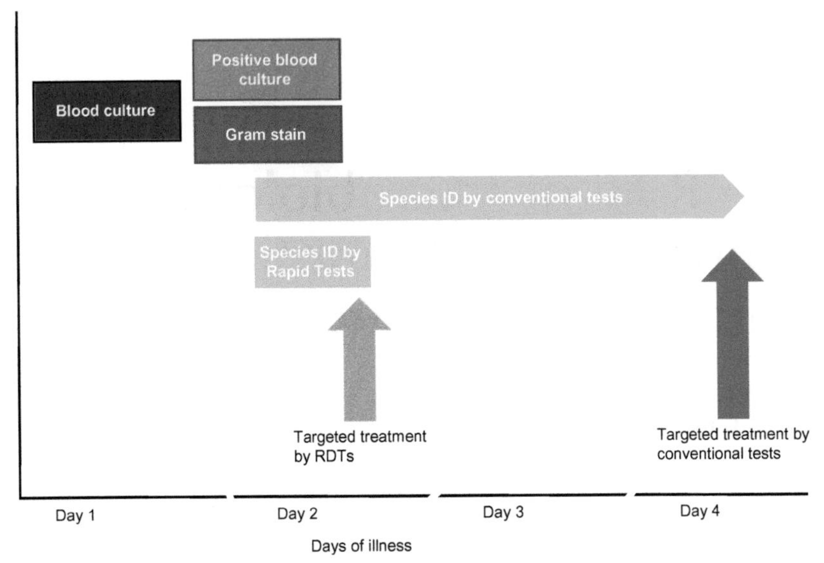

FIG. 1 Timing of targeted treatment by diagnosis with Rapid Diagnostic Tests (RDTs) vs Conventional tests.

microbiological methods. Numerous studies have shown the impact of early antimicrobial therapy on survival of the patient. Early results from microbiology laboratory by using rapid tests allow physicians to start life saving therapy for patients with serious infections (Fig. 1). The rapid tests not only benefit the individual patient but also improve AMS by decreasing antibiotic use, time to optimal antimicrobial therapy, length of stay, and costs [1,2].

Some most commonly used RDTs are matrix-assisted laser desorption-ionization time of flight (MALDI-TOF), quantitative polymerase chain reaction (qPCR) assays, peptide nucleic acid fluorescence in situ hybridization (PNA-FISH) assays, and multiplex nucleic acid assays.

MALDI-TOF system depends on analysis of intact proteins of microorganisms. Currently, two systems are cleared by the US Food and Drug Administration (FDA) for identification of cultured bacteria and yeasts. Although the list of cleared microorganisms in each system is different, both systems have good performance on identification of aerobic and anaerobic bacteria. MALDI-TOF has become a preferred method for identification of bacteria in clinical laboratories because of its easy to use protocol and over 90% accurate identification in genus and species level [3]. The performance of the system for identification of yeasts is good; however, for molds, it needs to be improved. The other limitation of MALDI-TOF systems is their low performance on antibiotic susceptibility testing. However, the application of MALDI-TOF on antibiotic susceptibility has been growing constantly. The current studies continue in two main approaches, including the detection of the modification at target sites like MRSA and VRE and the detection of degrading enzymes such as beta-lactamases and carbapenamases. The impact of pathogen identification via MALDI-TOF in patients with bacteremia and candidemia was demonstrated with reduced time to organism identification (from 84.0 to 55.9 h), decreased time to effective antibiotic therapy (from 30.1 to 20.4 h), and decreased fatality rate [4]. Moreover, early direct application of MALDI-TOF to positive blood cultures without waiting the subculture growth has shortened the optimal antimicrobial treatment time to 17.9 h compared with 36.1 h with classical identification [5].

Real-time PCR assays are usually developed for diagnosis of bloodstream infections. They accurately detect methicillin resistance in staphylococcal bacteremia after the detection of Gram-positive clustered cocci in blood culture, thereby improving early targeted antibiotic prescribing. In a recent randomized clinical study, diagnosis by real-time PCR assay revealed a decrease in the median time of reporting for methicillin resistance by 21.5 h, a decrease in the time to targeted antibiotic therapy by 20.5 h, and a decrease in unnecessary exposure to vancomycin in the patients with *S. aureus* bacteremia [6].

Multiplex nucleic acid assays can detect multiple pathogens in specimens. These systems identify the bacteria and some yeasts within 1.5 h from positive signaling in blood culture bottle. They also detect several resistance genes including *CTX-M, VIM, IMP, OXA, KPC, NDM, VanA*, and *mecA*. Several studies demonstrated the benefits of adding rapid multiplexed PCR-based identification systems to antimicrobial stewardship programs. In patients with bacteremia and fungemia, a decrease in time to specific antimicrobial therapy and a decrease in duration of broad-spectrum therapy were reported [7]. They also significantly enhance antimicrobial de-escalation (with multiplex PCR 21 h versus control 34 h) [8]. These assays reduced the length of hospital stay by 21.7 days and hospital costs by $60,729 in the patients with enterococcal bacteremia [9]. In a recent study, time to initiation of effective therapy was significantly reduced by over 24 h in the patients with enterococcal bacteremia as well [10].

PNA-FISH is another promising diagnostic technique that yields the results within 4 h after blood culture bottle growth signal. Pathogen-specific RNA targeted fluorescein-labeled probes are used. The PNA-FISH assay has a good performance in the accurate identification of *S. aureus*, enterococci, Gram-negative rods, and *Candida* species from blood cultures and cerebrospinal fluid cultures. These assays can provide the results at least one work day before MALDI-TOF mass spectrometry-based identification [11]. A recent study demonstrated the high impact of PNA-FISH assay along with AMS intervention with the outcomes of 3.8 days early pathogen identification, 1.7 days early initiation of treatment, and $415 cost savings [12].

TABLE 1 Impact of Rapid Diagnostic Tests (RDTs) in AMS

- Benefits of RDTs
  - Early pathogen identification
  - Early initiation of targeted therapy
  - Reduce broad-spectrum antibiotic therapy
- Limitations of RDTs
  - Antibiotic susceptibility testing
- Remaining Questions
  - The impact of RDTs on duration of therapy, fatality, and cost-effectiveness
  - The role of RDTs on nonblood tests and impact on AMS
  - The role of RDTs performed directly on specimens and impact on AMS

The important limitation of RDTs is their low performance on antibiotic susceptibility testing. Although some methods can identify certain resistance markers, these tests are still incapable of detecting full antimicrobial susceptibility profiles of pathogens. The currently available rapid sepsis panels need to be improved because they can only be performed after detection of growth signal in culture systems. The impact of these systems on clinical outcome of patients and fatality are still not well defined. The advantages and disadvantages of these tests are given in Table 1.

ANTIMICROBIAL SUSCEPTIBILITY TESTING

The rapid emergence of antimicrobial resistance (AMR) has led the clinical microbiology laboratory to take on a more critical role in promoting the rational use of antimicrobials to achieve optimal outcomes that is among the essential objectives of any given AMS program. Clinical microbiology laboratories perform antimicrobial susceptibility testing on causative agents of infections that enables the evidence-based and optimized treatment of individual patients. Collecting and organizing the cumulative susceptibility data also enable the creation of the antibiograms that reflect the institutional susceptibility patterns, thus providing sound evidence to make informed decisions on empirical treatment regimens. Furthermore, it is the role of a laboratory to engage in regional/national AMR surveillance networks that is extremely helpful in collecting data at a wider scale and enables to monitor susceptibility trends over periods of time. In case of an emerging resistance, the information obtained through surveillance studies can be used to formulate the strategies aimed at reducing or preventing any further development of resistance.

Performing of AMT, structured reporting of its results, and the use of cumulative susceptibility data have become essential in AMS programs since they have direct impact on improved patient outcomes, promote the rational use of antibiotics, and help tackle the emergence of AMR.

a. *Constructing and reporting AMT results*

The policies developed to fight against the emergence and spread of AMR should rely on evidence-based practices among which an important component is the administration of antimicrobials according to the documented presence of a causative organism that is accompanied by an AMT report. The emergence of resistance mechanisms in recent years rendered many broad-spectrum antimicrobials ineffective, and AMT become a necessity rather than an optional request of a physician. Furthermore, while inhibiting the growth of or killing the target pathogens, the broad-spectrum antimicrobials also have effects on the microbiota, promoting them to develop resistance. Due to these factors, AMT should be directed only toward the causative agent of infection, and the results of AMT should be reported to the clinician with caution. A policy to prevent the emergence of resistance should be adopted, which may include the selective reporting of susceptibility results.

Each laboratory is unique in its capacity, resources, level of experience or institutional needs, and the patient population being served. Therefore, the decision of which antimicrobials to test depends on each laboratory's specifications and cannot be generalized. The decision involves the opinions of infectious diseases specialist and the pharmacist and should also be in concordance with the hospital formulary. Generally, a laboratory defines 10–15 antimicrobial agents for routine testing against various organisms or organism groups, which is called antimicrobial panel or battery. Because the identity of the bacterial isolate is often not known at the time the AMT is performed, some drugs, which are inappropriate to report for that particular isolate, may be tested. These results, however, should be suppressed in the final report.

When reporting AMT results, the clinical microbiologist should consider the fact that the reports would be evaluated by many physicians other than infectious diseases specialists who may or may not have the required knowledge to correctly interpret the results, for example, the importance of the detected resistance mechanism, the need for implementing strict infection-control measures, or the clinical relevance of the organism isolated. For all these reasons, including brief comments in the AMT reports might help the physician in decision-making and identify the need for a consultation with an infectious diseases specialist.

b. *Performing AMT only for clinically relevant organisms*

Only the organisms that are likely to be the cause of an infection should be tested for antimicrobial susceptibility, which necessitates the differentiation between the normal flora that resides at the site of the infection and the actual organism causing the infection. Some important factors to be considered are to decide which bacterium or bacteria from a clinical specimen must be included in the AMT, such as the body site from which the organism was isolated, the presence of other bacteria and the quality of the specimen from which the organism was grown, the host's status, and the ability of the bacterial species to cause infection at the body site from which the specimen was obtained [13].

c. *AMT and reporting based on guidelines*

Each laboratory should follow standardized methods in the performance of AMT to be able to produce reliable and reproducible results. The standardized components of AMT include bacterial inoculum size, growth medium and the incubation conditions (atmosphere, temperature, and duration), and the concentrations of antimicrobials (disk potencies) to be tested. The procedural steps of each method must be followed strictly in order to obtain reproducible results.

The most frequently used guidelines are the Clinical and Laboratory Standards Institute (CLSI) and the European Committee on Antimicrobial Susceptibility Testing (EUCAST) guidelines. Both methods release yearly updates, and the laboratories should implement the recent edition in order to keep up to date with the new resistance mechanisms, developments in the methodology, and susceptibility testing of new antimicrobials, etc. In the recent years, the number of countries adopted the EUCAST guidelines has increased dramatically, and EUCAST has become the standard throughout Europe. Furthermore, countries such as Australia, Brazil, Canada, Iceland, Israel, Morocco, New Zealand, South Africa, and the United States have formed national antimicrobial susceptibility testing committees to promote the nationwide implementation of EUCAST methodology. Among the factors that enabled this wide and rapid endorsement of EUCAST methodology, the free availability of all documents on EUCAST website, the open and transparent decision-making process, and no industry involvement play an important role.

d. *The use of expert rules in AMT*

The performance of AMT and validation of the results is an important part of the clinical microbiology laboratory daily practice; however, the increased complexity in the AMR mechanisms and their clinical implications require an advanced level of knowledge to interpret the AMT results. When performing AMT, the laboratories use specific guidelines such as the CLSI or EUCAST, these guidelines publish breakpoint tables that allow the results to be categorized, as susceptible, intermediate, and resistant. Thus, the AMT results, whether disk diffusion zone diameters or MIC values, are reported to the clinicians in categories to which they correspond. However, by considering the genetic basis of the resistance mechanism that reflects itself in a certain phenotype and by interpreting the results of different classes of antimicrobials together rather than interpreting each antimicrobial individually, it is possible to produce a more clinically relevant

AMT result. This practice uses the expert rules that describe actions to be taken on the basis of specific AMT results. These actions might include recommendations on reporting, such as inferring susceptibility to other agents from results with one, suppression of results that may be inappropriate, and editing of results from susceptible to intermediate/resistant or from intermediate to resistant on the basis of an inferred resistance mechanism [14].

The expert rules also include the intrinsic resistances that describe the natural resistance characteristic of all or almost all isolates of the bacterial species, as opposed to acquired and/or mutational resistance and the unexpected resistance phenotypes that describe the resistances that have not yet been reported or reported only very rarely. Examples on expert rules in antimicrobial susceptibility testing are given in Table 2.

Expert rules, in general, have been developed to identify the underlying resistance mechanism and modify the results of AMT when clinically required. Interpretation of AMT results along with expert rules also serves as a quality control measure. Additionally, identification of the underlying resistance mechanisms helps to monitor the epidemiology of resistance mechanisms. As the resistance mechanisms constantly evolve and the laboratory methods developed for their detection become more complex, the clinical microbiologist should follow the latest developments and implement expert rules in the daily practice. This will not only increase the therapeutic success for the patients but also help prevent the emergence of resistance.

e. *Detection of important resistance mechanisms*

Detection of some specific resistance mechanisms is particularly important for the reporting of AMT results (determination of the susceptibility category) and also for infection control and public health. In order to help routine laboratories, EUCAST has published a special guideline in which the methods for the detection of important resistance mechanisms are being described [15]. Although the document includes seven important resistance mechanisms, some other clinically significant resistance mechanisms and their detection methods (e.g., inducible clindamycin resistance and low-level ciprofloxacin resistance in *Salmonella* spp.) are being described in EUCAST's breakpoint table [16].

EUCAST's detection of resistance mechanisms guideline briefly describes the resistance mechanism or specific resistance and the clinical and/or public health need for the detection of the resistance mechanism or specific resistance. Moreover, the laboratory methods recommended for the detection of the resistance mechanisms are summarized with an extensive list of references. The AMR mechanisms of clinical and/or epidemiological importance contained in the EUCAST guideline are given in Table 3.

TABLE 2 Examples on Expert Rules in Antimicrobial Susceptibility Testing

Expert Rule Type	Organism	Rule
Intrinsic resistances		Resistance to the following:
	Proteus mirabilis	Nitrofurantoin and colistin
	Serratia marcescens	Colistin
	Gram-positive bacteria	Aztreonam
	Enterococci	Fusidic acid
Exceptional resistance phenotypes		Susceptibility to the following:
	Streptococcus pyogenes	Penicillin
	Staphylococcus aureus	Vancomycin
Interpretive reading and expert rules		If... then...
	Staphylococcus spp.	If resistant to isoxazolyl-penicillins (as determined with oxacillin, cefoxitin, or by detection of *mecA*-gene or of PBP2a), then report as resistant to all β-lactams except those specifically licensed to treat infections caused by methicillin-resistant staphylococci owing to low affinity for PBP2a
	Haemophilus influenzae	If β-lactamase-positive, then report as resistant to ampicillin, amoxicillin, and piperacillin
	All organisms	If susceptible, intermediate, or resistant to erythromycin, then report the same category of susceptibility for azithromycin, clarithromycin, and roxithromycin
	Enterobacteriaceae	If resistant to ciprofloxacin, then report as resistant to all fluoroquinolones

It is important to note that some resistance mechanisms (e.g., extended-spectrum β-lactamases or carbapenemases in Enterobacteriaceae) do not always cause clinical resistance. However, the detection of resistance mechanism may be important for infection control and public health.

TABLE 3 Antimicrobial Resistance Mechanisms of Clinical and/or Epidemiological Importance

	Importance of Detection of Resistance Mechanism		
Resistance Mechanism	***Required for Antimicrobial Susceptibility Categorization***	***Infection Control***	***Public Health***
Carbapenemase-producing Enterobacteriaceae	No	Yes	Yes
Extended-spectrum β-lactamase-producing Enterobacteriaceae	No	Yes	Yes
Acquired AmpC β-lactamase-producing Enterobacteriaceae	No	Yes	Yes
Methicillin-resistant *Staphylococcus aureus* (MRSA)	Yes	Yes	Yes
Glycopeptide-nonsusceptible *Staphylococcus aureus*	Yes	Yes	Yes
Vancomycin resistant *Enterococcus faecium* and *Enterococcus faecalis*	Yes	Yes	Yes
Penicillin-nonsusceptible *Streptococcus pneumoniae*	Yes	No	Yes

f. *Selective reporting of AMT results*

Selective reporting (also called cascade or restricted reporting) is an active process to optimize the AMT report in a way to produce a more clinically relevant report. This includes reporting of the first-line antimicrobials that are indicated for the treatment of the infection, avoiding the susceptible results of broad-spectrum antimicrobials when the narrow spectrum option is susceptible, thereby promoting the rational use of antimicrobials and preventing the emergence of resistance. This approach helps to improve the clinical relevance of the reports by generating a personalized report considering the patient in question.

The objective of the AMT report should be to direct the physician to use the clinically effective, least toxic, and cost-effective option among available

TABLE 4 Basic Principles of Selective Reporting of AMT Results [18]

- Report susceptibilities to routinely used antibiotics
- Report susceptibility to the antibiotic that the prescriber has stated is in use
- Report all resistances for significant pathogens
- Always include susceptibility to non-β-lactams to cover penicillin allergy
- Whenever possible include antibiotics for oral therapy
- Select antibiotics reported to include those that have lowest risk of adverse events
- Take note of restrictions for special patient groups when reporting (e.g., tetracyclines not to be used in pregnancy or for children)

antimicrobials. The principles of selective reporting have been outlined by the CLSI in the M100 document [17]. In order to perform selective reporting, a laboratory should decide which agents to report routinely (first-line antimicrobials) and which might be reported selectively. This decision should be in consultation with the infection-control committee of the healthcare institution, and the principles of testing and reporting should be compliant with the local guidelines for antibiotic use and formulary.

It is important to state on the result sheet that the AMT results have been selectively reported, and further susceptibility results are available on request. Some basic principles of selective reporting of AMT results are given in Table 4.

g. *Surveillance*

Surveillance is defined as the ongoing and systematic collection, analysis, and interpretation of health data essential to the planning, implementation, and evaluation of public health practice [19]. In the field of AMR, the goal of the surveillance is to provide a valid description of the antimicrobial susceptibility of target bacterial pathogens in infections to the selected antimicrobial groups indicated for treatment of these infections. The sample of patients included in surveillance should aim to consist of a mix of patient types (e.g., pediatric, ICU, or neurosurgery patients) and infection types (e.g., community-acquired urosepsis and healthcare-associated bloodstream infections), in proportion to their occurrence in the total population [20]. The main purpose of surveillance is to detect changes in the susceptibility of target pathogens and thus to inform prescribers and other related parties of such changes. In case of an emerging resistance, the information obtained through surveillance studies can be used to formulate the strategies aimed at reducing or preventing any further development of resistance. Furthermore, the results of surveillance studies also allow the evaluation of strategies developed. Such strategies might include formulary changes, the development of policy guidelines, whether local, national, or international, and changes in prescribing practices and infection control [21].

The AMR surveillance activities can be undertaken locally at the institution level or in countries with an established network at the national level. Furthermore, currently there are the regional networks such as the European Antimicrobial Resistance Surveillance Network (EARS-Net) that includes the European Union countries, the regional network Central Asian and Eastern European Surveillance of Antimicrobial Resistance (CAESAR), and the global initiative the Global Antimicrobial Resistance Surveillance System (GLASS) that aims to collect data from the whole world. The Global Report on Surveillance published by the World Health Organization points out the possibility that patients in many places are treated for suspected bacterial infections in the absence of any information about the resistance situation in the local area, further emphasizing the need for more recent data at all levels to systematically monitor trends, to inform patient treatment guidelines, and to inform and evaluate containment efforts [22].

h. *Cumulative antibiogram to guide empirical treatment decisions*

At the postanalytic phase, the analysis of cumulative AMT results is important to monitor emerging trends in resistance at the local level. The monitoring of local resistance data is usually performed by preparation of an annual summary of susceptibility rates, which is also known as a cumulative antibiogram report. These institutional reports not only help to monitor the resistance trends but also are helpful to support clinical decision-making, infection-control interventions, and AMR containment strategies [23]. For the standardization of the methodology used in preparation and presentation of cumulative antibiograms, CLSI published a guideline with its recent edition being the M39-A4 [24]. Some key recommendations in this guideline are given in Table 5.

TABLE 5 Key Recommendations for Cumulative Antibiogram Preparation as Per CLSI M39-A4 Guideline

Frequency of reporting—analyze and present data at least annually
The number of isolates to include in a statistic—include only species with at least 30 isolates tested
The source of isolates—include diagnostic, not surveillance (or screening), isolates
Antimicrobial agents to analyze—include results only for drugs that are routinely tested
Elimination of repeat isolates—include the first isolate per patient in the period analyzed, irrespective of the body site from which the specimen was obtained or the antimicrobial susceptibility pattern
Presentation of percentages—calculate the percentage susceptible. Do not include the percentage of isolates with intermediate susceptibility

Although a cumulative antibiogram report may suffer from various limitations (e.g., culturing practices, patient population, specimen collection practices, laboratory AMT policies, and selective or sequential testing of antimicrobials), it is still considered as a useful tool in guiding initial empirical antimicrobial therapy decisions for the management of infections in patients for whom microbiological test data to target treatment do not yet exist. Its usefulness can further be improved by stratification of the results for select patient populations (e.g., inpatient vs outpatient), medical services (e.g., emergency vs intensive care unit), or different specimen types. The use of data contained in a cumulative antibiogram report requires special consideration since in these reports the antimicrobials with broad-spectrum activity exhibit in general higher percentage of susceptibility compared with their narrow-spectrum counterparts, which might favor the imprudent use of broad-spectrum antimicrobials. To circumvent this limitation, importance should be given to follow institutional or national algorithms to treat different types of infections that clearly describe first- and second-line antimicrobials.

COMMUNICATION BETWEEN AMS TEAM MEMBERS

The communication between the clinical microbiology laboratory and prescribing physicians is essential to achieve the goals of AMS. This communication may include structured meetings or direct conversations on patient's results or assessment of the results. The structured meeting on a weekly, biweekly, or monthly basis can be helpful for the improvement of AMS interventions. The other key element of AMS is reporting time of the results. Laboratories should contact to clinicians as quicker as possible after obtaining the results. Likewise, microbiologists should assist clinicians for the evaluation of results and limitations of methods.

REFERENCES

[1] Avdic E, Carroll KC. The role of the microbiology laboratory in antimicrobial stewardship programs. Infect Dis Clin North Am 2014;28(2):215–35.

[2] Goff DA, Jankowski C, Tenover FC. Using rapid diagnostic tests to optimize antimicrobial selection in antimicrobial stewardship programs. Pharmacotherapy 2012;32(8):677–87.

[3] Garner O, Mochon A, Branda J, Burnham CA, Bythrow M, Ferraro M, *et al.* Multi-centre evaluation of mass spectrometric identification of anaerobic bacteria using the VITEK(R) MS system. Clin Microbiol Infect 2014;20(4):335–9.

[4] Huang AM, Newton D, Kunapuli A, Gandhi TN, Washer LL, Isip J, *et al.* Impact of rapid organism identification via matrix-assisted laser desorption/ionization time-of-flight combined with antimicrobial stewardship team intervention in adult patients with bacteremia and candidemia. Clin Infect Dis 2013;57(9):1237–45.

[5] Verroken A, Defourny L, le Polain de Waroux O, Belkhir L, Laterre PF, Delmee M, *et al.* Clinical impact of MALDI-TOF MS identification and rapid susceptibility testing on adequate antimicrobial treatment in sepsis with positive blood cultures. PLoS One 2016;11(5), e0156299.

[6] Emonet S, Charles PG, Harbarth S, Stewardson AJ, Renzi G, Uckay I, *et al.* Rapid molecular determination of methicillin resistance in staphylococcal bacteraemia improves early targeted antibiotic prescribing: a randomized clinical trial. Clin Microbiol Infect 2016;22: e9-946.e15.

[7] MacVane SH, Nolte FS. Benefits of adding a rapid PCR-based blood culture identification panel to an established antimicrobial stewardship program. J Clin Microbiol 2016;54:2455–63.

[8] Banerjee R, Teng CB, Cunningham SA, Ihde SM, Steckelberg JM, Moriarty JP, *et al.* Randomized trial of rapid multiplex polymerase chain reaction-based blood culture identification and susceptibility testing. Clin Infect Dis 2015;61(7):1071–80.

[9] Sango A, McCarter YS, Johnson D, Ferreira J, Guzman N, Jankowski CA. Stewardship approach for optimizing antimicrobial therapy through use of a rapid microarray assay on blood cultures positive for Enterococcus species. J Clin Microbiol 2013;51(12): 4008–11.

[10] MacVane SH, Hurst JM, Boger MS, Gnann Jr JW. Impact of a rapid multiplex polymerase chain reaction blood culture identification technology on outcomes in patients with vancomycin-resistant Enterococcal bacteremia. Infect Dis (Lond) 2016;48(10):732–7.

[11] Calderaro A, Martinelli M, Motta F, Larini S, Arcangeletti MC, Medici MC, *et al.* Comparison of peptide nucleic acid fluorescence in situ hybridization assays with culture-based matrix-assisted laser desorption/ionization-time of flight mass spectrometry for the identification of bacteria and yeasts from blood cultures and cerebrospinal fluid cultures. Clin Microbiol Infect 2014;20(8):O468–75.

[12] Heil EL, Daniels LM, Long DM, Rodino KG, Weber DJ, Miller MB. Impact of a rapid peptide nucleic acid fluorescence in situ hybridization assay on treatment of Candida infections. Am J Health Syst Pharm 2012;69(21):1910–4.

[13] Karatuna O. Quality assurance in antimicrobial susceptibility testing. In: Akyar I, editor. Latest research into quality control. Rijeka: InTech; 2012. p. 413–33.

[14] Leclerq R, Cantón R, Brown DFJ, Giske CG, Heisig P, MacGowan AP, *et al.* EUCAST expert rules in antimicrobial susceptibility testing. Clin Microbiol Infect 2013;19:141–60.

[15] Giske CG, Martínez-Martínez L, Cantón R, Stefani S, Skov R, Glupczynski Y, *et al.* EUCAST Guidelines for detection of resistance mechanisms and specific resistances of clinical and/or epidemiological importance, Version 1.0. 2013. http://www.eucast.org/fileadmin/src/media/PDFs/EUCAST_files/Resistance_mechanisms/EUCAST_detection_of_resistance_mechanisms_v1.0_20131211.pdf. Accessed: 16 August 2016.

[16] The European Committee on Antimicrobial Susceptibility Testing. Breakpoint tables for interpretation of MICs and zone diameters. Version 6.0, 2016. http://www.eucast.org.

[17] CLSI . Performance standards for antimicrobial susceptibility testing. 26th ed. CLSI supplement M100S. Wayne, PA: Clinical and Laboratory Standards Institute; 2016.

[18] British Society for Antimicrobial Chemotherapy (BSAC), Guidance on reporting antimicrobial susceptibility. http://bsac.org.uk/wp-content/uploads/2012/02/AST-testing-and-Reporting-guidance-v1-Final.pdf. Accessed: 16 August 2016.

[19] Klaucke DN, Buehler JW, Thacker SB, Parrish RG, Trowbrigde FL, Berkelmann RL, *et al.* Guidelines for evaluating surveillance systems. MMWR 1988;37(S-5):1–18.

[20] World Health Organization (WHO). Central Asian and Eastern European Surveillance of Antimicrobial Resistance, Annual Report 2014. 2015, WHO. http://www.euro.who.int/__data/assets/pdf_file/0006/285405/CAESAR-Surveillance-Antimicrobial-Resistance2014.pdf. Accessed 16 August 2016.

[21] Bax R, Bywater R, Cornaglia G, Goossens H, Hunter P, Isham V, *et al.* Surveillance of antimicrobial resistance–what, how and whither? Clin Microbiol Infect 2001;7:316–25.

[22] World Health Organization (WHO). Antimicrobial Resistance Global Report on Surveillance, 2014. 2015, WHO. http://apps.who.int/iris/bitstream/10665/112642/1/9789241564748_eng.pdf. Accessed 16 August 2016.

[23] Hindler JF, Stelling J. Analysis and presentation of cumulative antibiograms: a new concensus guideline from the Clinical and Laboratory Standards Institute. Clin Infect Dis 2007;44:867–73.

[24] CLSI. Analysis and presentation of cumulative antimicrobial susceptibility test data; approved guideline—fourth edition. CLSI document M39-A4. Wayne, PA: Clinical and Laboratory Standards Institute; 2014.

Chapter 10

The Role of Pharmacists

Philip Howard
Leeds Teaching Hospitals NHS Trust, Leeds, United Kingdom

INTRODUCTION

Pharmacists are considered to be experts in medicines and are probably the ideal choice to oversee antimicrobial stewardship programs within hospitals and in primary care [1]. They oversee medicine optimization in all healthcare sectors. Table 1 summarizes the roles that pharmacists can play in AMS.

HOSPITAL PHARMACY

Within hospitals, the pharmacy is responsible for purchasing, distribution, or dispensing of medicines to inpatients and to outpatients or emergency department attendees in many countries. Most hospitals in the developed world have a clinical (or ward) pharmacy service. Clinical pharmacists review the prescribing of doctors to ensure safety and efficacy and to optimize long-term medicines that patients are taking prior to admission [2].

ANTIMICROBIAL STEWARDSHIP PROGRAMS

Most countries either have mandated or are in the process of introducing national antimicrobial stewardship guidance [3] that requires hospitals to have an AMS program [1]. Local ASPs usually run within the hospital's infection prevention and control (IPC) program. The pharmacy department in most hospitals usually run the drug and therapeutics committee (DTC) and therefore is ideally placed to provide professional secretariat support to ASPs for committee meetings and the day-to-day activities of the program [4]. Most hospitals have a policy for the use of medicines, so the antimicrobial stewardship policy can be written by the AMS pharmacist as part of the hospitals overarching medicine policies. Within the AMS program, the AMS pharmacist often leads the development of systems for the control of antimicrobial supply such as restricted or protected antibiotics and intravenous to oral switch guidance.

Antimicrobial Stewardship. http://dx.doi.org/10.1016/B978-0-12-810477-4.00010-6

TABLE 1 Roles That Pharmacists Can Play in AMS

AMS Strategy	Medical Lead	Pharmacy Lead
AMS Committee	AMS Chair—better medical engagement	AMS Prof Secretary—Good at organizing committees
Guidelines and policies	Diagnosis, investigations, non antimicrobial treatment, local drug choice	Drug dosing, processes, e.g., IV to oral therapy, AMS policy, new antibiotic review
Audit and feedback	Feedback to difficult audiences	Tools, doing and feedback
Education	AMS ward rounds - diagnosis and investigations	Antibiotic related, e-learning
Surveillance	Antimicrobial resistance	Antimicrobial usage
Individual patient advice	Treatment failures Telephone support	Dose optimization (TDM), ITU infusions, OPAT management
Miscellaneous		Formulary and restriction IT systems: web, Apps Patient safety (incidents, systems, prescriptions), communication, shortages

The introduction of new antimicrobials into a hospital or local health economy formulary is through the DTC process. The AMS pharmacist can work with infectious disease specialists to identify the potential place in therapy within both its licensed indications and potential off-label uses. The AMS pharmacist will usually produce the review of the evidence to support the introduction of the new antimicrobial, including its cost-effectiveness and advantages over existing therapies.

One of the key elements of any ASP is development of comprehensive antimicrobial treatment and prophylaxis guidelines. These are either tailored from national or international guidelines based on local AMR patterns or are developed locally [5]. Recommendations regarding diagnosis and non-antimicrobial management are best developed in conjunction with microbiology or infectious diseases. Recommendations regarding antimicrobial dosing in special populations should be developed by the pharmacy. Effective dissemination of guidelines and keeping them up to date are challenges. Some AMS programs have introduced smartphone "apps" to enhance uptake of guidelines by prescribers [6]. For hospitals with electronic prescribing, building protocols into the system based on local hospital guidelines may improve adherence. Mandating the reason for the antibiotic on all prescriptions (paper or electronic) may help monitoring of adherence to guidelines.

One of the challenges of an ASP is identification of patients to visit. Follow-up of restricted or protected antibiotic supply or patients with positive blood cultures is relatively easy, but in the absence of electronic prescribing, targeting of other patients is more difficult [7]. Electronic prescribing is still not the standard practice in most hospitals [3], so many AMS teams rely on clinical pharmacists to identify patients where intervention may be required. Few hospitals have managed to introduce data warehousing (linking prescribed antibiotics to pathology susceptibility data) to monitor drug-bug mismatch [3], but where this is present, there can be improvements in patient outcomes [8,9]. It is important that the pharmacy team introducing electronic prescribing ensures that the system can integrate with the pathology system. Many hospitals rely on the clinical pharmacists to review antimicrobial prescribing against local guidelines and to confirm with prescribers where this is not a justified reason. Compliance to antimicrobial guidelines can reduce patient mortality [10]. Predefined referral criteria for use by clinical pharmacists to decide which patients should be seen by the AMS team can improve consistency in referral patterns [11]. De-escalation based on culture results can reduce patient mortality [10] and is a key activity of the nonspecialist clinical pharmacist. A switch from IV to oral therapy can decrease inpatient length of stay, save money (as IV antibiotics are usually more expensive than oral) and save nursing time [12]. In some US states, some pharmacists are authorized to undertake IV to oral therapy in patients meeting predefined criteria [13]. Elsewhere, pharmacists can drive the IV to oral switch via the use of "prompt" stickers on the drug chart [14]. The alternative where patients need to be continued on IV antimicrobials because of deep-seated infections but are otherwise well is referral into outpatient parenteral antibiotic therapy (OPAT). The primary role of the pharmacist in OPAT is arranging appropriate antibiotic supplies for home use [15] but can also include assessing patients, optimizing antibiotic choice, monitoring blood levels of antimicrobials with narrow therapeutic window (e.g., glycopeptides and aminoglycosides), or monitoring for anticipated adverse events (e.g., bone marrow suppression associated with linezolid). The establishment of OPAT services can often be challenging, but the involvement of the pharmacist in preparing business cases can improve the success of implementation [16].

Restriction of antibiotics has been shown to reduce the use of broad-spectrum antimicrobials, but AMS programs should monitor for increased in use of unrestricted agents [10]. Pharmacy departments should ensure that restriction does not have negative effect on delaying treatment of patients with sepsis while authorization is sought [17]. Follow-up for authorization is key for hospitals that use second dose only restriction.

THERAPEUTIC DRUG MONITORING AND DOSE OPTIMISATION

The monitoring of blood levels for antimicrobials with narrow therapeutic spectrum has been shown to reduce the risk of nephrotoxicity [10], and this should be a key activity of the clinical pharmacist or the AMS pharmacist at the ward level [18]. The development of specific dosing tools or prescriptions can also help achieve optimum levels. There is increasing evidence that administering beta-lactam antibiotics as continual infusions in patients in critical care can improve patient outcomes [19]. The role of the pharmacist in intensive care or other high-risk populations is especially as important as optimizing antimicrobial therapy based on organ dysfunction, obesity, and cystic fibrosis [20].

AUDIT AND FEEDBACK

The use of prospective audit and feedback has been described as an effective intervention [21], but the quality from the published literature is poor. Audit and feedback is a key task of the AMS pharmacist, which includes developing and disseminating standard audit tools to allow efficient audit. Ideally, these should allow self-audit and collect patient outcomes or surrogates where possible. Quality indicators have been developed to help standardization to allow comparison between European hospitals as part of the DRIVE-AB project [22]. Effective feedback is essential. For difficult audiences, the AMS pharmacist should get the support of the lead AMS physician or infection expert that works with that specialty. Audits that are done often in small numbers that use run charts to demonstrate progress will probably be more effective in achieving change rather than a large annual audit.

SURVEILLANCE OF ANTIMICROBIAL CONSUMPTION AND LINKING TO RESISTANCE

Most antimicrobials are issued through the pharmacy computer system. In order to demonstrate improvement in antimicrobial usage, there needs to be a standardized approach to quantity measurement. The DRIVE-AB project has recommended that at least two quantity metrics are used for inpatients [22]. The choice will depend on the data that are available to the hospital. Hospitals without electronic prescribing will generally use defined daily doses, but length of therapy may be a better measure where it does exist. Where possible, linking antimicrobial consumption to resistance will be most useful for hospitals for monitoring. Many countries have identified key drug-bug combinations for monitoring and reporting at a national level [23] and developed online databases for interrogation at a national level (e.g., Scottish Antimicrobial Prescribing Group Report on Antimicrobial Use and Resistance in Humans; http://www.isdscotland.org/Health-Topics/Prescribing-and-Medicines/SAPG/AMR-Annual-Report/).

PROCUREMENT, MEDICINES PREPARATION AND MANAGING SHORTAGES

The main role of the pharmacy department in any hospital is to ensure a robust supply of antimicrobials at the best possible price. However, shortages of antibiotics are an increasing problem over the past 20 years. Between 2001 and 2013 in the United States, 148 antibacterial drugs were in short supply, with 22% experiencing multiple shortage periods [24]. These included antimicrobials such as carbapenems and colistin used for multidrug-resistant pathogens. More than half of US physicians reported that shortages had negatively affected patient outcomes because the alternative drugs were less effective, more toxic, or more costly in a 2011 survey [25]. The United States has a live medicine shortages website (www.ashp.org/shortages) to help pharmacists manage the shortages. This gives the reason for the shortage, available products, estimated resupply dates, implications for patient care, safety and alternative agents, and management. In Europe, a survey of more than 600 hospital pharmacists from 36 countries reported that antimicrobials were the agents most affected by shortages [26]. The Cooperation in Science and Technology has established a 4-year-work program to investigate European medicine shortages and to develop some solutions (www.cost.eu/COST_Actions/ca/CA15105). Where shortages occur, pharmacists should work with infectious disease physicians to develop strategies for alternate treatment that maximize patient outcome but minimize harm.

The availability of ready-to-use IV antimicrobials will save nursing time [27] and decrease the opportunity for errors to be made in preparation of IV antimicrobials. Some countries have national web-based injectable medicine guides (e.g., Wales Medusa) that summarize available information on compatibility and rate the risks of local preparation.

RISK MANAGEMENT

With up to a third of patients being on antimicrobials at any one time in a hospital, it is not surprising that antimicrobials are the most common class of medications associated with errors [28]. Antimicrobial-related medication errors need to be investigated as part of the ASP, and specific evidence-based interventions implemented to minimize future occurrence [29]. Inadvertent prescription and administration in patients with claimed antimicrobial allergy is common [30] and with occasional devastating results [30]. However, inappropriate antibiotic allergy labeling reduces available antimicrobial options for infection treatment and prophylaxis and is associated with higher incidence of *C. difficile*, MRSA, and VRE infections [31]. There is a role for pharmacy to assess claimed antimicrobial allergy on patient admission to hospital [32].

EDUCATION OF HEALTHCARE, STAFF AND PATIENTS

Education on prudent antimicrobial prescribing should start early in the undergraduate curriculum, preferably in the third year of undergraduate training in medicine and correspondent level in nonmedical curricula of pharmacy, dentistry, midwifery, nursing, and veterinary medicine to reach all health professionals [33]. While the principles of AMS are generally taught to undergraduate doctors, nurses, pharmacists, and vets in the United Kingdom, there is wide variation in the components of AMS delivered [34]. Education should probably be delivered by the expert in that element of AMS [33]. In the United Kingdom, antimicrobial prescribing and stewardship competencies have been developed for prescribers [35]. There have also been some competencies developed for AMS pharmacist in the United Kingdom [36]. Education on AMS for postregistrant healthcare professionals should ideally be tailored to their current training stage. Passive education alone is not likely to be effective (lectures or printed educational materials), unlike active training such as small group training delivered as part of an educational clinical ward round with a member of the AMS team [33] or interactive online learning [37].

PRIMARY CARE PHARMACY

The majority of antibiotic prescribing occurs in primary care, but the implementation of antimicrobial stewardship in this setting is limited [38]. The range of infections seen in primary care is small and often guided by national primary care guidelines [39]. While primary care physicians might be willing to accept supportive measures, they are against restrictions to their prescribing freedom [40]. The role of the AMS pharmacist in primary care is growing [41] and includes the localization of national guidelines, delivering education and training to doctors and prescribing nurses and pharmacists, academic detailing of general practitioners or family practices, and coordination of public health campaigns for world antibiotic awareness week.

COMMUNITY PHARMACY

Community pharmacies are often the first point of call for patients seeking advice for infections or self-care treatments for minor ailments. They can perform the triage role and direct patients to their family doctor or to an urgent care facility. Importantly, community pharmacists should say: "I am sending you to the doctor for a further opinion," but not mention antibiotics, as this generates expectation in the patient and makes it difficult for the doctor not to prescribe them. Community pharmacists can check prescribing against local guidelines by asking the patient what type of infection they are for. They can also play an important role in counseling the patient to complete the course and return any unused antibiotics back for destruction and warn of the dangers of sharing antibiotics.

Community pharmacies run health promotion campaigns to reduce the need for antibiotics by promoting good hygiene, vaccination, and self-care of viral infections. They can also raise awareness of sepsis and the benefits of early treatment. Some countries run minor ailment schemes where community pharmacies can provide symptomatic relief for no charge, thereby avoiding opportunities for patients to ask for antibiotics. Targeted point-of-care group A streptococcus antigen testing in community pharmacies has demonstrated opportunities for decreasing antibiotic prescribing rates for sore throats [42].

REFERENCES

[1] FIP. Fighting antimicrobial resistance: contribution of pharmacists. The Hague: International Pharmaceutical Federation; 2015.

[2] NICE_guidelines. Medicines optimisation: the safe and effective use of medicines to enable the best possible outcomes. Manchester 2015.

[3] Howard P, Pulcini C, Levy Hara G, West RM, Gould IM, Harbarth S, *et al.* An international cross-sectional survey of antimicrobial stewardship programmes in hospitals. J Antimicrob Chemother 2015;70(4):1245–55. PubMed PMID: 25527272.

[4] Gilchrist M, Wade P, Ashiru-Oredope D, Howard P, Sneddon J, Whitney L, *et al.* Antimicrobial Stewardship from policy to practice: experiences from UK Antimicrobial Pharmacists. Infect Dis Ther 2015;4(Suppl. 1):51–64. PubMed PMID: 26362295, Pubmed Central PMCID: 4569645.

[5] Trivedi KK, Dumartin C, Gilchrist M, Wade P, Howard P. Identifying best practices across three countries: hospital antimicrobial stewardship in the United Kingdom, France, and the United States. Clin Infect Dis 2014;59(Suppl. 3):S170–8. PubMed PMID: 25261544.

[6] Charani E, Kyratsis Y, Lawson W, Wickens H, Brannigan ET, Moore LS, *et al.* An analysis of the development and implementation of a smartphone application for the delivery of antimicrobial prescribing policy: lessons learnt. J Antimicrob Chemother 2013;68(4):960–7. Pubmed Central PMCID: Pmc3594497. Epub 2012/12/22. J Antimicrob Chemother Eng.

[7] Carreno JJ, Kenney RM, Bloome M, McDonnell J, Rodriguez J, Weinmann A, *et al.* Evaluation of pharmacy generalists performing antimicrobial stewardship services. Am J Health Syst Pharm 2015;72(15):1298–303. PubMed PMID: 26195656.

[8] Hermsen ED, VanSchooneveld TC, Sayles H, Rupp ME. Implementation of a clinical decision support system for antimicrobial stewardship. Infect Control Hosp Epidemiol 2012;33(4):412–5. PubMed PMID: 22418640.

[9] Forrest GN, Van Schooneveld TC, Kullar R, Schulz LT, Duong P, Postelnick M. Use of electronic health records and clinical decision support systems for Antimicrobial Stewardship. Clin Infect Dis 2014;59(Suppl. 3):S122–33.

[10] Schuts EC, Hulscher ME, Mouton JW, Verduin CM, Stuart JW, Overdiek HW, *et al.* Current evidence on hospital antimicrobial stewardship objectives: a systematic review and meta-analysis. Lancet Infect Dis 2016;16(7):847–56. PubMed PMID: 26947617.

[11] ASHP statement on the pharmacist's role in antimicrobial stewardship and infection prevention and control. Am J Health Syst Pharm 2010;67(7):575–7. PubMed PMID: 20237387. Epub 2010/03/20. eng.

[12] Mertz D, Koller M, Haller P, Lampert ML, Plagge H, Hug B, *et al.* Outcomes of early switching from intravenous to oral antibiotics on medical wards. J Antimicrob Chemother 2009;64(1):188–99. PubMed PMID: 19401304, Pubmed Central PMCID: 2692500.

[13] Kuti JL, Le TN, Nightingale CH, Nicolau DP, Quintiliani R. Pharmacoeconomics of a pharmacist-managed program for automatically converting levofloxacin route from i.v. to oral. Am J Health Syst Pharm 2002;59(22):2209–15. PubMed PMID: 12455304.

[14] Lesprit P, Landelle C, Girou E, Brun-Buisson C. Reassessment of intravenous antibiotic therapy using a reminder or direct counselling. J Antimicrob Chemother 2010;65(4):789–95. PubMed PMID: 20139143, Epub 2010/02/09. eng.

[15] Gilchrist M, Seaton RA. Outpatient parenteral antimicrobial therapy and antimicrobial stewardship: challenges and checklists. J Antimicrob Chemother 2015;70(4):965–70. PubMed PMID: 25538169.

[16] British Society for Antimicrobial C. BSAC Out Patient Antimicrobial Therapy Initiative 2016. Available from: http://www.e-opat.com/.

[17] Messina AP, van den Bergh D, Goff DA. Antimicrobial stewardship with pharmacist intervention improves timeliness of antimicrobials across thirty-three hospitals in South Africa. Inf Dis Ther 2015;4(Suppl. 1):5–14. PubMed PMID: 26362291, Pubmed Central PMCID: 4569642.

[18] Avent ML, Vaska VL, Rogers BA, Cheng AC, Van Hal SJ, Holmes NE, *et al.* Vancomycin therapeutics and monitoring: a contemporary approach. Intern Med J 2013;43(2):110–9.

[19] Roberts JA, Abdul-Aziz MH, Davis JS, Dulhunty JM, Cotta MO, Myburgh J, *et al.* Continuous versus Intermittent Beta-lactam Infusion in Severe Sepsis: a meta-analysis of individual patient data from randomized trials. Am J Respir Crit Care Med 2016;14. PubMed PMID: 26974879.

[20] Huttner A, Harbarth S, Hope WW, Lipman J, Roberts JA. Therapeutic drug monitoring of the beta-lactam antibiotics: what is the evidence and which patients should we be using it for? J Antimicrob Chemother 2015;70(12):3178–83. PubMed PMID: 26188037.

[21] Davey P, Peden C, Charani E, Marwick C, Michie S. Time for action-Improving the design and reporting of behaviour change interventions for antimicrobial stewardship in hospitals: Early findings from a systematic review. Int J Antimicrob Agents 2015;45(3):203–12. PubMed PMID: 25630430.

[22] DRIVE-AB. Quality indicators and quantity metrics of antibiotic use (DRIVE-AB WP1A). Innovative Medicines Initiative Joint Undertaking under grant agreement no. 115618 [Driving re-investment in R&D and responsible antibiotic use—DRIVE-AB—www.drive-ab.eu]; 2016.

[23] ESPAUR. English surveillance programme for antimicrobial utilisation and resistance (ESPAUR) 2010 to 2014. 2015 Report. Nov 2015. London: Public Health England; 2015.

[24] Quadri F, Mazer-Amirshahi M, Fox ER, Hawley KL, Pines JM, Zocchi MS, *et al.* Antibacterial drug shortages from 2001 to 2013: implications for clinical practice. Clin Infect Dis 2015;60(12):1737–42. PubMed PMID: 25908680.

[25] Gundlapalli AV, Beekmann SE, Graham DR, Polgreen PM. Infectious Diseases Society of America's Emerging Infections N. Perspectives and concerns regarding antimicrobial agent shortages among infectious disease specialists. Diagn Microbiol Infect Dis 2013;75(3):256–9 PubMed PMID: 23305775.

[26] Torjesen I. Drug shortages: it's time for Europe to act. Pharm J 2015;294:7847. Epub 31-01-2015.

[27] Mertz D, Plagge H, Bassetti S, Battegay M, Widmer AF. How much money can be saved by applying intravenous antibiotics once instead of several times a day? Infection 2010;38(6):479–82. PubMed PMID: 20981469.

[28] Lewis PJ, Dornan T, Taylor D, Tully MP, Wass V, Ashcroft DM. Prevalence, incidence and nature of prescribing errors in hospital inpatients: a systematic review. Drug Saf 2009;32(5):379–89. PubMed PMID: 19419233. eng.

[29] Keers RN, Williams SD, Cooke J, Walsh T, Ashcroft DM. Impact of interventions designed to reduce medication administration errors in hospitals: a systematic review. Drug Saf 2014;37(5):317–32. PubMed PMID: 24760475.

[30] Dworzynski K, Ardern-Jones M, Nasser S. Guideline Development Group; National Institute for Health and Care Excellence. Diagnosis and management of drug allergy in adults, children and young people: summary of NICE guidance. BMJ 2014;349:g4852. PubMed PMID: 25186447.

[31] Macy E, Contreras R. Health care use and serious infection prevalence associated with penicillin "allergy" in hospitalized patients: a cohort study. J Allergy Clin Immunol 2014;133(3):790–6. PubMed PMID: 24188976.

[32] Chen JR, Tarver SA, Alvarez KS, Tran T, Khan DA. A proactive approach to penicillin allergy testing in hospitalized patients. J Allergy Clin Immunol Pract 2016. PubMed PMID: 27888034.

[33] Pulcini C, Gyssens IC. How to educate prescribers in antimicrobial stewardship practices. Virulence 2013;4(2):192–202. PubMed PMID: 23361336, Pubmed Central PMCID: 3654620, Epub 2013/01/31. eng.

[34] Castro-Sanchez E, Drumright LN, Gharbi M, Farrell S, Holmes AH. Mapping Antimicrobial Stewardship in Undergraduate Medical, Dental, Pharmacy, Nursing and Veterinary Education in the United Kingdom. PLoS ONE 2016;11(2):e0150056. PubMed PMID: 26928009, Pubmed Central PMCID: 4771156.

[35] Public_Health_England. Antimicrobial prescribing and stewardship competencies; 2013.

[36] Sneddon J, Gilchrist M, Wickens H. Development of an expert professional curriculum for antimicrobial pharmacists in the UK. J Antimicrob Chemother 2015;70(5):1277–80.

[37] Rocha-Pereira N, Lafferty N, Nathwani D. Educating healthcare professionals in antimicrobial stewardship: can online-learning solutions help? J Antimicrob Chemother 2015;70(12):3175–7. PubMed PMID: 26429566.

[38] Ashiru-Oredope D, Budd EL, Bhattacharya A, Din N, McNulty CA, Micallef C, *et al.* Implementation of antimicrobial stewardship interventions recommended by national toolkits in primary and secondary healthcare sectors in England: TARGET and Start Smart Then Focus. J Antimicrob Chemother 2016;10. PubMed PMID: 26869693.

[39] McNulty CAM. European Antibiotic Awareness Day 2012: general practitioners encouraged to TARGET antibiotics through guidance, education and tools. J Antimicrob Chemother 2012;67(11):2543–6.

[40] Giry M, Pulcini C, Rabaud C, Boivin JM, Mauffrey V, Birge J. Acceptability of antibiotic stewardship measures in primary care. Med Mal Infect 2016;46(6):276–84. PubMed PMID: 27056661.

[41] Harris DJ. Initiatives to improve appropriate antibiotic prescribing in primary care. J Antimicrob Chemother 2013;68(11):2424–7. PubMed PMID: 24030546.

[42] Thornley T, Marshall G, Howard P, Wilson AP. A feasibility service evaluation of screening and treatment of group A streptococcal pharyngitis in community pharmacies. J Antimicrob Chemother 2016;71(11):3293–9. PubMed PMID: 27439523.

Chapter 11

The Roles of Nurses in Antimicrobial Stewardship

Oliver J. Dyar* and Céline Pulcini**
**Karolinska Institutet, Stockholm, Sweden*
***Université de Lorraine and Nancy University Hospital, Nancy, France*

INTRODUCTION

Nursing and midwifery staff make up a large proportion of the global healthcare workforce: there are between two and six nurses for every doctor in high-income countries [1]. The roles of nurses vary between and within countries, creating different patterns of responsibility and autonomy. A growing number of nurses with prescribing rights are now working in community and inpatient settings in many countries.

During a hospital admission, patients will have more contact with nursing staff than with any other healthcare worker. From triage to discharge (Fig. 1), patients encounter nurses in many different roles. Nurses are central coordinators and communicators in hospital settings, both with other healthcare workers and with patients, their relatives and their caregivers. Nurses are also often responsible for organizing movements of patients within hospitals and managing transitions from inpatient to outpatient care settings. All of these activities have the potential to influence antibiotic management at the level of the individual patient.

In the community setting, nurses are responsible for the day-to-day care of patients in long term care facilities. Nurses also work in primary care centers, running clinics for patients with acute illnesses and chronic conditions.

Many of the routine activities that nurses engage in are central to stewardship efforts, such as taking of cultures, administering of antibiotics, routinely assessing patients' input and output, catheter sites and skin integrity, and educating of patients, relatives, and caregivers. These contributions have largely lacked recognition and formal integration within antimicrobial stewardship (AMS) programs [2], and consequently, there has been little assessment of the impact that nurses have on AMS efforts. Similarly, few attempts have been made to improve this impact. Research is needed to help develop

Antimicrobial Stewardship. http://dx.doi.org/10.1016/B978-0-12-810477-4.00011-8

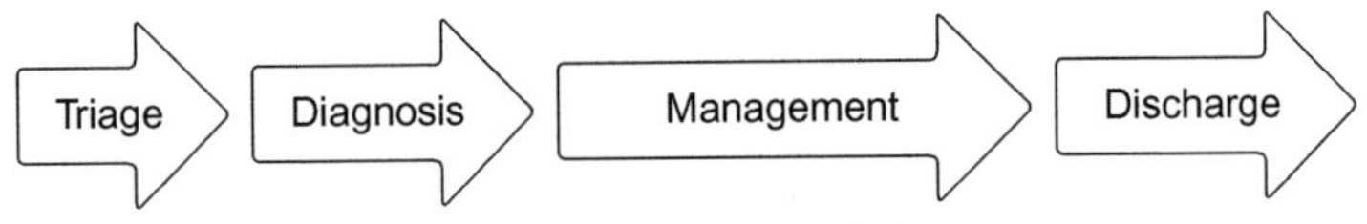

FIG. 1 Traditional journey of a patient through the hospital.

relevant structure, process, and outcome indicators for the various roles of nurses. Furthermore, undergraduate training and continuing professional development (CPD) efforts need to emphasize the importance of these stewardship-related activities.

This lack of recognition applies both to traditional stakeholders in AMS and nurses themselves: despite daily conducting many tasks relevant to the aims of AMS, fewer than one in five nurses in a recent study in Scotland were even aware of the term "antimicrobial stewardship." [3] Nurse executives and nurse educators working within individual healthcare institutions can help improve this awareness [4]. Nursing staff usually rotate less frequently between locations during their careers, compared with other healthcare workers; nurses are in a strong position to contribute to organizational memory for antimicrobial management [5]. Recently, senior nurses in the United Kingdom have been identified as core members of hospital AMS teams [6].

In this chapter, we will divide discussion of the specific roles of nurses into three parts: nurses with prescribing rights, nonprescribing staff nurses, and nurses working within an AMS team.

NURSES WITH PRESCRIBING RIGHTS

In some countries, nurses can acquire the right to prescribe medications through additional postgraduate training courses. These roles and specific prescribing rights of *nurse prescribers* and *nurse practitioners* are highly heterogeneous, even within individual countries. In hospitals, nurse prescribers often perform roles previously undertaken by junior doctors [7]. There are close to 190,000 nurse practitioners working in the United States and 65,000 nurse prescribers in the United Kingdom [8,9]. Nurse practitioners in the United States commonly prescribe antibiotics, with 90% of respondents in a survey prescribing between 1 and 15 antibiotics per week [10]. In Scotland, nurse prescribers were responsible for 4% of all antibiotic prescriptions in primary care in 2013 [9].

Antimicrobial therapy featured in the curricula of 95% of nurse practitioner programs in the Unites States, but the duration was generally limited to less than 10 h [11]. The availability of continuing education is important for nurses who have acquired prescribing rights; there may be a lack of established CPD programs when compared with physicians and pharmacists. An additional challenge is that the individual practices of nurse prescribers may be more isolated and subject to less review than other prescriber groups.

Nurse practitioners and doctors in a US hospital had similar knowledge, attitudes, and beliefs concerning antibiotic use and resistance [12]. Little data are available on the quality of antibiotic prescribing by nurse prescribers, but most studies indicate similar appropriateness to physicians [13,14]. It is likely that interventions that have successfully improved antimicrobial prescribing among physicians, such as individual audit and feedback (see Chapter 4), will have similar benefits when used for nurse prescribers. However, interventions for nurse prescribers may need to be more targeted at individuals than they are for doctors: out of 2500 nurse prescribers in Scotland, only 40% wrote any prescriptions for antibiotics in 2013, with 10% of the nurse prescribers responsible for 56% of the antibiotics prescribed [9]. These figures reflect the diversity of roles that nurse prescribers are engaged in; interventions may be best targeted toward nurses working in roles with high frequencies of antibiotic prescribing, such as minor illness clinics in primary care.

NONPRESCRIBING STAFF NURSES IN HOSPITALS, IN THE OUTPATIENT SETTING AND IN LONG TERM CARE FACILITIES

Nonprescribing staff nurses have been involved in infection control and prevention activities for a long time and are involved daily in many activities related to AMS. It is therefore surprising that nurses are not currently included in many formal AMS guidelines. Furthermore, there has been little systematic examination of the activities that nurses undertake and of how strengthening these could complement other stewardship efforts.

What Could Be Done?

Nurses spend more time with patients in inpatient settings than any other healthcare provider. Nurses collect and submit microbiological samples, administer prescribed antibiotic therapy, and monitor its effects. Nurses also have a potentially important reflective role to play regarding antimicrobial use: they work alongside many prescribers and can notice differences in their prescribing behaviors.

We have listed practical examples of AMS activities that could be performed by nonprescribing nurses in Table 1 (a nonexhaustive list) [2,5]. These activities should be adapted to the healthcare setting (hospital, primary care, and long term care facility) and to the local/national context. Nurses should be informed about the AMS policy in their workplace. This can help them to contribute actively and consistently to the AMS activities and can ensure they feel involved and recognized.

TABLE 1 Practical Examples of AMS Activities That Could Be Performed by Nonprescribing Nurses

Aspect of Management	Potential Nurse Contribution
Patient admission	• Document accurate allergy history • Medication reconciliation
Microbiological sampling	• Early and appropriate microbiological sampling, before giving antibiotics • Avoidance of unnecessary sampling (e.g. smelly or cloudy urine in the absence of clinical information suggesting infection)
Antibiotic treatment management	• Ensure timely and appropriate administration of antibiotics, in accordance with the prescribed treatment plan; record doses and timing in the patient chart • Help document the indication and duration of the treatment in the medical record • Prompt review of antibiotics around day 3 and day 7 by prescribers (including IV-oral switch) • Help comply with surgical prophylaxis quality indicators
Daily clinical progress monitoring	• Monitor progress and report • Timely information of doctors regarding any microbiological or therapeutic drug monitoring result
Patient safety and quality monitoring	• Monitor and report adverse events • Monitor and report any change in patient condition
Patient (and families) education	• Situations where an antibiotic is not needed • Main principles of responsible antibiotic use and antibiotic resistance • Therapeutic education: advice on how to take the antibiotic treatment (timing, interactions with food, etc.) and the adverse events and encourage medication compliance • Education on returning leftover antibiotics back to the pharmacy
Patient discharge	• Engage nurses in decision-making regarding a patient's suitability for outpatient parenteral antibiotic therapy (OPAT) • Medication reconciliation, including information on antibiotic prescriptions

Pathways can also be developed for specific clinical scenarios. As an example, some Californian emergency departments stipulated that the triage nurse would initiate a diagnostic work-up and notify the nurse and attending physician in charge whenever a patient had two or more systemic inflammatory response syndrome criteria together with suspected infection and signs of hypoperfusion; after the protocol was implemented, there were improvements in serum lactate measurements and in timely administration of antibiotics [15]. In another US study, outpatients with uncomplicated acute respiratory infections were offered telephone consultation with a nurse, instead of a physician visit [16]; the nurses confirmed the diagnosis, educated the patient about the self-limited nature of the disease, and advised on self-care including nonprescription analgesics and decongestants. This led to fewer unnecessary antibiotic prescriptions and medical visits, and lower healthcare costs.

What Needs to Be Addressed?

For nurses to be efficiently involved in AMS activities, some barriers need to be addressed.

First, a clear definition of nurses' input and responsibilities is useful to limit tensions between nurses and prescribers. Speaking about "antibiotic management", rather than "antibiotic prescribing", might be a simple way to acknowledge the existing contributions of nonprescribing healthcare professionals in the whole process and to reinforce the need for teamwork [5].

Second, educating nurses and ensuring they are aware of their unique roles in stewardship is paramount. A global survey of AMS activities in hospitals found that nurses received less education than doctors, both at induction to a new role and within mandatory training [17]. Nurses also need to have access to—and be supported by—information, such as practical booklets or computerized decision-support systems on how to prepare and administer antibiotics as well as on possible side effects and drug-drug interactions [18,19]. Designing a way to allow nurses to easily identify antibiotics among other medications is a first step. Table 2 summarizes educational needs for nurses participating in AMS.

Finally, the usual barriers and facilitators to implementation of a change need to be addressed and contextualized for nurses working in diverse roles and settings (see Chapter 4).

NURSES WORKING WITHIN AN AMS TEAM

A 2012 international survey including 660 hospitals worldwide showed that nurses were quite often involved in AMS teams (mean of 6 h/week), mostly outside Europe, North America, and Oceania [17].

Nurses who are part of AMS teams have additional roles compared with nonprescribing staff nurses. They are in a perfect position to educate and train the nursing staff about best practices in AMS and to help implement AMS strategies. They can also participate in audits and point prevalence surveys, and help monitor antibiotic use and resistance, among other activities of AMS teams.

TABLE 2 Education for Nurses in Antimicrobial Stewardship

Educational Need	Description
Microbiology diagnostics	– Understanding and skills on how to obtain specimens – Understanding of laboratory processes – Basic principles to interpret microbiology results
Pathophysiology and pharmacotherapy of infection	Understanding the basic principles of: – Streamlining/de-escalation – IV to oral switching
Clinical knowledge and skills	– Recognizing subtle signs of infection – Differentiating between colonization and infection
Communication skills	Improving confidence in asking prescribers about infection and antibiotic treatment

(Adapted from Olans RD, Nicholas PK, Hanley D, DeMaria A. Defining a role for nursing education in staff nurse participation in antimicrobial stewardship. J Contin Educ Nurs 2015;46:318–321.)

CONCLUSIONS AND PERSPECTIVES

The extent to which nurses can and do contribute to AMS programs is currently poorly acknowledged and studied. As the most consistent providers of care at the bedside, and with medication chart review being part of their routine professional practice, many nurses are in an ideal position to enhance antibiotic management through multidisciplinary collaboration.

Given the diversity of roles and activities undertaken by nurses, it is necessary to gain a contextual understanding of the barriers and facilitators to their contributions to AMS and of how this aspect of nursing can be developed in the future.

REFERENCES

1. The World Bank. World development indicators 2014, http://databank.worldbank.org/data; 2014 [Accessed 29 June 2016].
2. Olans RN, Olans RD, DeMaria A. The critical role of the staff nurse in antimicrobial stewardship-unrecognized, but already there. Clin Infect Dis 2016;62:84–9.
3. McGregor W, Brailey A, Walker G, Bayne G, Sneddon J, McEwen J. Assessing knowledge of antimicrobial stewardship. Nurs Times 2015;111:15–7.
4. Manning ML, Giannuzzi D. Keeping patients safe: antibiotic resistance and the role of nurse executives in antibiotic stewardship. J Nurs Adm 2015;45:67–9.
5. Edwards R, Drumright L, Kiernan M, Holmes A. Covering more territory to fight resistance: considering nurses' role in antimicrobial stewardship. J Infect Prev 2011;12:6–10.

6. Ladenheim D, Rosembert D, Hallam C, Micallef C. Antimicrobial stewardship: the role of the nurse. Nurs Stand 2013;28:46–9.
7. Inkster T, Marek A, Khanna N. Improving antimicrobial prescribing by targeting clinical nurse practitioners. J Hosp Infect 2010;76:85–6.
8. Manning ML. The urgent need for nurse practitioners to lead antimicrobial stewardship in ambulatory health care. J Am Assoc Nurse Pract 2014;26:411–3.
9. Ness V, Malcolm W, McGivern G, Reilly J. Growth in nurse prescribing of antibiotics: the Scottish experience 2007–13. J Antimicrob Chemother 2015;70:3384–9.
10. Goolsby MJ. 2004 AANP National Nurse Practitioner Sample Survey, Part II: nurse practitioner prescribing. J Am Acad Nurse Pract 2005;17:506–11.
11. Sym D, Brennan CW, Hart AM, Larson E. Characteristics of nurse practitioner curricula in the United States related to antimicrobial prescribing and resistance. J Am Acad Nurse Pract 2007;19:477–85.
12. Abbo L, Smith L, Pereyra M, Wyckoff M, Hooton TM. Nurse practitioners' attitudes, perceptions, and knowledge about antimicrobial stewardship. J Nurs Pract 2012;8:370–6.
13. Ladd E. The use of antibiotics for viral upper respiratory tract infections: an analysis of nurse practitioner and physician prescribing practices in ambulatory care, 1997–2001. J Am Acad Nurse Pract 2005;17:416–24.
14. Lenz ER, Mundinger MO, Kane RL, Hopkins SC, Lin SX. Primary care outcomes in patients treated by nurse practitioners or physicians: two-year follow-up. Med Care Res Rev 2004;61:332–51.
15. Bruce HR, Maiden J, Fedullo PF, Kim SC. Impact of nurse-initiated ED sepsis protocol on compliance with sepsis bundles, time to initial antibiotic administration, and in-hospital mortality. J Emerg Nurs 2015;41:130–7.
16. Pittenger K, Williams BL, Mecklenburg RS, Blackmore CC. Improving acute respiratory infection care through nurse phone care and academic detailing of physicians. J Am Board Fam Med 2015;28:195–204.
17. Howard P, Pulcini C, Levy Hara G, West RM, Gould IM, Harbarth S, *et al.* An international cross-sectional survey of antimicrobial stewardship programmes in hospitals. J Antimicrob Chemother 2015;70:1245–55.
18. Wentzel J, van Drie-Pierik R, Nijdam L, Geesing J, Sanderman R, van Gemert-Pijnen JEWC. Antibiotic information application offers nurses quick support. Am J Infect Control 2016;44:677–84.
19. Olans RD, Nicholas PK, Hanley D, DeMaria A. Defining a role for nursing education in staff nurse participation in antimicrobial stewardship. J Contin Educ Nurs 2015;46:318–21.

Chapter 12

Antifungal Stewardship

Ozlem K. Azap* and Önder Ergönül**
*Başkent University, Ankara, Turkey
**Koç University, Istanbul, Turkey

INTRODUCTION

Invasive fungal infections are commonly detected because of increasing number of immunocompromised patients. New diagnostic tools and increasing number of antifungal drug choices require clear decision-making process. In a study performed at a tertiary center, antifungal prescriptions were inappropriate in 40% of prophylaxis prescriptions, 78.6% of empirical prescriptions, 50% of preemptive prescriptions, and 25% of tailored therapy prescriptions; overall 57% of antifungal prescriptions were inappropriate [1]. Antifungal drug use was decreased by 50% in intensive care unit (ICU) after implementation of antifungal stewardship (AFS) [2]. It is well-known that delay in antifungal treatment in certain clinical situations results in higher mortality [3]. The term "antimicrobial stewardship" generally indicates "antibiotic stewardship," that is, dealing with antibiotics, not antifungals or antivirals.

BASIC PRINCIPLES OF ANTIFUNGAL STEWARDSHIP

The aim of implementing an AFS is to optimize the use of antifungal drugs to achieve the best outcomes while minimizing adverse events and the emergence of resistance. AFS is less established than antibacterial stewardship because of a narrower and more complex evidence base along with only a few number of available drugs.

There are some differences between AFS and antibacterial stewardship (Table 1).

The first step of implementing an AFS is obviously the establishment of the team. The members of the stewardship team were outlined in recently published reviews and it was emphasized that a multidisciplinary approach is essential [4,5]. The team optimally should include an infectious disease (ID) specialist, a clinical pharmacist with ID training, a clinical microbiologist, an

Antimicrobial Stewardship. http://dx.doi.org/10.1016/B978-0-12-810477-4.00012-X

TABLE 1 Differences Between Antibacterial and Antifungal Stewardships

	Antibacterial	Antifungal
Diagnosis	CRP Procalcitonin Culture: Earlier	Beta-D-glucan Galactomannan Computerized tomography Culture: Not very early Difficult if seated deeply
Resistance reports	Set	Improving
Prophylactic measures	Defined	Not clearly defined
Tailored treatment	Developed	Not well developed
Consensus in treatment	Better	Needs to be improved

information system specialist, an infection control professional, and a hospital epidemiologist. These members should work closely with infection control committee and pharmacy and be supported by hospital administration and medical staff leadership [4]. In addition to good collaboration, it is stated that team members should be seen as local authorities and opinion leaders [5].

The two core strategies should be used together: the first prospective audit and feedback and the second formulary restriction and preauthorization. Both strategies should be supported by educational programs. Institutional guidelines regarding candidiasis, aspergillosis, and other fungal diseases if needed should be prepared. Local epidemiological data of the institution, that is, the distribution of the causative agents, antifungal susceptibility data, the diagnostic tests readily used in the laboratory, and the available antifungal drugs should be taken into consideration while preparing the guideline. Optimization of empirical therapy, that is, de-escalation in many instances, after laboratory results are obtained is of paramount importance and should be strongly encouraged in the guidelines.

The impact of the stewardship programs should be monitored by process measures and outcome measures determined by the stewardship team. Process measures such as length of therapy, modification of therapy according to the laboratory results, switch to oral therapy, and outcome measures such as length of hospital stay, recurrence rates, and 30-day-mortality rates are recommended in the IDSA Guideline about the implementation of antimicrobial stewardship [6].

There are many guidelines published to help physicians in the management of invasive fungal disease (IFD) [7–11]. Generally, adherence to recommendations is poor resulting in huge amount of antifungal misuse. AFS programs, with a multidisciplinary approach, focus on the appropriate use of antifungals, resulting in higher clinical and microbiological success with lower resistance, toxicity, and cost [1,2,12–14]. Two steps are of paramount importance in the management of fungal infections because of long treatment durations and high costs; the first one is when to START, and the second is when to STOP. Rapid diagnostic tests and cultures and imaging modalities aid both at the initial phase and at the end of therapy. Meticulous clinical follow-up of the patient offers the best aid in evaluating the response to treatment.

There are only few antifungal drugs that should be used cautiously for the patients' well-being and also to avoid resistance. Patients with IFD often have comorbid diseases such as renal or hepatic insufficiency. Therapeutic drug monitoring is an important part of the management although they cannot be performed in many institutions. Drug-drug interactions should be taken into consideration while prescribing antifungal drugs.

INVASIVE CANDIDIASIS

Invasive candidiasis is the most common IFD, and *Candida* spp. are the fourth most common cause of bloodstream infection [15]. There are 15 *Candida* species that cause human disease, but 90% of invasive candidiasis is caused by the most common five species: *C. albicans*, *C. glabrata*, *C. tropicalis*, *C. parapsilosis*, and *C. krusei* [8]. *C. albicans* is the most commonly isolated species, but the proportion of nonalbicans *Candida* spp. has reached to 50% and is increasing gradually (Table 2).

Candida infections involving only oropharynx, esophagus, and vagina are not considered to be invasive disease; they are called mucosal candida infections. This clinical entity is not included in this chapter.

Invasive candidiasis is seen in three forms:

- Candidemia in the absence of deep-seated candidiasis
- Candidemia associated with deep-seated candidiasis
- Deep-seated candidiasis in the absence of candidemia [16].

Candidemia incidence was reported to be 6.9 per 1000 ICU patients, and 7.5% of ICU patients received antifungal therapy. Candidemia increases mortality rates in the range of approximately 50% [7].

DIAGNOSIS (TABLE 3)

Use of Cultures for the Diagnosis of Invasive Candidiasis

Sensitivity of blood culture for the diagnosis of invasive candidiasis is nearly 50% [16]. The incubation period should be at least 5 days [17]. Sensitivity varies depending on the species and the system used. Because the

TABLE 2 *Candida* **spp. at a Glance**

	Features	Antifungal Resistance
C. albicans	Responsible for about 50% of candidemia	Fluconazole 1%–2%
C. parapsilosis	Biofilm Skin contamination Fatality rate is lower than *C. albicans* More common in Southern Europe	MIC of echinocandins is high
C. glabrata	More common in elderly and HIV-positive patients	Dose-related resistance for azoles
C. tropicalis	More common among cancer patients	Less resistance to fluconazole
C. krusei	Less common Fatality rate is higher than *C. albicans*	Resistance to fluconazole Echinocandins considered

performance of blood culture for the diagnosis of candidemia is not high, it cannot be considered as an early diagnostic technique.

Is There a Useful Biomarker or a Molecular Diagnostic Tool for the Diagnosis of Invasive Candidiasis?

The β-1,3-D-glucan (*BDG*) is present in many fungal species, that is, it is a panfungal test. BDG has a poor specificity and potential of false positivity for the diagnosis of invasive candidiasis [8]. BDG test should mainly be used to rule out invasive candidiasis [17].

Candida antigen and anti-Candida antibody detection is used more commonly in Europe than in the United States. Antigen detection is limited by rapid clearance and has limited value in immunosuppressed patients. Mannan/antimannan antibody detection maybe useful for the detection of hepatosplenic (chronic disseminated) candidiasis [8].

Polymerase Chain Reaction (*PCR*) tools are not readily useful in clinical practice because of standardization and validation difficulties [17]. In Europe, a whole blood, multiplex real-time PCR assay (SeptiFast, Roche) that detects six fungi (*C. albicans*, *C. glabrata*, *C. parapsilosis*, *C. tropicalis*, *C. krusei*, and *Aspergillus fumigatus*) along with 19 bacteria has been studied, and

TABLE 3 Diagnosis of Candida Disease According to ESCMID Candida Disease Diagnosis Guideline 2012

Disease	Specimen	Test	Recommendation
Candidemia	Blood Serum	Blood culture	Essential investigation[a]
		Mannan/antimannan	Recommended
		Β-D-Glucan	Recommended
		Other antibodies	No recommendation[b]
		SeptiFast PCT kit	No recommendation
		In-house PCR	No recommendation
Invasive candidiasis	Blood Serum	Blood culture	Essential investigation
		Mannan/antimannan	No recommendation
		Β-D-Glucan	Recommended
		SeptiFast PCT kit	No recommendation
		In-house PCR	No recommendation
	Tissue and sterile body fluids	Direct microscopy and histopathology	Essential investigation
		Culture	Essential investigation
		Immunohistochemistry	No recommendation
		Tissue PCR	No recommendation
		In situ hybridization	No recommendation
Chronic disseminated candidiasis	Blood Serum	Blood culture	Essential investigation
		Mannan/antimannan	Recommended
		Β-D-Glucan	Recommended
		SeptiFast PCT kit	No recommendation
		In-house PCR	No recommendation

Continued

TABLE 3 Diagnosis of Candida Disease According to ESCMID Candida Disease Diagnosis Guideline 2012—Cont'd

Disease	Specimen	Test	Recommendation
	Tissue and sterile body fluids	Direct microscopy and histopathology	Essential investigation
		Culture	Essential investigation
		Immunohistochemistry	No recommendation
		Tissue PCR	No recommendation
		In situ hybridization	No recommendation

[a]*It must be done if possible.*
[b]*No data.*
(Adapted from Cuenca-Estrella M, Verweij PE, Arendrup MC, Arikan-Akdagli S, Bille J, Donnelly JP, *et al.* ESCMID* guideline for the diagnosis and management of Candida diseases 2012: diagnostic procedures. Clin Microbiol Infect 2012;18(Suppl. 7):9–18.)

sensitivity of the test was found to be 94% [8]. The role of PCR in testing samples other than the blood has not been established yet.

What Should Be Done for Detection of Distant Foci?

All patients with candidemia should undergo a dilated funduscopic examination performed by an ophthalmologist within the first week after initiation of antifungal therapy [7,8]. Recent data demonstrate that 16% of patients with candidemia have ocular involvement, some of which will develop sight-threatening endophthalmitis [8]. For neutropenic patients, dilated funduscopic examination should be performed within the first week after recovery from neutropenia because ophthalmologic findings of choroid and vitreal infection maybe minimal until recovery from neutropenia [8].

Transesophageal echocardiography is recommended in addition to funduscopic examination in ESCMID guidelines [7]. It is emphasized that a recent observational study found infectious endocarditis in 8.3% of patients with candidemia [7]. In IDSA guidelines, it is stated that endocarditis should be suspected when blood cultures are persistently positive, when a patient with candidemia has persistent fever despite appropriate treatment or when a new heart murmur, heart failure, or embolic phenomena occur in the setting of candidemia [8].

Follow-up cultures should be performed every day or every other day both to determine a distant focus and to show the exact time of clearance of candidemia [7,8].

Is There a "Pre-Emptive" Modality for the Treatment of Invasive Candidiasis?

There is growing body of evidence in the field of both prevention and treatment of IFD including invasive candidiasis, but still, there are challenging issues in the management of these infections [18].

Preemptive (diagnosis-driven) treatment of candidiasis is defined as therapy triggered by microbiological evidence of candidiasis without proof of invasive fungal infection in ESCMID guideline. There is no "preemptive/diagnosis-driven treatment" section in IDSA guideline. A recently published trial aimed at determining the role of preemptive therapy for invasive candidiasis following gastrointestinal surgery for intraabdominal infections was unable to provide evidence for benefits of preemptive therapy with echinocandins. Two main limitations were reported; one was too late administration of the drug; the other was the low number of invasive candidiasis cases [19].

Are Antifungal Susceptibility Tests Helpful for the Management of Invasive Candidiasis?

Intensive efforts have resulted in standardized antifungal susceptibility methods, one established by CLSI and the other by EUCAST [20]. Interpretive breakpoints for susceptibility take into account the MIC and pharmacokinetic/pharmacodynamic data and animal model data. They are reported for each species. Breakpoints have been established for the five most common *Candida* species, namely, *C. albicans*, *C. glabrata*, *C. tropicalis*, *C. parapsilosis*, and *C. krusei*. The susceptibility of *Candida* to the currently available antifungal agents is generally predictable if the causative agent was identified [8]. Antifungal resistance among *C. albicans* isolates is an emerging issue but still rare. Recent data point out triazole resistance in *C. glabrata* strains, which necessitates antifungal susceptibility results [8].

How to Choose an Antifungal Drug

There are mainly three classes of antifungals: azoles (fluconazole, itraconazole, voriconazole, and posaconazole), amphotericin B, and echinocandins (caspofungin, anidulafungin, and micafungin). When deciding which drug to choose among this narrow armamentarium, two main issues should be taken into consideration: the status of the host (immunocompromised patient, ICU patient, total parenteral nutrition (TPN) receiving patient, etc.) and the local data regarding the distribution of the *Candida* isolates and antifungal susceptibility rates.

For candidemia in both neutropenic and nonneutropenic patients, one of the echinocandins seems to be the first choice [7,8], although this recommendation has been recently challenged by an observational study in which fluconazole was similarly effective. Step-down therapy with fluconazole should be prescribed if the causative agent is reported as susceptible and only after the patient is clinically stable [7,8].

Lipid formulation of amphotericin B is a reasonable alternative if there is intolerance or resistance problem [7,8].

Voriconazole offers little advantage over fluconazole in cases other than resistance problem [8].

C. parapsilosis: Treat With What?

All three echinocandins are used safely in invasive candidiasis. Treatment of *C. parapsilosis* infections with echinocandins is controversial because minimal inhibitory concentrations are higher than other *Candida* species. In echinocandin trials, no statistically important difference was reported regarding *C. parapsilosis*. However, there are numerically higher numbers of persistent candidemia episodes in cases due to *C. parapsilosis* [7]. Fluconazole is still the first choice for fungemia due to *C. parapsilosis* [7].

What is the Duration of Therapy?

Candidemia should be treated for 14 days after the end of candidemia in uncomplicated cases. In cases of metastatic foci, duration of therapy should be individualized [7,8].

When to Switch to Oral Therapy?

Switching to oral drug can be considered after 10 days of intravenous treatment [7].

What About Intravascular Catheters?

Central venous catheters should be removed as soon as possible in nonneutropenic patients, but because gastrointestinal tract is the main source of candidemia in neutropenic patients, catheter removal should be considered on individual basis [8].

Approach to Candiduria

The presence of *Candida* in urine does not always mean infection; yet it generally reflects colonization [21]. Asymptomatic candiduria should not be treated.

Patients at high-risk for dissemination should receive antifungal drugs. High-risk group includes neutropenic patients and patients who will undergo urologic interventions [8].

Removal of the catheter alone results in the clearance of candiduria in 35% of the cases [21]. Fluconazole and amphotericin B with or without flucytosine maybe used for candida pyelonephritis [7].

Echinocandins do not achieve high concentrations in urine so they are rarely used for the treatment of urinary tract infections due to *Candida* spp. Anyway, some cases have been treated successfully with caspofungin [7].

Candida Growth in Respiratory Specimens

An autopsy study, clearly, has demonstrated that pneumonia due to candida strains is extremely rare [21]. Candida growth from respiratory specimens should be evaluated just as a colonization—not an infection [8].

Candida Colonization: Does It Matter?

Candida colonization is one of the well-known factors predisposing to invasive candidiasis. 5%–15% of patients were found to be colonized with *Candida* spp. at admission to ICU, and this ratio reaches to 50%–80% in ICU unit. Recently, it was reported that invasive candidiasis was caused by the colonizing strain in 95% of the cases [22]. There are indices, namely, colonization index, corrected colonization index, and candida score, attempting to describe the status of the patient regarding candida colonization. This may help physicians to determine the high-risk patients both for prophylaxis and early treatment.

What Are the Differences in the Management of Candidiasis Among Neutropenic and Nonneutropenic Patients?

There are two main differences in the management of patients with invasive candidiasis among neutropenic: the first one is about catheter removal. Because candidemia in neutropenic patients results from an endogenous source, catheter removal should be evaluated in the individual basis. However, in nonneutropenic patients, catheter is generally the source of candidemia, and catheter removal is recommended for management [8].

The second issue is the timing of ophthalmologic examination for the detection of distant foci. Neutropenic patients should undergo funduscopic examination within one week after the neutrophil recovery to enable to demonstrate choroid and vitreal infection. Nonneutropenic patients should undergo funduscopic examination within one week after the diagnosis of candidemia [8].

Evidence-based principles of antifungal therapy in invasive candidiasis were presented in Table 4.

Antifungal Prophylaxis

ESCMID recommendation: Patients who had recent abdominal surgery *and* recurrent gastrointestinal perforations or anastomotic leakages should receive fluconazole 400 mg/day [7].

TABLE 4 Evidence-Based Principles of Antifungal Therapy in Invasive Candidiasis [8]

1. When to start?
Empirical therapy
Risk factors (candida colonization) AND fever
Tailored therapy
ANY candidal growth from the blood, tissue, or sterile body fluids
Urinary candidal growth along with urinary tract infection symptoms
Histopathologic diagnosis of candida in tissue samples
2. How to follow-up?
All patients
Blood cultures should be drawn every day OR every other day
Transesophageal or at least transthoracic echocardiography
For nonneutropenic patients
Catheter should be removed because the source is probably the catheter
Funduscopic examination within 1 week of antifungal therapy
For neutropenic patients
Catheter removal should be considered because the source may also be the endogenous flora of the patient
Funduscopic examination within the first week after neutrophil recovery
3. When to stop?
For candidemia WITHOUT distant foci
14 days after the last negative blood culture
For candidemia WITH distant foci
Endophthalmitis: 6 weeks
Endocarditis: Valve replacement if possible and *at least* 6 weeks after surgery

IDSA recommendation: Fluconazole, 800 mg (12 mg/kg) loading dose and then 400 mg (6 mg/kg), could be used in high-risk patients in adult ICUs with a high rate (>5%) of invasive candidiasis [8].

INVASIVE ASPERGILLOSIS

Aspergillosis is the second most common type of IFD with increasing incidence among hematologic malignancy patients reaching rates of 25% [23]. There are many uncertainties and controversies in the field of aspergillosis mainly due to the heterogeneous group of patients such as allo-hematopoietic stem cell transplantation (HSCT) patients, autologous HSCT patients, and solid organ transplantation (SOT) patients.

Empirical treatment is described as "fever driven" in which the aim is to reduce the incidence of IA and certainly to reduce fungal mortality rate. Preemptive treatment is described as "diagnosis driven" and is usually defined by positive Galactomannan (GM) testing along with thorax CT (low-dose) findings. Patients receiving chemotherapy and who have fever and neutropenia for longer than 96 h despite receiving antibacterial drugs for more than 96 h are candidates for fever-driven approach.

Patients with hematologic malignancies—mostly acute leukemia or myelodysplastic syndromes requiring induction or consolidation chemotherapy—and patients receiving immunosuppressive therapy for allogeneic HSCT, particularly those with graft versus host disease, have high-risk for IA and are candidates for diagnostic-driven approach.

There is no difference between the mortality outcomes of "fever-driven" and "diagnosis-driven" approaches [24].

Empirical antifungal drugs (started based on fever-driven approach) should be DISCONTINUED if no infiltrates are seen after leukocyte recovery. The duration of tailored therapy is long, and there are only weak data to support any recommendation. The basic requirement for DISCONTINUATION of therapy is the complete resolution of radiological findings in addition to the resolution of clinical and microbiological findings. An important point is the discrimination of targeted/salvage therapy and secondary prophylaxis. Switching the therapy to oral form maybe considered in stable patients with close follow-up. Patients should be followed closely via onset of fever, symptoms, and radiological imaging after discontinuation of therapy.

DIAGNOSIS

The Role of "Conventional" Microbiology in the Diagnosis of IA: Microscopy, Stains, and Culture

Direct microscopy provides an important diagnostic benefit that is greater than culture alone. An important advantage of microscopy is the rapid availability of results. Prompt diagnosis is of pivotal importance for the management of IFD because delay in diagnosis may be lethal.

Microscopy can distinguish whether an infection is caused by a septate mold that affects the choice of antifungal treatment modality [25].

Optical brighteners are recommended on all samples from immunocompromised patients.

Cultures should be performed via inoculating the material on sabouraud dextrose agar (SDA), brain-heart infusion (BHI), and potato dextrose agar (PDA) and incubating at 30°C and 37°C for 72 h. Isolation of aspergillus from respiratory samples can suggest infection, allergy, colonization, or environmental contamination. Quantitative cultures are not discriminative for infection and colonization.

Sensitivity of culture and microscopy of BAL in IA is 50% in high-risk patients with hematologic disease. Sensitivity of fungal culture is increased in patients with advanced disease [25].

Histopathologic examination basically includes hematoxylin and eosin (HE), Gomori's methenamine silver stain (GMS), and periodic acid-Schiff (PAS) and is an essential investigation. There are also monoclonal antibodies such as WF-AF-1 (specific for *A. fumigatus*, *A. flavus*, and *A. niger*) used for immunohistochemistry tests.

Is There a Useful Biomarker or a Molecular Diagnostic Tool for the Diagnosis of IA?

Galactomannan (*GM*) in the blood is a suitable biomarker for the diagnosis of IA in neutropenic patients. Highest test accuracy is obtained by two consecutive samples with an optical density (OD) index of $\geq$0.5. GM screening of serum (two times per week) from patients with hematologic malignancies at high-risk for IA should be considered in those *not receiving antimold prophylaxis*; GM index values lower than 0.5 enables IA to be excluded in neutropenic patients [11,25].

Clinical follow-up and high-resolution computed tomography should be evaluated together along with the GM positivity.

GM has lower sensitivity in nonneutropenic patients. It is not recommended for screening in SOT recipients [11].

GM assay should be completed soon after delivery to the laboratory. Short- or long-term storage of serum at 4°C should be avoided. Testing of positive/negative serum and bronchoalveolar lavage (BAL) fluid pools showed no decline in GM index over 11 months at −20°C.

False positivity of GM may result from ingestion of ice pops, transfusions, antibiotics, and Plasma-Lyte infusion. Tazocin is no longer responsible for false positive results.

Cross reactivity of GM maybe due to histoplasmosis, fusariosis, penicilliosis, and trichosporonosis.

GM in BAL is a good tool to diagnose pulmonary aspergillosis. Optimal cutoff is between 0.5 and 1.0 [11]. BAL GM index lower than 0.5 rules out pulmonary aspergillosis [25].

There is no validated cutoff for GM in CSF for the diagnosis of cerebral aspergillosis [25].

BDG is used to diagnose IFD, not specific for IA. Fungitell is used in the United States and Europe. Specificity limits its value in the diagnosis of IA. BDG screening of serum from patients at high-risk of IFD (IA in this setting) should be considered; a negative result has a high negative predictive value, enabling the exclusion of IFD (IA in this setting) [11,25].

PCR screening of serum for aspergillus from patients at high-risk of IA disease should be considered, but there is still no consensus for the routine use of nucleic acid testing in clinical specimens [11].

It is mentioned that a negative result has a high negative predictive value enabling the exclusion of IA. Two positive tests of aspergillus PCR from the blood are required to confirm diagnosis of IA (because of significantly higher specificity) [25]. The combination of two independent detection methods, for example, GM ELISA and PCR, is superior to PCR only for the diagnosis of IA [11].

The Role of Radiology in the Diagnosis of IA

HSCT patients, SOT patients, leukemic patients, or profoundly neutropenic patients (<500 neutrophil/mL) who have a new cough, chest pain, or hemoptysis should undergo CT scan of thorax regardless of chest radiograph results. Routine use of contrast during a thorax CT scan for the diagnosis of invasive pulmonary aspergillosis is not recommended. CT scan with contrast is recommended when a nodule or a mass is located close to a large blood vessel [11,25].

Unresolved temperature after 5 days of antibiotic ± antifungal therapy, abnormal chest X-ray, microscopic evidence of fungal hyphae or positive culture of *Aspergillus* spp., and GM or BDG positivity are the other indications for a CT scan [25].

Dense, well-circumscribed lesions with or without a halo sign, air-crescent sign, and cavity are typical signs for IA as mentioned by the European Organization for the Research and Treatment of Cancer-Mycoses Study Group (EORTC-MSG) [26]. IA also may present with atypical CT signs; interdisciplinary image interpretation along with clinical findings and other diagnostic tests is essential in this setting.

Antifungal Susceptibility Tests for the Management of IA

For filamentous fungi, the first document was published in 2002 by CLSI, and the second edition of "Reference Method for Broth Dilution Antifungal

Susceptibility Testing of Filamentous Fungi; Approved Standard" was published in 2008 [27]. EUCAST also has published breakpoint standards [28]. Routine use of susceptibility tests in clinical practice is not common yet. Evidence-based principles of antifungal therapy of IA was summarized in Table 5.

Primary Prophylaxis

There are some differences between the recommendations of ECIL 5 and EFISG about primary prophylaxis. Herein recommendations of EFISG will be included. The groups recommended are:

- Hematologic malignancies with profound and prolonged neutropenia (e.g., acute leukemia)
- Allogeneic HSCT (until neutrophil recovery)

TABLE 5 **Evidence-Based Principles of Antifungal Therapy in Invasive Aspergillosis [11]**

When to start?
Empirical therapy
High-risk patients (HSCT recipients and patients treated for AML) with prolonged neutropenia and fever despite broad-spectrum antibacterial therapy
Preemptive therapy
Serum galactomannan OD index of $\geq$0.5 in two consecutive samples
BAL galactomannan OD index of $\geq$1
Typical features of invasive pulmonary aspergillosis on CT imaging
Tailored therapy
Histopathologic, cytopathologic, direct microscopic examination, or culture of a specimen obtained from a normally sterile site
When to stop?
Empirical therapy
If no infiltrates are seen after leukocyte recovery
Preemptive therapy
It is defined by whether a definite diagnosis is achieved or not
Tailored therapy
Complete resolution of radiological findings in addition to the resolution of clinical and microbiologic findings

- Allogeneic HSCT with moderate/severe GvHD and/or intensified immunosuppression

Posaconazole (200 mg tid, po) seems to be the first choice for primary antifungal prophylaxis.

Primary antifungal prophylaxis is not recommended for hematologic malignancies other than acute leukemia and for autologous HSCT patients.

Secondary Prophylaxis

Secondary prophylaxis is the treatment strategy to prevent recurrence of IA in patients at risk for IA recurrence because of a history of IA and entering a subsequent risk period of immunosuppression (6). The groups recommended are:

- Allogeneic HCT
- Chemotherapy resulting in severe neutropenia (i.e., <500/μL for at least 7 days)
- GvHD
- T-cell suppressing therapy (e.g., >1 mg/kg prednisolone equivalent corticosteroids, etc)

The choice of antifungal drug depends on the drug that was effective for that patient, that is, voriconazole or any other antimold active drug.

Ongoing therapies for IA during risk periods are not considered as secondary prophylaxis.

REFERENCES

[1] Valerio M, Rodriguez-Gonzalez CG, Munoz P, Caliz B, Sanjurjo M, Bouza E, *et al.* Evaluation of antifungal use in a tertiary care institution: antifungal stewardship urgently needed. J Antimicrob Chemother 2014;69:1993–9.

[2] Valerio M, Munoz P, Rodriguez CG, Caliz B, Padilla B, Fernandez-Cruz A, *et al.* Antifungal stewardship in a tertiary-care institution: a bedside intervention. Clin Microbiol Infect 2015;21:492.e1–9.

[3] Morrell M, Fraser VJ, Kollef MH. Delaying the empiric treatment of candida bloodstream infection until positive blood culture results are obtained: a potential risk factor for hospital mortality. Antimicrob Agents Chemother 2005;49:3640–5.

[4] Pfaller MA, Castanheira M. Nosocomial candidiasis: antifungal stewardship and the importance of rapid diagnosis. Med Mycol 2016;54:1–22.

[5] Agrawal S, Barnes R, Brüggemann RJ, Rautemaa-Richardson R, Warris A. The role of the multidisciplinary team in antifungal stewardship. J Antimicrob Chemother 2016;71:37–42.

[6] Barlam TF, Cosgrove SE, Abbo LM, MacDougall C, Schuetz AN, Septimus EJ, *et al.* Implementing an Antibiotic Stewardship Program: guidelines by the Infectious Diseases Society of America and the Society for Healthcare Epidemiology of America. Clin Infect Dis 2016;62: e51–77.

[7] Cornely OA, Bassetti M, Calandra T, Garbino J, Kullberg BJ, Lortholary O, *et al.* ESCMID* guideline for the diagnosis and management of Candida diseases 2012: non-neutropenic adult patients. Clin Microbiol Infect 2012;18(Suppl. 7):19–37.

[8] Pappas PG, Kauffman CA, Andes DR, Clancy CJ, Marr KA, Ostrosky-Zeichner L, *et al.* Clinical practice guideline for the management of candidiasis: 2016 update by the Infectious Diseases Society of America. Clin Infect Dis 2016;62:e1–e50.

[9] Dimopoulos G, Antonopoulou A, Armaganidis A, Vincent JL. How to select an antifungal agent in critically ill patients. J Crit Care 2013;28:717–27.

[10] Munoz P, Valerio M, Vena A, Bouza E. Antifungal stewardship in daily practice and health economic implications. Mycoses 2015;58(Suppl. 2):14–25.

[11] Patterson TF, Thompson 3rd GR, Denning DW, Fishman JA, Hadley S, Herbrecht R, *et al.* Practice guidelines for the diagnosis and management of aspergillosis: 2016 Update by the Infectious Diseases Society of America. Clin Infect Dis 2016;63:e1–e60.

[12] Mondain V, Lieutier F, Hasseine L, Gari-Toussaint M, Poiree M, Lions C, *et al.* A 6-year antifungal stewardship programme in a teaching hospital. Infection 2013;41:621–8.

[13] Micallef C, Aliyu SH, Santos R, Brown NM, Rosembert D, Enoch DA. Introduction of an antifungal stewardship programme targeting high-cost antifungals at a tertiary hospital in Cambridge, England. J Antimicrob Chemother 2015;70:1908–11.

[14] Leroux S, Ullmann AJ. Management and diagnostic guidelines for fungal diseases in infectious diseases and clinical microbiology: critical appraisal. Clin Microbiol Infect 2013;19:1115–21.

[15] Guinea J. Global trends in the distribution of Candida species causing candidemia. Clin Microbiol Infect 2014;20(Suppl. 6):5–10.

[16] Clancy CJ, Nguyen MH. Finding the "missing 50%" of invasive candidiasis: how nonculture diagnostics will improve understanding of disease spectrum and transform patient care. Clin Infect Dis 2013;56:1284–92.

[17] Cuenca-Estrella M, Verweij PE, Arendrup MC, Arikan-Akdagli S, Bille J, Donnelly JP, *et al.* ESCMID* guideline for the diagnosis and management of Candida diseases 2012: diagnostic procedures. Clin Microbiol Infect 2012;18(Suppl. 7):9–18.

[18] Knitsch W, Vincent JL, Utzolino S, Francois B, Dinya T, Dimopoulos G, *et al.* A randomized, placebo-controlled trial of preemptive antifungal therapy for the prevention of invasive candidiasis following gastrointestinal surgery for intra-abdominal infections. Clin Infect Dis 2015;61:1671–8.

[19] Kullberg BJ, Arendrup MC. Invasive candidiasis. N Engl J Med 2015;373:1445–56.

[20] Kauffman CA. Candiduria. Clin Infect Dis 2005;41(Suppl. 6):S371–6.

[21] Meersseman W, Lagrou K, Spriet I, Maertens J, Verbeken E, Peetermans WE, *et al.* Significance of the isolation of Candida species from airway samples in critically ill patients: a prospective, autopsy study. Intensive Care Med 2009;35:1526–31.

[22] Klingspor L, Tortorano AM, Peman J, Willinger B, Hamal P, Sendid B, *et al.* Invasive Candida infections in surgical patients in intensive care units: a prospective, multicentre survey initiated by the European Confederation of Medical Mycology (ECMM) (2006-2008). Clin Microbiol Infect 2015;21:87.e1–87.e10.

[23] Pagano L, Caira M, Candoni A, Offidani M, Fianchi L, Martino B, *et al.* The epidemiology of fungal infections in patients with hematologic malignancies: the SEIFEM-2004 study. Haematologica 2006;91:1068–75.

[24] Freemantle N, Tharmanathan P, Herbrecht R. Systematic review and mixed treatment comparison of randomized evidence for empirical, pre-emptive and directed treatment strategies for invasive mould disease. J Antimicrob Chemother 2011;66(Suppl. 1):i25–35.

[25] Schelenz S, Barnes RA, Barton RC, Cleverley JR, Lucas SB, Kibbler CC, *et al.* British Society for Medical Mycology best practice recommendations for the diagnosis of serious fungal diseases. Lancet Infect Dis 2015;15:461–74.

[26] De Pauw B, Walsh TJ, Donnelly JP, Stevens DA, Edwards JE, Calandra T, *et al.* Revised definitions of invasive fungal disease from the European Organization for Research and Treatment of Cancer/Invasive Fungal Infections Cooperative Group and the National Institute of Allergy and Infectious Diseases Mycoses Study Group (EORTC/MSG) Consensus Group. Clin Infect Dis 2008;46:1813–21.

[27] Reference method for broth dilution antifungal susceptibility testing of yeasts. 3rd informational supplement. Wayne, PA: Clinical and Laboratory Standards Institute; 2008.

[28] EUCAST. European Committee on Antimicrobial Susceptibility Testing Antifungal Agents, 2015.

Section C

AMS in Specific Clinical Settings

Chapter 13

Optimising Prescribing for Acute Respiratory Tract Infections as an Example for Antimicrobial Stewardship in Primary Care

Femke Böhmer, Anja Wollny and Attila Altiner
Rostock University Medical Center, Rostock, Germany

The worldwide overuse of antibiotics in primary care is one of several major causes for rising global antibiotic resistance. Antibiotics being available in many countries for direct purchase without a valid prescription further worsen the problem [1]. In this chapter, we describe different approaches of antimicrobial stewardship (AMS) in primary care aiming at improving the antibiotic prescribing quality. Whereas in hospital AMS also uses a so-called front-end approach to stewardship, which basically means to restrict prescriptive authority, for primary care, such an approach is not feasible. Thus, a back-end approach to stewardship is used focusing on de-escalation and modification of antibiotic prescribing. De-escalation includes changing a broad-spectrum antibiotic to one with narrower coverage, changing from combination therapy to monotherapy or—most importantly—not using antibiotic therapy altogether [2].

ANTIBIOTICS IN PRIMARY CARE

Besides the apparent threat of bacterial resistance, inadequate antibiotic use poses risks for the individual patient: side effects occur in up to every fourth patient treated with an antibiotic. It is estimated that up to 20% of all outpatient emergency department consultations due to drug side effects are caused by antibiotics [3]. Nonindicated antibiotic therapy therefore also causes relevant avoidable costs in health-care. More than three quarters of the overall increase in global antibiotic consumption between 2000 and 2010 was

Antimicrobial Stewardship. http://dx.doi.org/10.1016/B978-0-12-810477-4.00013-1

attributable to BRICS countries (Brazil, Russia, India, China, and South Africa), which largely exceeds their increase in global population. To prevent further spread of antimicrobial resistance, AMS programs are not only needed in the industrialized part of the world but also needed for primary care in emerging and developing countries [4]. However, most research on AMS in primary care has been done (or at least published) in industrialized countries and in regard to acute respiratory tract infections (ARTIs). ARTIs are among the most frequent consultation reasons in primary care and by far the most common reason for antibiotic prescriptions. ARTIs are mostly caused by viral pathogens and are usually self-limiting in otherwise healthy children and adults [5].

The treatment of the overwhelming majority of ARTIs in primary care is not a medically challenging task and should aim at helping patients to get appropriate symptomatic relieve. Antibiotics should only be prescribed for severe cases (e.g., pneumonia) of ARTI or for patients with relevant comorbidities. Analyzing the reasons for inappropriate antibiotic prescribing for ARTI is of special interest because it may also shed some light into general phenomena of inadequate prescribing.

Geographic variations in antibiotic prescriptions can be observed and cannot be explained by epidemiological reasons. Medico-cultural phenomena have been discussed as the range of antibiotic prescriptions for ARTI across different European countries reaches from 30% up to 80% of ARTI patients [1]. The highest rates of antibiotic consumption in Europe can be observed in Mediterranean countries (e.g., Spain, Greece, and Italy). The United Kingdom and Ireland show midlevel antibiotic prescription rates (50%–60%) in primary care, with narrow-spectrum antibiotics being the most prescribed substances. The Netherlands, the Scandinavian Countries, and Germany rank among the relatively low-prescribing countries. However, German primary care physicians are unexplainably fond of newer broad-spectrum antibiotics like fluoroquinolones for the treatment of ARTI. Both on a population level and for the individual patient, the link between country-specific antibiotic consumption and bacterial resistance rates has been proved.

BARRIERS TO RATIONAL USE OF ANTIBIOTICS

Even though a consensus exists that antibiotic prescription rates based on rational medical decision-making for ARTI in primary care should not exceed 10%–15% of ARTI cases, even low-prescribing countries do not reach such low-prescribing levels. Multiple studies showed that nearly every primary care physician *knows* that antibiotics are usually not necessary for the treatment of ARTI and other self-limiting infections and is aware of the problems of increasing bacterial resistance caused by nonindicated antibiotic prescribing [6]. Research identified reasons for antibiotic prescribing *despite better knowledge* (Table 1). Relevant explanations can be found at the level of

TABLE 1 Barriers to Rational Use of Antibiotics in Primary Care

Consultation level	Doctor-patient communication Doctors' side – Overestimation of patients' wish for antibiotics on doctor side – Irrational feeling of greater patient safety when prescribing antibiotics Patients' side – Miscommunication of consultation expectations
Community level	Over-the-counter availability of antibiotics Low health literacy
Governance level	Lack of applicable evidence-based guidelines Limited access to meaningful antibiotic surveillance data

doctor-patient communication. Primary care physicians seem to misinterpret and overestimate patient expectations toward antibiotic prescriptions and thus perceive a pressure to prescribe antibiotics, irrespectively of the health-care setting [7,8].

Patients wish for a plausible diagnosis, a truthful prognosis, and guidance on symptom management. The actual desire to be prescribed an antibiotic ranks far behind [9]. Nevertheless, patients are often worried about their ARTI symptoms and communicate this anxiety both verbally and nonverbally. This can create a potentially uneasy atmosphere in the consultation, which then leads to an antibiotic prescription in a partially unconscious effort by the physician to avoid conflict. In this way, irrational prescribing of antibiotics nourishes a vicious circle of medicalization of a self-limiting illness in which patients learn that antibiotics seem to be necessary for their symptoms and then expect them the next time they experience them. A misguided impression of increased patient safety *just in case* can furthermore explain why some physicians prescribe nonindicated antibiotics.

Time constraints though seem not to play a substantial role in nonindicated antibiotic prescribing as no relation between consultation length and the likelihood of an antibiotic prescription can be observed. Research also shows that patient satisfaction does not decrease if antibiotics are sparsely prescribed for ARTI [10,11].

APPROACHES TO AMS IN PRIMARY CARE

Beginning from the early 1990s, interventions were developed aiming at lowering prescription rates and improving the choice of antimicrobial substances in primary

TABLE 2 Ways to Facilitate AMS in Primary Care

Concept	Goal	Setting	Technique
Doctor-patient communication	Fostering shared decision-making	Low antibiotic prescribing	– Communication skills training to overcome miscommunication between patients' expectations and doctors' perceptions – Decision support systems or operationalized communication techniques like "delayed prescribing" – Patient education (informational leaflets and public campaigns) to empower patients to engage in shared decision-making
Point-of-care testing	Supporting shared decision-making	Medium and high antibiotic prescribing	– Ruling out severe infection – Communicating harmlessness of ARTI to patients
Feedback	Facilitating self-assessment	High prescribers	– Most effective if targeted at high prescribers – Should contain precise information, possible solutions, specific targets, and action plans – Should be delivered regularly and may be augmented by audit

care. The early interventional approaches however could not demonstrate sustained lowering in inappropriate prescribing, probably as they primarily targeted the physicians' knowledge level [1]. More elaborate interventional concepts were then developed and tested in high-quality trials. Although it seems unquestioned that the design and distribution of educational interventions should be adapted to the local medical context, interventional approaches have been developed that universally demonstrated the ability to optimize antibiotic prescribing (Table 2). Sufficient medium strength evidence exists for communication skills training and laboratory testing. Low strength evidence exists for interventions including provider and/or patient education, dissemination of guidelines, delayed prescribing, and computerized clinical decision support [12]. The potential of community pharmacists in primary care AMS is currently being discussed. By educating the public on the problems associated with antibiotic overuse and offering symptom management in ARTI, community pharmacists could play an important role for primary care AMS in the future.

Communication

As nonbiomedical aspects play an important part in antibiotic prescribing, a better understanding within the consultation combined with an improved patient-centered approach helps to reduce unnecessary antibiotic prescriptions. Although providing patient information *alone* has not shown to be effective, most interventions make use of information materials to support doctor-patient communication. Targeted information can improve patients' understanding of therapy options and choices and even initialize a process of shared decision-making in which the primary care physician and the patients communicate together using the best available evidence regarding the use of antibiotics for ARTI. Patients are ideally enabled to consider the possible positive and negative consequences of taking antibiotics for ARTI with the aim to reach consensus between doctor and patient about the best course of action. Several studies demonstrated that fostering shared decision-making can significantly reduce antibiotic prescriptions [13].

Delayed prescribing describes a slightly different communication strategy to reduce unnecessary antibiotics. Primarily in situations in which consent cannot be reached (e.g., patient/parent is still worried after receiving suitable information), the patient can pick up an antibiotic prescription after a short waiting period (24–48 h), usually without seeing the doctor again. As most ARTI are self-limiting patients may choose not to pick up the antibiotic prescription, especially when adequate symptom relief can be provided. Limited data suggest that delayed prescribing strategies reduce antimicrobial prescriptions and are safe [14,15].

Lab-Tests and Biomarkers

Due to the relatively low pretest probability for relevant bacterial infections, most laboratory tests do not have the diagnostic accuracy to sufficiently differentiate between bacterial and viral infection in ARTI in a primary care settings. However, assessment of inflammatory biomarkers can help to rule out *severe infections* irrespective of bacterial or viral origin and thus help to reduce unnecessary antibiotic prescriptions. Most research in this field has been done on point-of-care (POC) tests. POC tests have the advantage over conventional lab-tests that results are available to be discussed within the actual consultation.

Thus, a negative result can give reassurance for both patients and physician that antibiotic treatment is not indicated. Most studies have so far investigated C-reactive protein (CRP) and procalcitonin in regard to POC tests [12,16,17]. POC tests seem to have the most potential to optimize antibiotic prescribing in medium- to high-prescribing settings. The same can be said regarding the rapid strep test (RST), which is a rapid antigen detection test that is used to assist in the diagnosis of bacterial pharyngitis caused by group A streptococci (GAS). Mainly because of the very low incidence of rheumatic fever and post-streptococcal glomerulonephritis in industrialized countries, some debate is going on whether bacterial pharyngitis needs to be treated with antibiotics in many cases, which in consequence would very much limit the use of the RST in primary care. Theoretically, huge potential exists for optimizing the treatment in urinary tract infections. Although antibiotic treatment is the current standard for uncomplicated and of course complicated urinary tract infections, the empirical choice of substance is often not targeted. Thus, advanced POC tests that would deliver some form of antimicrobial susceptibility testing without the need of a classic culture could be a real game changer.

Systematic Prescribing Feedback

Systematic prescribing feedback can help to lower antibiotic prescribing [18]. Newer studies show that feedback is most effective in high-prescribing individuals. If confronted with the own high-prescribing behavior in comparison with a suitable better performing peer group, a behavioral modification can be started, mainly by creating problem awareness. However, to be effective, prescribing feedback needs to be individualized, targeted, systematic, precise, and timely. Generalized nontargeted prescribing feedback did not correlate with better antibiotic prescribing [12]. So far, it remains unclear if prescribing feedback has any influences on patient outcomes.

The advantage of prescribing feedback is that it can be upscaled with relative ease as compared with other more resource-intensive interventions like communication training and thus allows to address a large number of physicians. However, a precondition for a working feedback program is a near real-time access to routine prescribing data to be analyzed and used to create individualized feedback reports. These organizational, technical, and regulatory standards are currently only met by a small number of countries with well-developed primary care systems.

GOVERNANCE AND THE PUBLIC

In the light of rising antibacterial resistance, the World Health Organization (WHO) describes the imminent dangers of a relapse to the preantibiotic era and urges its member states to develop AMS strategies that include the primary care sector [19]. Ideally, such AMS strategies should include public information campaigns, education for health-care providers, local surveillance of antibiotic resistance, and research grants to generate evidence needed to continually improve AMS. Many countries have instituted some form of national governmental or nongovernmental programs to encourage appropriate use of antibiotics. Some programs are the UK five-year antimicrobial resistance strategy, the German DART 2020 strategy, the Australian National Antimicrobial Resistance Strategy 2015–19, or the Indian Chennai Declaration, just to name a few. But substantial funding is necessary to build and maintain meaningful programs. As the effect of large-scale campaigns is very difficult to measure—especially in terms of cost-effectiveness—sustained funding unfortunately remains a controversial political issue.

CONCLUSIONS

Evidence exists that antibiotic prescribing can be optimized with the help of tailored educational interventions in primary care. Limited evidence shows that a reduction in antibiotic use does not adversely affect patient outcomes in primary care. Albeit interventions work the better they are tailored to the actual health-care setting, generalizable elements for improving antibiotic prescribing quality can be named. The most promising approaches are currently communication training, use of laboratory tests—preferably POC tests—and systematized feedback on provider prescribing.

The present-day challenge is now to close the apparent gap between the generated evidence and transferring successful approaches into sustainable local, regional, national, and international AMS policies in primary care [20].

REFERENCES

[1] Coenen S, Ferech M, Haaijer-Ruskamp FM, Butler CC, Vander Stichele RH, Verheij TJ, *et al.* European Surveillance of Antimicrobial Consumption (ESAC): quality indicators for outpatient antibiotic use in Europe. Qual Saf Health Care 2007;16:440–5.

[2] Doron S, Davidson LE. Antimicrobial stewardship. Mayo Clin Proc 2011;86:1113–23.

[3] Harris A, Hicks L, Qaseem A. Appropriate antibiotic use for acute respiratory tract infection in adults. Ann Intern Med 2016;164(6):425–34.

[4] Van Boeckel TP, Gandra S, Ashok A, Caudron Q, Grenfell BT, Levin SA, *et al.* Global antibiotic consumption 2000 to 2010: an analysis of national pharmaceutical sales data. Lancet Infect Dis 2014;14(8):742–50.

[5] Altiner A, Bell J, Duerden M, Essack S, Kozlov R, Noonan L, *et al.* More action, less resistance: report of the 2014 summit of the Global Respiratory Infection Partnership. Int J Pharm Pract 2015;23(5):370–7.

[6] Wigton RS, Darr CA, Corbett KK, Nickol DR, Gonzales R. How do community practitioners decide whether to prescribe antibiotics for acute respiratory tract infections? J Gen Intern Med 2008;23:1615–20.

[7] Butler C, Rollnick S, Maggs-Rapport F, Pill RM, Stott NCH. Understanding the culture of prescribing: a qualitative study of general practitioners' and patients' perceptions of antibiotics for sore throats. BMJ 1998;317:637–42.

[8] Kumar S, Little P, Britten N. Why do general practitioners prescribe antibiotics for sore throat? Grounded theory interview study. BMJ 2003;18:138–41.

[9] van Driel ML, de Sutter A, Deveugele M, Peersman W, Butler CC, de Meyere M, *et al.* Are sore throat patients who hope for antibiotics actually asking for pain relief? Ann Fam Med 2006;4:494–9.

[10] Macfarlane J, Holmes W, Macfarlane R, Britten N. Influence of patients' expectations on antibiotic management of acute lower respiratory tract infection in general practice: questionnaire study. BMJ 1997;315:1211–4.

[11] Grigoryan L, Burgerhof JG, Degener JE, Deschepper R, Lundborg CS, Monnet DL, *et al.* Self-Medication with Antibiotics and Resistance (SAR) Consortium. Determinants of self-medication with antibiotics in Europe: the impact of beliefs, country wealth and the healthcare system. J Antimicrob Chemother 2008;61:1172–9.

[12] Drekonja D, Filice G, Greer N, Olson A, MacDonald R, Rutks I, *et al.* Antimicrobial stewardship programs in outpatient settings: a systematic review. Washington, DC: Department of Veterans Affairs; 2014.

[13] Coxeter P, Del Mar CB, McGregor L, Beller EM, Hoffmann TC. Interventions to facilitate shared decision making to address antibiotic use for acute respiratory infections in primary care. Cochrane Database Syst Rev 2015;11:CD010907.

[14] Colvin J, Gumaste M, Blake N, Adams M, Byrne J, Smucny J. Is delayed antibiotic prescribing a good strategy for managing acute cough? J Fam Pract 2001;50:625–8.

[15] Spurling GK, Del Mar CB, Dooley L, Foxlee R. Delayed antibiotics for respiratory infections. Cochrane Database Syst Rev 2007;18:CD004417.

[16] Cals JW, Butler CC, Hopstaken RM, Hood K, Dinant GJ. Effect of point of care testing for C reactive protein and training in communication skills on antibiotic use in lower respiratory tract infections: cluster randomised trial. BMJ 2009;338:b1374.

[17] Minnaard MC, van de Pol AC, Hopstaken RM, van Delft S, Broekhuizen BD, Verheij TJ, *et al.* C-reactive protein point-of-care testing and associated antibiotic prescribing. Fam Pract 2016;33(4):408–13.

[18] Arnold SR, Straus SE. Interventions to improve antibiotic prescribing practices in ambulatory care. Cochrane Database Syst Rev 2005;4:CD003539.

[19] World Health Organization Regional Committee for Europe. European strategic action plan on antibiotic resistance, http://www.euro.who.int/__data/assets/pdf_file/0008/147734/wd14E_AntibioticResistance_111380.pdf; 2011 [Accessed 3 October 2016].

[20] Campbell M, Fitzpatrick R, Haines A, Kinmonth AL, Sandercock P, Spiegelhalter D, *et al.* Framework for design and evaluation of complex interventions to improve health. BMJ 2000;321(7262):694–6.

Chapter 14

Antimicrobial Stewardship: What to Tell the Patients and the General Public

Vera Vlahović-Palčevski*,**
*University Hospital Rijeka, Rijeka, Croatia
**University of Rijeka Medical Faculty, Rijeka, Croatia

BACKGROUND

Antimicrobial stewardship refers to coordinated interventions designed to improve and measure the appropriate use of antimicrobials [1]. Interventions were initially intended for organizations, prescribers, and health and social care practitioners, including general practitioners (GPs), nurses, and pharmacists. However, it has been well documented that the knowledge and attitudes of both the prescribing physician and the patient influence the prescribing of antibiotics [2].

In a WHO publication issued for the World Health Day 2011, it is stated that in order to promote responsible use of antimicrobials, education strategies need to target both health-care practitioners and patients to effectively minimize unnecessary prescription of antimicrobials and reduce demand for these medicines [3].

The Australian Commission on Safety and Quality in Health Care has issued a national Antimicrobial Stewardship Clinical Care Standard, which consists of nine quality statements that describe the best quality clinical care where antimicrobial therapy is being considered. Applicable to both hospital and general practitioners, the clinical care standard provides clear guidelines for patients so that they understand how they might be best managed for their infection and to make informed treatment decisions in partnership with their clinician [4].

Patients can reduce unnecessary exposure to antibiotics by accepting evidence-based medicine advice when health-care providers indicate they are not useful for their illnesses, for example, for acute viral infections.

Antimicrobial Stewardship. http://dx.doi.org/10.1016/B978-0-12-810477-4.00014-3

Similarly, patients should be informed when an antimicrobial is necessary to treat their condition and understand why it is important to take an antimicrobial as prescribed.

STUDIES OF PATIENT SATISFACTION

An association between antibiotic prescription and patient satisfaction has been demonstrated in a number of studies.

To determine whether antibiotic prescribing had an impact on patient satisfaction, Ashworth and colleagues assessed results from the 2012 General Practice Patient Survey that included data from nearly all general practices in England ($n = 7800$). Doctors and practices with the highest volumes of antimicrobial prescribing were associated with the highest rates of overall satisfaction compared with all other prescriptions, including antidepressants, hypnotics, antipsychotics, and low-cost statins [5].

In a US survey between 2012 and 2013, more than half of interviewed health-care providers said they thought their patients expected an antimicrobial to treat a viral illness; however, only 26% of all consumers said they anticipated a prescription. Forty-two percent of consumers said they expected reassurance from their physician instead of a prescription. The results also revealed cultural differences in antimicrobial knowledge and attitudes. Hispanic consumers were more likely to believe that antimicrobials could treat cold symptoms and were less aware of resistance problems. Although 41% of Hispanic consumers expected an antimicrobial during an office visit, they were more likely to expect recommendations for symptom relief (58%). These findings indicated that provider counseling, not an antimicrobial prescription, is critical for consumer satisfaction [6].

A latent class analysis evaluating knowledge and attitudes about antimicrobial use and resistance was performed among Swedish general population to identify which groups within population are in particular need of knowledge/attitude improvement. The questionnaire was sent to 2500 randomly selected individuals aged 18–74. With a response rate of 57%, the results showed that men, younger, and more educated people were more knowledgeable, but men had less restrictive attitude toward antimicrobials [7].

KNOWLEDGE ON ANTIMICROBIALS AND RESISTANCE

In a systematic review of 24 cross-sectional studies investigating general populations' knowledge about antimicrobial use and resistance, it was found that around 50% of the sample population did not know that antimicrobial drugs are not useful for viral infections, and 27% did not know that misuse of these drugs could lead to resistance. The authors concluded that in order to improve knowledge on antimicrobials and resistance, it is advisable to strengthen initiatives in the community (i.e., campaigns) to control inappropriate demand for antimicrobials and to push physicians to inform patients of the importance of correct behavior concerning antimicrobial use [8].

In another systematic review of public's knowledge and beliefs about antimicrobial resistance, it was shown that the general population has an incomplete understanding of antimicrobial resistance and misperception about it and its causes and does not believe they contribute to its development. In conclusion, the authors suggest taking this into account when designing interventions to change the public's beliefs about how they can contribute to tackling this global issue. Public health campaigns could address resistance issues by providing information about how bacteria develop resistance, emphasizing that individual antibiotic use increases individuals' risk from resistance, and by highlighting that resistance is reversible if antibiotic use is minimized. At individual level, clinicians could use strategies such as shared decision-making when antibiotic prescribing is concerned [9].

Recently, the Wellcome Trust conducted a qualitative research about people's relationship with antibiotics. The aim was to get a deeper understanding of how people think and feel about antibiotics, their understanding of the resistance issue, and the language they use around this area—how they talk about it and what words they use.

The key findings in patients' words are presented in Box 1:

As a result, it was concluded that the current language needs to change, AMR is meaningless, and "antibiotic resistance" does not take people to the

BOX 1 How People Think and Feel About Antibiotics

Getting antibiotics means you have got a "real" illness; it is "proof" you are ill.

Antibiotics make you better and mean the trip to the doctor was "worth it;" so many see them as the "perfect solution" and will take them without hesitation.

- Some are more reluctant; this often stems from understanding the impact they have on your body ("good bacteria") and/or preferring "natural" options.
- Antimicrobial resistance (AMR) means nothing to people and resistance is only on the radar of a few.
- The concept is very hard to understand, and nearly everyone assumes it is the person not the bacteria becoming resistant—though when it is understood it has impact.
- Making it feel relevant and real and part of "my world" is vital, and the "ways in" most regularly used (cost/ deaths) do not achieve this; specific bugs have more impact.
- The language of "antibiotic resistance" and "superbugs" does not help; the challenge is to find simple, clear language that focuses on illness and implications.

Source: *Taken from Exploring the consumer perspective on antimicrobial resistance, Wellcome Trust, June 2015. Retrieved from https://wellcome.ac.uk/sites/default/files/exploring-consumer-perspective-on-antimicrobial-resistance-wellcome-jun15.pdf.*

right place; the focus of the resistance "story" for the general public needs to shift away from macrofactors such as number of deaths and cost to the economy and epidemics/pandemics; there is a need for a communication campaign for the public, which makes the issue feel real and relevant, so that the tide of opinion is behind taking action; doctors (and dentists) are key—while more research may be needed, it appears there is a need for a behavior change program for doctors, which provides clear guidelines and targets around when to prescribe antibiotics and advice on how to manage patients. Evidence of interventions to deliver effective behavior change is required urgently [10].

CAMPAIGNS

A number of initiatives have been going on in order to spread the messages about the risks associated with inappropriate use of antimicrobials.

The World Health Organization is leading a global campaign "Antibiotics: Handle with Care" calling on individuals, governments, health, and agriculture professionals to take action to address this urgent problem. The World Antibiotic Awareness Week was designated for the first time on November 16–22, 2015. It aims to increase awareness of global antibiotic resistance and to encourage best practices among the general public, health workers, and policy makers to avoid further emergence and spread of antibiotic resistance [11].

In the United States, CDC's Get Smart: Know When Antibiotics Work program works to make sure antibiotics are prescribed only when they are needed and used as they should be. Get Smart About Antibiotics Week since 2008 has been an annual effort to coordinate the work of CDC's Get Smart: Know When Antibiotics Work campaign, state-based appropriate antibiotic use campaigns, nonprofit partners, and for-profit partners during a 1-week observance of antibiotic resistance and the importance of appropriate antibiotic use [12].

The Australian Government is providing funding to NPS MedicineWise to run an antimicrobial resistance awareness and education campaign targeting both consumers and health professionals, with a particular focus on general practitioners as the main prescribers of antibiotics in Australia. The 5-year campaign commenced in February 2012, with the aim of reducing current antibiotic prescribing rates by 25% in general practice [13].

The European ECDC has launched the European Antibiotic Awareness Day (EAAD), an annual European public health initiative that takes place on 18 November every year to emphasize the importance of taking antibiotics responsibly by putting an end to unnecessary use of antibiotics and encouraging people to follow their doctor's instructions on how to take antibiotics in the appropriate way. The EAAD has been observed since 2008. The ECDC Campaign Communication Materials for the European Antibiotic Awareness Day aim to support the activity carried out by European national health authorities so as to achieve a comprehensive and consistent communication campaign across Europe with regard to the rational use of antibiotics. A European-wide campaign compliments national strategies where they exist and encourage campaigns in countries where they do not exist [14].

There are many other national or regional activities in different countries and continents with different degrees of comprehensiveness but carrying the same message [15].

MEASURING THE IMPACT OF INTERVENTIONS

As with all antimicrobial stewardship interventions, assessment of interventions to raise awareness about the importance of improvement of antimicrobial drug use may be conducted at various levels. Short-term effects of awareness interventions concern changes in the knowledge, attitudes, and beliefs about antibiotics. Midterm effects concern changes in the quantity and quality of antibiotic use, and long-term affects concern changes in the outcomes like resistance and patient outcomes. To date, studies presenting the effects of awareness interventions have evaluated only short- and midterm effects.

A systematic review of the evidence of effectiveness and cost-effectiveness of changing the public's risk-related behaviors pertaining to antimicrobial use has found that educational interventions were more likely to lead to improvements in knowledge regarding when it is appropriate to use antimicrobials rather than improvements in knowledge of antimicrobial resistance. Direct contact types of educational interventions (e.g., interventions given by GPs, teachers, and researchers and face-to-face to patients and students) are consistently more effective than mass media-type interventions [16].

A meta-analysis review was undertaken to examine the effects of mass media campaigns on changes in behavior, knowledge, and self-efficacy in general public. Although the analysis did not capture any antibiotic campaigns, other health-related campaigns (smoking, exercise, cardiovascular disease, and cancer) have provided a standard by which future campaigns might measure their success. The 5% benchmark for behavior change provides critical input when conducting power analysis. This gain needs to be weighed against the cost of campaign. When predicting an impact of a campaign, target behavior (e.g., cessation vs adoption), evaluation design, and target audience have to be taken into account [17].

Beliefs relevant to health behaviors include perceptions of risk and efficacy. Stronger perception of risk such as severity of infection caused by resistant bacteria motivates people to act. Efficacy perception determines whether people's risk-induced motivation to act results in engaging in the recommended actions. People tend to engage in actions that they believe will work. Patients (and sometimes physicians) have an inaccurate perception of what antimicrobials actually can treat, which results in misuse [18].

In a review assessing outcomes of public campaigns aimed at improving the use of antibiotics in outpatients in high-income countries, Huttner and colleagues have described campaigns of various degrees of comprehensiveness and cost. A total of 22 public education campaigns at national or regional levels between 1990 and 2007 were identified and the characteristics and outcomes evaluated. The campaigns were distributed in Europe (14), North America (3), Oceania (2), and Israel (1). In the Unites States, the Get Smart program included more than 30 different regional campaigns. In most cases, the campaigns were part of a national strategy to reduce antimicrobial use. All campaigns focused mainly on respiratory tract infections and education was mostly symptom-oriented. The intensity of the campaigns varied widely, from simple use of Internet distribution channels to expensive mass-media campaigns.

The results of most campaigns have shown positive effect on the use of antibiotics. The greatest effect had multifaceted campaigns conducted repeatedly over several years. However, the impact on AMR could not be assessed from the data available [15].

Antibiotic awareness campaigns in England using predominantly posters and leaflets had little or no impact on knowledge, behavior, or prescription rates. On the contrary, campaigns that in addition to printed educational materials used mass media and social marketing techniques have led to reduction in antibiotic use and changes in professional and public attitudes [19].

A systematic review evaluating effectiveness of information leaflets used for informing patients about common infections during consultations in general practice suggests that the use of information leaflets in general practice consultations is effective in reducing antibiotic prescription by GPs and actual antibiotic use by patients and their intention to reconsult for future similar episodes of illness. Included studies mostly focused on the immediate effects of leaflets on reconsultation rates. It was concluded that the use of patient information leaflets for common infections in general practice should be encouraged but their contributing role in multifaceted interventions targeting management of common infections in primary care needs to further exploration [20].

CAMPAIGN MESSAGES

In September 2015, NICE has published a draft guideline for consultation on antimicrobial stewardship—changing risk-related behaviors in the general population, which proposes interventions to change public and professional behaviors. It covers interventions to change people's behavior to help reduce antimicrobial resistance and stop the spread of resistant microorganisms. This includes making people aware of the importance of using antimicrobials correctly and the dangers associated with their overuse and misuse. The guideline includes measures to prevent and control infection that can stop people needing antimicrobials or spreading infection to others [21].

To develop effective interventions to promote antibiotic stewardship, a strong understanding of cultures, health-care systems, and antimicrobial use and resistance rates across countries is necessary.

A person-centered approach using latent class analysis identifies three different profiles based on their antibiotic stewardship behaviors: stewards, stockers, and demanders. This stratification suggests three different types of campaign goals: to encourage stewards to follow through on their intentions, to encourage stockers to dispose of their antibiotics, and to influence demanders to accept medical advice when an antibiotic is not indicated and to dispose of their leftover antibiotics.

The demanders' class provides an important target for campaign designers. Demanders could be targeted with message to increase perceived severity of antibiotic-resistant infections and correcting their inaccurate perceptions of response efficacy. Much more messages have been addressing to increase perceived severity, and less attention has been paid to construct messages that correct inaccurate perceptions of response efficacy. In addition, messages should offer effective treatment or prevention alternatives. Stockers' class needs different approaches. Interestingly, they seem to be opinion leaders with the higher knowledge about antibiotics. They like to be a source of health advice, but they may also be holding onto health resources such as leftover antibiotics. Thus, it may be critical to target these mavens first in order to prevent them spreading misinformation and allow them to spread accurate messages about the efficacy of antibiotics [18].

A study was performed in the United Kingdom about clinicians' and parents' perceptions of communication within consultations for respiratory tract infections in children 3 months to 12 years and what influence clinician-to-clinician communication had on parents' understanding of antibiotic treatment. For that purpose, 60 consultations in six primary care practices were videorecorded. Clinicians commonly told parents that antibiotics are not effective against viruses, but it did not have much impact on parents' beliefs about antibiotics. They believed that antibiotics were needed for more severe infections and it was supported by the way clinicians accompanied viral diagnoses with problem minimizing language and antibiotic prescriptions with more problem-oriented language. Antibiotic prescriptions tended to confirm patients' beliefs about what indicated illness severity [22].

A study on antibiotic prescribing and consumption in the community for urinary tract infection (UTI) from the perspective of the GP and community member performed in Ireland demonstrated how qualitative research can identify the interacting processes, which are instrumental to the decision to prescribe or consume an antibiotic. The authors identified at least three patient profiles: the young professional ("quick fixers"), the young mothers ("advice seekers"), and the mature patient ("experienced consulters"). Each type of patient can be satisfied differently. The quick fixers prioritize their personal health, adopt a low involvement approach, and are satisfied to receive their antibiotic prescription; the advice seekers adopt a higher involvement perspective, discussing different treatment options for their illness, not necessarily an antibiotic, and the experienced consulters have experienced a UTI and antibiotic treatment in the past reinforcing the norm and expectations of treatment. They concluded that behavioral interventions should focus on improving the quality of antibiotic prescribing for UTIs by encouraging GPs to reflect on their current antibiotic prescribing practices, including when they prescribe and what antibiotics they choose, supporting a dialogue between the GP and the patient within the consultation about the positive and negative aspects of antibiotic treatment for UTI, and integrating behavioral change messages into routine care without elongating the consultation [23].

A variety of patient educational initiatives have been shown to effectively improve patient knowledge of appropriate antibiotic use [14]. Most educational efforts target adult patients. An EU project, e-Bug, aims to develop and disseminate, across Europe, a junior and senior school educational resource for teachers covering microbes, hygiene, antibiotics, and prevention of infection. An educational pack containing fun lesson plans and activities is accompanied by a web site hosting the lesson plans and complementary games for young people and their families to play in the classroom or at home. The resources have been translated, adapted for and disseminated to schools across countries in Europe, and endorsed by the relevant government departments of health and education [24,25].

Education about antimicrobials and resistance is a critical element in the multifaceted approach required to ensure appropriate antimicrobial use and ultimately slowing antibiotic resistance.

Patients and general public can play an important role in slowing antibiotic resistance by engaging in behaviors to prevent infections, such as receiving recommended vaccines and practicing proper personal hygiene in their daily lives. Antimicrobial stewardship programs intended for general public and patients should address these aspects as well.

REFERENCES

[1] Society for Healthcare Epidemiology of America, Infectious Diseases Society of America, Pediatric Infectious Diseases Society. Policy Statement on Antimicrobial Stewardship by the Society for Healthcare Epidemiology of America (SHEA), the Infectious Diseases Society of America (IDSA), and the Pediatric Infectious Diseases Society (PIDS). Infect Control Hosp Epidemiol 2012;33(4):322–7.

[2] Hwang TJ, Gibbs KA, Podolsky SH, Linder JA. Antimicrobial stewardship and public knowledge of antibiotics. Lancet Infect Dis 2015;15(9):1000–1.

[3] Policy package to combat antimicrobial resistance. WHO, http://www.reactgroup.org/uploads/publications/other-publications/WHO-Policy-Package-to-Combat-Antimicrobial-Resistance.pdf [Accessed 01 June 2016].

[4] Australian Commission on Safety and Quality in Health Care. Antimicrobial Stewardship Clinical Care Standard, http://www.safetyandquality.gov.au/publications/antimicrobial-stewardship-clinical-care-standard/ [Accessed 30 August 2016].

[5] Ashworth M, White P, Jongsma H, Schofield P, Armstrong D. Antibiotic prescribing and patient satisfaction in primary care in England: cross-sectional analysis of national patient survey data and prescribing data. Br J Gen Pract 2016;66(642):e40–6.

[6] Watkins LKF, Sanchez GV, Albert AP, Roberts RM, Hicks LA. Knowledge and attitudes regarding antibiotic use among adult consumers, adult hispanic consumers, and health care providers—United States, 2012–2013. MMWR Morb Mortal Wkly Rep 2015;64(28):767–70.

[7] Vallin M, Polyzoi M, Marrone G, Rosales-Klintz S, Tegmark Wisell K, Stålsby Lundborg C. Knowledge and attitudes towards antibiotic use and resistance—a latent class analysis of a Swedish population-based sample. PLoS ONE 2016;11(4):e0152160.

[8] Gualano MR, Gili R, Scaioli G, Bert F, Siliquini R. General population's knowledge and attitudes about antibiotics: a systematic review and meta-analysis. Pharmacoepidemiol Drug Saf 2015;24(1):2–10.

[9] McCullough AR, Rathbone J, Parekh S, Hoffmann TC, Del Mar CB. Not in my backyard: a systematic review of clinicians' knowledge and beliefs about antibiotic resistance. J Antimicrob Chemother 2015;70(9):2465–73.

[10] Wellcome Trust and Good Business. Exploring the consumer perspective on antimicrobial resistance. http://wellcomelibrary.org/media/b24978000/0/40db9838-5dda-4cf1-b0ff-c8b2feff02cc.pdf [Accessed 23 August 2016].

[11] World Antibiotic Awareness Week. WHO, http://www.who.int/mediacentre/events/2015/world-antibiotic-awareness-week/event/en/ [Accessed 01 June 2016].

[12] Get smart: now when antibiotics work. CDC, http://www.cdc.gov/getsmart/community/ [Accessed 01 June 2016].

[13] Antimicrobial Resistance. Australian Government Department of Health, http://www.health.gov.au/internet/main/publishing.nsf/Content/ohp-amr.htm [Accessed 01 June 2016].

[14] European Antibiotic Awareness Day. ECDC, http://ecdc.europa.eu/en/eaad/antibiotics-plan-campaign [Accessed 01 June 2016].

[15] Huttner B, Goossens H, Verheij T, Harbarth S. Characteristics and outcomes of public campaigns aimed at improving the use of antibiotics in outpatients in high-income countries. Lancet Infect Dis 2010;10(1):17–31.

[16] King S, Exley J, Taylor J, Kruithof K, Larkin J, Pardal M. Antimicrobial stewardship: the effectiveness of educational interventions to change risk-related behaviours in the general population: a systematic review. Santa Monica, CA: RAND Corporation; 2015.

[17] Anker AE, Feeley TH, McCracken B, Lagoe CA. Measuring the effectiveness of mass-mediated health campaigns through meta-analysis. J Health Commun 2016;21(4):439–56.

[18] Smith RA, Quesnell M, Glick L, Hackman N, M'Ikanatha NM. Preparing for antibiotic resistance campaigns: a person-centered approach to audience segmentation. J Health Commun 2015;20(12):1433–40.

[19] Ashiru-Oredope D, Hopkins S. Antimicrobial resistance: moving from professional engagement to public action. J Antimicrob Chemother 2015;70(11):2927–30.

[20] de Bont EG, Alink M, Falkenberg FC, Dinant G, Cals JW. Patient information leaflets to reduce antibiotic use and reconsultation rates in general practice: a systematic review. BMJ Open 2015;5(6):e007612.

[21] Antimicrobial stewardship—changing risk related behaviours in the general population. NICE guideline, https://www.nice.org.uk/guidance/gid-phg89/resources/antimicrobial-resistance-changing-riskrelated-behaviours-in-the-general-population-full-guideline2.

[22] Cabral C, Ingram J, Lucas PJ, Redmond NM, Kai J, Hay AD, *et al.* Influence of clinical communication on parents' antibiotic expectations for children with respiratory tract infections. Ann Fam Med 2016;14(2):141–7.

[23] Duane S, Domegan C, Callan A, Galvin S, Cormican M, Bennett K, *et al.* Using qualitative insights to change practice: exploring the culture of antibiotic prescribing and consumption for urinary tract infections. BMJ Open 2016;6(1):e008894.

[24] e-bug. www.e-bug.eu/.

[25] e-Bug: educating children and young people on hygiene, the spread of infection and antibiotics. Perspect Public Health 2016;136:192–3.

Chapter 15

Antimicrobial Stewardship in Long-Term Care Facilities

Céline Pulcini
Université de Lorraine and Nancy University Hospital, Nancy, France

INTRODUCTION

The number of elderly people living in long-term care facilities (LTCFs) is increasing worldwide; in Europe, around 1% of the population lives in such facilities.

Antibiotic use is high in LTCFs compared with the primary care setting, and bacterial resistance prevalence is also much higher. Antimicrobial stewardship (AMS) is thus a necessity in LTCFs, but quite surprisingly, AMS in LTCFs has been quite overlooked so far compared with hospitals and primary care settings.

Some recent literature reviews have nicely addressed AMS in LTCFs [1–6], and we will provide here a practical overview of the topic, based on these reviews, focusing on the challenges and the possible AMS strategies.

ANTIBIOTIC USE IN LTCFS

The prevalence of antibiotic use is very high in LTCFs: 3%–15% of residents receive an antibiotic on any given day in point-prevalence surveys, 80% of residents with a suspected infection are prescribed antibiotics, and 50%–80% of residents receive at least one antibiotic course per year. There are also huge (5- to 10-fold) variations of antibiotic use between LTCFs, suggesting that overuse is frequent.

The types of antibiotics that are prescribed in LTCFs are quite similar to primary care patterns, except for higher use of parenteral route (10%–20% in France, e.g., mostly ceftriaxone). The main motives for antibiotic use are suspected urinary tract infections (UTI, 32%–66%), respiratory tract infections (RTI, 15%–36%), and skin and soft tissue infections (SSTI, 13%–18%).

As in any other setting, around one-third of antibiotic prescriptions are appropriate, whereas one-third are unnecessary and one-third are inappropriate

Antimicrobial Stewardship. http://dx.doi.org/10.1016/B978-0-12-810477-4.00015-5

(i.e., an antibiotic is indicated, but the choice of the antibiotic, the dose, the duration, etc. are inappropriate). Excessive durations of treatment are the leading cause of inappropriate prescriptions in LTCFs, with treatments longer than 7 days in half of the cases in a Canadian study [7]; in this study, duration was dependent on the prescriber, not the patients' characteristics.

BACTERIAL RESISTANCE IN LTCFS

The prevalence of multidrug-resistant bacteria in LTCFs is very high and sometimes higher than in hospitals. Residents are often colonized for several months, since they accumulate many risk factors (high level of transmission, antibiotic use, wounds, catheters, etc.). LTCFs are therefore considered as a multidrug-resistant bacteria "reservoir."

The incidence of *Clostridium difficile* infections is also much higher than in primary care settings, for the same reasons.

ANTIBIOTIC PRESCRIBING: A CHALLENGE IN LTCFS

As shown by van Buul *et al.* in their qualitative study conducted in LTCFs in the Netherlands [8], the physician considering prescribing an antibiotic in a resident living in a LTCF is influenced by and will take into account many factors, in particular:

- Assessment of the current and past medical history, consideration of advance care plans for residents in palliative care
- Utilization of diagnostic tests
- Perception of the risks of treatment and the risks of nontreatment
- Influence of others: colleagues, patient/family, and nursing staff
- Influence of the environment/culture

We will discuss here the main reasons explaining why antibiotic prescribing is particularly challenging in LTCFs.

Difficult Decision-Making Process: High Level of Diagnostic Uncertainty

Making an accurate diagnosis of infection is particularly challenging in residents of LTCFs, even for the most expert clinician. Getting relevant clinical information is difficult, since residents frequently suffer from hearing loss, dementia, aphasia, etc. Clinical findings are often atypical and nonspecific, and elderly patients with multiple comorbidities can develop infections without running a fever.

Investigations are also of limited added value to the clinician. First, there is usually a lack of onsite diagnostic facilities (biology, imaging, etc.), and transporting elderly residents to perform investigations is not easy. Second, getting good quality samples/investigations is difficult: for example, residents are often incontinent and it is complicated to get a good quality urine sample; chest X-rays are often of poor quality since residents do not comply with the instructions. Finally, even when a culture comes back positive (which is often the case, the samples having a high risk of being contaminated and colonization being very frequent in LTCFs residents), making the difference between colonization and infection is tricky, in light of the frequently atypical clinical findings and difficulties in getting a clear history from the patient.

For all these reasons, diagnostic uncertainty is the rule rather than the exception in patients suspected of having an infection in LTCFs.

Healthcare Organization and Culture

The way LTCFs are organized is very different from one country to another, and there is also considerable heterogeneity within the same country. Understanding the organization and culture of the LTCF is crucial to design any AMS program (see Chapter 4).

Regarding medical staff, LTCFs are usually served by multiple doctors (to the extreme, each resident can be attended to by his/her own general practitioner). There is usually a lack of onsite doctors to provide immediate clinical assessment, explaining that around half of antibiotics are prescribed over the phone. Locum staff is also quite frequently involved in patient care. All this can lead to some degree of unfamiliarity with patients, which makes clinical assessment of residents even more difficult.

Nurses are the cornerstone of care in LTCFs, and doctors rely on the information they provide to prescribe antibiotics. However, rapid staff turnover, shortage of nurses, and lack of training on infection are frequent issues among nursing staff.

Finally, doctors sometimes prescribe antibiotics to avoid hospitalization or a revisit.

Lack of Local Resistance and Antibiotic Use Data

These data are available in less than 20% of European LTCFs, and this leads to a lack of awareness from healthcare professionals and administrators regarding the problem of antibiotic overuse and high prevalence of multidrug-resistant bacteria.

High Prevalence of Bacterial Colonization

Residents in LTCFs are often colonized with bacteria, mostly in wounds and in urine (100% of long-term catheterized patients have positive urine cultures; 25%–50% of noncatheterized women and 15%–40% of noncatheterized men have asymptomatic bacteriuria).

Microbiological samples thus often come back positive in elderly residents living in LTCFs. It is then up to the healthcare professional to decide if this positive result corresponds to a true infection or is simply reflecting colonization, based on the clinical findings; it has been shown in the literature that a positive culture result is a powerful driver for unnecessary antibiotic use. Treatment of asymptomatic bacteriuria confers no benefit in elderly residents, with or without a urinary catheter.

Antibiotic Use and End-of-Life Care

Antibiotics are largely prescribed in residents living their last days in LTCFs (mostly for suspicion of pneumonia in demented patients). The positive impact of such a treatment on patients' outcomes (mostly comfort) is not proven, and most experts agree that antibiotics are not needed in that situation. Advance care plans might be helpful to limit these unnecessary prescriptions.

Patients' and Families' Expectations

Healthcare professionals working in LTCFs face the same problems as in primary care practice regarding potential patients' and families' expectations for antibiotics.

Guidelines

Infection guidelines are often not available in LTCFs. When they are, they rarely include best practice recommendations for investigations (microbiology, imaging, etc.) and both diagnostic and therapeutic recommendations for antibiotic prescriptions.

Moreover, some clinicians argue that the available guidelines are not applicable to the older LTCF population, which is supposed to be somewhat different and more "frail." This usually leads to overprescription of antibiotics, the risks of nontreatment being perceived as far more important than the risks of treatment by the clinician ("better safe than sorry" concept).

Lack of Awareness of the Bacterial Resistance Problem

There is a general lack of awareness of the bacterial resistance problem, and this is even more the case in LTCFs. The impact of bacterial resistance is overlooked, all the more since residents living in LTCFs have quite a short life expectancy. In LTCFs, antimicrobial stewardship is not a priority compared with other competing issues, such as palliative care, pain, and nutrition management.

POTENTIAL STRATEGIES TO IMPROVE ANTIBIOTIC USE IN LTCFS

Specific evidence-based guidelines regarding prudent antibiotic use in LTCFs are lacking. The 2016 Infectious Diseases Society of America (IDSA) guidelines on AMS mention the topic but without much detail [9]. Meanwhile, the best strategy is probably to depart from current guidelines on AMS/general principles of AMS programs (see Chapters 2 and 4) and to adapt them to the LTCF specificities, as done by the US Centers for Disease Control and Prevention (CDC, Core elements of antibiotic stewardship for nursing homes, available at http://www.cdc.gov/longtermcare/index.html); qualitative studies might also be useful [10].

The lack of LTCFs-specific evidence-based AMS guidelines is not so surprising, as there are only around 20 published interventional studies in LTCFs aiming at promoting responsible antibiotic use so far. Mostly were conducted in the United States and Canada and only a few in Europe.

We will highlight here a selection of practical suggestions, but generic components of AMS programs also apply to LTCFs (availability of guidelines, audit, feedback, etc.).

Where to Start?

You need to target situations where antibiotic misuse is frequent (Table 1) and where improving prescribing will be easier. You then follow a stepwise approach and tackle progressively situations that are more challenging.

In order to change the healthcare professionals' behavior, you need of course to motivate them, but changing the system is also a powerful tool (e.g., no reporting of susceptibility testing in urines coming from a long-term catheterized patient [11]).

TABLE 1 Situations Where Room for Improvement is Usually Large in LTCFs
• Unnecessary antibiotic prophylaxis (especially for UTI) • Unnecessary antibiotic treatments for colonization (especially unnecessary treatments of asymptomatic bacteriuria in residents with or without a urinary catheter) • Absence of diagnostic and therapeutic guidelines • Excessive use of broad-spectrum antibiotics (such as cephalosporins and quinolones) • Unnecessary topical antibiotic treatments • Excessive durations of treatment

Limit Antibiotic Prescriptions Without a Clinical Examination

Education

Mandatory and regular training of physicians and nurses on prudent antibiotic use and infection management is crucial.

Patients and their families must also be informed on the benefits and risks of antibiotics and on the situations when an antibiotic is not needed.

Microbiological Investigations

Unnecessary samples (e.g., urine cultures or wound swabs) will drive unnecessary antibiotic use if the culture comes back positive. Having clear indications for performing microbiological investigations is thus very important for the nursing staff. Ideally, these investigations should be prescribed by a physician, after a clinical examination, but it is not always possible to do so.

Reassess Antibiotic Prescriptions Around Day 3

Given the high level of diagnosis uncertainty when prescribing an antibiotic in a LTCF resident, reassessing the prescription around day 3, taking into account the clinical evolution and the microbiological results is quite logical. A "day 3 bundle" has been validated and might be useful (Table 2) [12]. The nurses can participate in this reassessment process: an English cluster-randomized control study showed that asking nurses completing a

TABLE 2 Day 3 Bundle [12]

Four process measures should be documented in the medical notes 24–96 h after the antibiotic course is started, thus around day 3 of treatment:

1. Is there an antibiotic plan (name, dose, route, interval of administration, and planned duration)?
2. Is there a review of the diagnosis?
3. If positive microbiological results are available, is there any adaptation of the antibiotic treatment, for example, streamlining (=de-escalation) or discontinuation?
4. If the patient was initially started on intravenous (iv) antibiotic therapy, is the possibility of iv-oral switch documented?

A fifth measure is made up of the grouping of the preceding four measures to form a care bundle. A care bundle is a grouping of best practices with respect to a disease process that individually improve care, but when applied together result in substantially greater improvement. Their application favors an "all or none" approach as opposed to piecemeal measures. Bundles are dichotomous, so compliance is assessed in a simple yes/no measure.

5. The "day 3 bundle" is deemed completed only if all preceding four measures are completed.

checklist at initiation and 48–72 h after starting antibiotic treatment led to a significant decrease in antibiotic use [13].

Role of the Microbiology Laboratory

The microbiologist plays a major role in improving antibiotic use, particularly by improving reporting of the results. Educational messages on the report might be useful (e.g., for a positive urine culture, stating that only clinical findings can differentiate between colonization and infection). Selective reporting of susceptibility testing results is also a powerful tool (see Chapter 9) and is recommended in the 2016 IDSA guidelines [9].

Rapid and Point-of-Care Diagnostic Tests

No interventional studies have been conducted in LTCFs to assess the impact of point-of-care diagnostic tests or biomarkers (e.g., C-reactive protein, procalcitonin, and influenza test) on antibiotic use, to the best of our knowledge. Research on that topic is clearly needed.

Innovative Strategies Need to be Tested

Strategies that have proved useful in hospitals or primary care settings should also be tested and evaluated in LTCFs: infection champions, AMS teams, and infectious disease specialist advice available on the phone or using telemedicine, computerized decision support systems, etc.

Process and outcome indicators are of course needed to monitor these interventions and adapt the AMS program (see Chapters 3 and 4, and Ref. [14]).

Regulatory Measures

At a national level, regulatory measures are a very powerful trigger for change. Adding AMS in the list of mandatory missions of LTCF medical coordinators is one way to go. Integrating AMS in existing LTCFs quality and/or safety and/or infection prevention and control programs may also help and allows sharing resources; one might also consider integrating LTCFs' AMS programs into existing primary care or hospital-based AMS programs. Finally, making AMS programs mandatory components of certification/accreditation of LTCFs has of course a huge and quick impact.

CONCLUSIONS AND PERSPECTIVES

AMS in LTCFs is currently vastly overlooked, but it deserves our full attention since elderly residents receive many antibiotics and are at high risk of being infected or colonized with multidrug-resistant bacteria. Research is also needed to help identify the most successful interventions.

REFERENCES

[1] Morrill HJ, Caffrey AR, Jump RL, Dosa D, LaPlante KL. Antimicrobial stewardship in long-term care facilities: a call to action. J Am Med Dir Assoc 2016;17:183.e1–183.e16.
[2] Dyar OJ, Pagani L, Pulcini C. Strategies and challenges of antimicrobial stewardship in long-term care facilities. Clin Microbiol Infect 2015;21:10–9.
[3] Crnich CJ, Jump R, Trautner B, Sloane PD, Mody L. Optimizing antibiotic stewardship in nursing homes: a narrative review and recommendations for improvement. Drugs Aging 2015;32:699–716.
[4] Fleming A, Bradley C, Cullinan S, Byrne S. Antibiotic prescribing in long-term care facilities: a meta-synthesis of qualitative research. Drugs Aging 2015;32:295–303.
[5] Nicolle LE. Antimicrobial stewardship in long term care facilities: what is effective? Antimicrob Resist Infect Control 2014;3:6.
[6] Rhee SM, Stone ND. Antimicrobial stewardship in long-term care facilities. Infect Dis Clin North Am 2014;28:237–46.
[7] Daneman N, Gruneir A, Bronskill SE, Newman A, Fischer HD, Rochon PA, *et al.* Prolonged antibiotic treatment in long-term care: role of the prescriber. JAMA Intern Med 2013;173:673–82.
[8] van Buul LW, van der Steen JT, Doncker SM, Achterberg WP, Schellevis FG, Veenhuizen RB, *et al.* Factors influencing antibiotic prescribing in long-term care facilities: a qualitative in-depth study. BMC Geriatr 2014;14:136.
[9] Barlam TF, Cosgrove SE, Abbo LM, MacDougall C, Schuetz AN, Septimus EJ, *et al.* Implementing an antibiotic stewardship program: guidelines by the Infectious Diseases Society of America and the Society for Healthcare Epidemiology of America. Clin Infect Dis 2016;62:e51–77.
[10] Lim CJ, Kwong M, Stuart RL, Buising KL, Friedman ND, Bennett N, *et al.* Antimicrobial stewardship in residential aged care facilities: need and readiness assessment. BMC Infect Dis 2014;14:410.
[11] Leis JA, Rebick GW, Daneman N, Gold WL, Poutanen SM, Lo P, *et al.* Reducing antimicrobial therapy for asymptomatic bacteriuria among noncatheterized inpatients: a proof-of-concept study. Clin Infect Dis 2014;58:980–3.
[12] Pulcini C, Defres S, Aggarwal I, Nathwani D, Davey P. Design of a 'day 3 bundle' to improve the reassessment of inpatient empirical antibiotic prescriptions. J Antimicrob Chemother 2008;61:1384–8.
[13] Fleet E, Gopal Rao G, Patel B, Cookson B, Charlett A, Bowman C, *et al.* Impact of implementation of a novel antimicrobial stewardship tool on antibiotic use in nursing homes: a prospective cluster randomized control pilot study. J Antimicrob Chemother 2014;69:2265–73.
[14] Mylotte JM. Antimicrobial stewardship in long-term care: metrics and risk adjustment. J Am Med Dir Assoc 2016;17:672.e13–8.

Chapter 16

Antimicrobial Stewardship in ICU

Jeroen Schouten* and Jan De Waele**
**Radboud University Medical Center, and Canisius Wilhelmina Ziekenhuis, Nijmegen, The Netherlands*
***Ghent University Hospital, Ghent, Belgium*

INTRODUCTION

Management of infections is an important issue in many healthcare settings, but severe infections are most prevalent, and antimicrobial use is most abundant at the intensive care unit (ICU). Not surprisingly, antimicrobial resistance (AMR) has emerged primarily in the intensive care setting, where multiple facilitators for the development of resistance are present: high antibiotic pressure, loss of physiological barriers, and high transmission risk [1].

DEFINITION OF AMS IN THE ICU, RATIONALE FOR STEWARDSHIP. WHO DOES WHAT?

Many interventions and programs have been designed to improve appropriate antimicrobial use in terms of choice of drugs, dosing, timing, de-escalation, and discontinuation. Such interventions are collectively known as antimicrobial stewardship programs (ASPs). An ICU ASP can be thought of as a menu of interventions that is adapted and customized to fit the infrastructure and organization of ICUs [2,3].

The ICU is—more than any department in the hospital—a place where specialists work together to provide most optimal patient care. This is especially a challenge in the treatment of patients with infections as infectious disease physicians (IDP), microbiologists, and clinical pharmacists, relying on their own expertise, all advise the ICU physician on the optimal use of antibiotics.

An ICU ASP may exhibit some very ICU-specific goals and strategies, but an ICU is still located within the walls of a hospital, and a large part of its admissions come through the wards. Resistance patterns in ICU mimic those in the wards, and antibiotic use patterns are usually similar. ICU physicians

Antimicrobial Stewardship. http://dx.doi.org/10.1016/B978-0-12-810477-4.00016-7

should thus actively be involved in hospital antibiotic stewardship teams and responsibilities of infectious disease physicians, microbiologists, and pharmacists in ICU should be clearly defined in an ASP.

Influencing the use of antibiotics in ICUs can be a challenging path for ID physicians, clinical pharmacists, and clinical microbiologists. Recent evolution in the organization of ICUs, increasingly reverting from an open to a closed format (where intensivists are primary responsible for patient care and provide 24/7 cover), may contribute to reluctance of ICU physicians to accept outside interference.

Some factors may aid in establishing a successful relationship with the ICU team: try to engage in collaborative research and to ensure that dedicated ASP team members are involved in the ICU (not a different one every day!) and provide data regularly on AMR and antibiotic use patterns. Participating in bedside rounds may be a good opportunity to teach other ASP team members on a particular topic.

APPROPRIATE DIAGNOSTICS IN SUSPECTED BACTERIAL INFECTIONS IN ICU PATIENTS

Accurately diagnosing infection—a prerequisite to the appropriate use of antibiotics—is a challenging task in critically ill patients. Often, signs and symptoms of inflammation rather than infection itself are used to screen for patients with infection. Many conditions encountered in the ICU however cause inflammation, and these will often misguide the clinician [4,5].

Because of this, it is evident that signs of inflammation cannot be the only tools to guide antibiotic decision-making, and specific signs and symptoms of the infection, such as pneumonia or abdominal infection, should be sought after. This search for an anatomical source of infection is important, to select the appropriate antibiotic and determine the need for controlling the source of the infection. Physical examination and directed use of imaging are key elements in this process; obtaining samples for microbiological diagnosis is crucial, in some cases with the use of rapid techniques. However, usually microbiological confirmation will generally come late and has only a limited role at the start of antibiotic therapy.

The threshold to initiate antibiotic therapy is often low, as there is pressure on clinicians not to delay antibiotic therapy in order to improve outcome. As a result, many patients receive antibiotic therapy for noninfectious causes; apart from the futility of the treatment and the impact on the microbiome of the patient, this may also delay establishing an accurate diagnosis and appropriate therapy.

SELECTION OF INITIAL ANTIBIOTIC THERAPY AND TIMELINESS

Initial appropriate antibiotic therapy is highly desirable yet challenging as the microbiology results will only become available at a later stage. Empirical therapy should therefore cover the expected pathogens in a particular patient. This may be determined by many factors such as site of infection, local ecology, previous antibiotic exposure, and length of stay in the hospital, and known colonization status with multidrug-resistant (MDR) pathogens can aid in determining the appropriate empirical choice of therapy.

Traditionally, the focus of initial antibiotic therapy has been on timing and spectrum, and guidelines advocate the initiation of broad-spectrum antibiotic therapy within the hour of hypotension in patients with septic shock. Whereas the latter is indeed crucial, the role of delays in antibiotic therapy is more debated. Whereas several studies have suggested that mortality increases with the hour that antibiotics are delayed, other studies could not confirm this, and indeed, the timing of antibiotic therapy could be a surrogate marker for overall care.

Implementation of international guidelines should be carefully considered as these are often focused on what is relevant in a particular country; what may be appropriate in the United States may not be a correct empirical choice in China or the Netherlands. These guidelines offer a hint of what could be an appropriate choice in a particular setting but do not necessarily constitute the best therapy all elements considered.

PHARMACOKINETIC-PHARMACODYNAMICS-OPTIMIZED ANTIMICROBIAL THERAPY

When considering optimizing antibiotic therapy, it is important to acknowledge the determinants of antibiotic efficacy, which are the patient, the pathogen, and the drug. In critically ill patients, these may differ considerably from outpatients or patients in the general ward. This results in a need for a different approach in prescribing antibiotics [6] (see Chapter 7).

The host. The changed physiology in the host will fundamentally change the pharmacokinetics of the antibiotic administered. Increases in the volume of distribution, in drug clearance, and in protein binding are the most pronounced changes described and particularly relevant for hydrophilic antibiotics. Drug clearance from the circulation and especially increased clearance, augmented renal clearance, defined as a clearance of 130 mL/min or higher, are frequent and may lead to lower concentrations of renally cleared antibiotics.

The pathogen. As the causative microorganism is unknown at the start of empirical therapy; this will only impact the later stages of antibiotic therapy. In the early stage of treatment, it is appropriate to consider a worst-case scenario when it comes to identification and susceptibility of the pathogen; in practice, the epidemiological cutoffs of antibiotic susceptibility may be used, targeting the least susceptible pathogen for which the antibiotic would be appropriate. Close collaboration with the microbiology laboratory is essential to get an early identification and susceptibility at a later stage. Rapid diagnostic tests can sometimes help (see Chapter 6).

The antibiotic. Although the drug is the only consistent and well-known component, there are significant differences between drugs when it comes to the pharmacokinetics and pharmacodynamics. Although often not considered in noncritically ill patients where these changes have been accounted for in the recommended dose, the changed physiology of the ICU patients and the interventions such as renal replacement therapy, combined with the increase in antibiotic resistance, require a more advanced approach in the ICU.

PK/PD Optimized Therapy—Putting the Pieces of the Puzzle Together

PK/PD-optimized therapy refers to the optimized dosing and infusion strategy and can be applied for most antibiotics in the majority of patients. In a PK/PD-optimized therapy (Fig. 1), all of the three determinants of antibiotic therapy are considered. A stepwise approach for this is advised and consists of (1) selection of the PK/PD target for the antibiotic, (2) front loading at the start of therapy, and (3) individualized maintenance dosing.

Step 1. Selecting the PK/PD Target

Depending on the antibiotic used, the PK/PD target will be different. Some antibiotics such as the beta-lactam antibiotics are time-dependent antibiotics, which means that antibiotic efficacy is determined by the duration for which the antibiotic concentration is kept above the MIC. In vitro data found that this is between 40% and 60% of the time of the dosing interval to achieve bacteriostasis and in critically ill patients up to 100% of the time

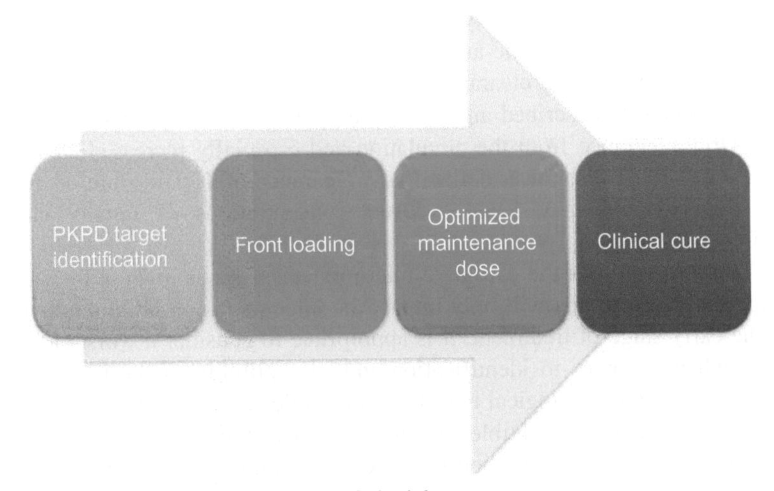

FIG. 1 Practical application of PK/PD optimized therapy.

for 1–4 times the MIC has been advocated. Aminoglycosides are different (concentration-dependent) and require high enough peak concentrations, with optimal efficacy at 8–10 times the MIC. Efficacy of other categories such as glycopeptides or fluoroquinolones will be determined by the AUC_{24}/MIC. Determining this target will guide the clinician in selecting the proper loading and maintenance dose, as well as the appropriate infusion strategy.

Step 2. Front Loading

Because of the changes in the physiology, a fitting loading dose is required to achieve sufficient concentrations from the first hours of therapy. This is particularly important when prolonged infusion strategies are used (see step 3) but now has been applied in the standard dosing schemes of many newly developed antibiotics; its use should however not be limited to new antibiotics. Furthermore, also in patients with acute or chronic renal insufficiency, the loading dose should not be reduced; as it is only the clearance from the circulation that is affected, only the subsequent dose should be adapted to the kidney function.

To further refine therapy and determine the optimal loading dose, the use of population pharmacokinetic models can be used. Software packages are available that make this process simpler.

Step 3. Optimized Maintenance Dose

Finally, also the maintenance dose should be optimized in terms of dose and method of administration. For beta-lactam antibiotics, given that the time>-MIC is the PK/PD determinant, the use of prolonged infusion (either extended or continuous infusion) results in improved antibiotic exposure; in some patients, this may not yet be enough, and even higher doses are required to maintain sufficient concentrations. To further refine therapy, therapeutic drug monitoring (TDM) can be helpful. Similar strategies can be used for other antibiotics (Table 1).

This PK/PD-optimized therapy may also allow using antibiotics for XDR or PDR pathogens. Using optimized infusion strategies and higher doses, combined with TDM when available, can avoid the need for more toxic combinations. These combinations should only be considered in situations where no other antibiotics to which the pathogen is susceptible are available [7–9].

ANTIMICROBIAL THERAPY DE-ESCALATION

The definition of antibiotic de-escalation is challenging. First, many different interpretations have been used in the literature: therapy targeted at the causative pathogen; using a narrower spectrum antibiotic, withdrawing redundant or unnecessary antibiotics (e.g., for MRSA or VRE); reducing the amount of antibiotics. Second, "narrowing the spectrum" seems an easy concept, but it is not: determining which antibiotic has the least ecological impact proves

TABLE 1 Practical Examples of PKPD Optimized Therapy in the ICU

Antibiotic	PKPD Target Selection	Front Loading	Optimized Maintenance Dose
Penicillins carbapenems	Intermittent and extended infusion: >50%fT>MIC, up to 100% for maximal effect Continuous infusion: 100 fT>4–5xMIC	Yes	Prolonged infusion after initial dose, higher maintenance dose in augmented renal clearance
Cefalosporins	Intermittent and extended infusion: >70%fT>MIC, up to 100% for maximal effect Continuous infusion: 100 fT>4–5xMIC	Yes	Prolonged infusion after initial dose, higher maintenance dose in augmented renal clearance
Aminoglycoside	C_{max}/MIC>8	No	Based on TDM (Cmax)
Flurorquinolones	$AUC_{0\text{-}24}$>35–250	Yes	Higher doses may be required
Vancomycin	$AUC_{0\text{-}24}$/MIC 400–600	Yes	Continuous infusion, higher doses in augmented renal clearance, TDM guided

much more difficult than it seems. Different groups of experts have used consensus procedures to establish a definition and draw up a list to order antibiotics in a range of "ecological impact on the development of resistance." No consensus could be reached. Rather than trying to define de-escalation in depth, it has been advocated to broaden the scope to a more general concept of "review of empirical therapy at day 3." This concept suggests a more customized and holistic approach including decisions like "stop therapy if no evidence for infection" and "optimize dosing specific to PK/PD profile."

De-escalation is advocated as an important intervention of a stewardship approach in the ICU. The underlying concept of customizing empirically started broad-spectrum antibiotic therapy to narrow-spectrum therapy targeted at the causative pathogen is suggestive of reducing the emergence of AMR.

However, the evidence supporting this claim is scarce. Most retrospective studies show that de-escalation does not negatively influence patient outcomes (such as mortality and ICU length of stay). The few prospective studies that have been done show no negative influence on ICU stay or mortality but do suggest a higher rate of reinfection and an increase of the total use of antibiotics during subsequent ICU and hospital stay. All considered data are reassuring that de-escalation is probably safe in the ICU.

With increasing evidence that shortening of therapy (often to 5 days) for most ICU patients with an infection is safe (see below), the complexity of performing de-escalation in daily practice should be weighed against the benefit of reducing broad-spectrum ecological exposure for only a few (1–2) days [10–12].

SHORTENING ANTIMICROBIAL TREATMENT DURATION

Until recently, there has been little attention to the decision-making in the final phase of antibiotic therapy. As it is likely that the duration of antibiotic therapy is an important determinant of acquiring MDR, limiting antibiotic exposure at the end of the therapy is probably as important as making the correct empirical choice.

Although many guidelines recommend a duration of 7 days for many common ICU infections, most recommendations are based on expert advice and in real life the duration of therapy is often longer. Therefore, the duration of antibiotic therapy is certainly an attractive target if we want to reduce antibiotic exposure in critically ill patients.

There are several reasons why physicians may be reluctant to stop antibiotic therapy. Often, the patient may be improving but not yet fully recovered from organ dysfunction; traditional biomarkers do not offer much support to guide antibiotic therapy, and also cultures from samples collected in the days after the start of antibiotic therapy may suggest persisting infection leading to prolongation of antibiotic therapy. Clinical criteria to discontinue antibiotic therapy are not helpful either so we are often in the dark as to when to stop antibiotic therapy.

Procalcitonin (PCT), a novel biomarker that has been introduced initially to guide antibiotic therapy initiation, seems to have more value in aiding antibiotic discontinuation decision-making (see Chapter 6). A recent pragmatic, multicenter study from the Netherlands found that PCT-guided therapy can even further reduce antibiotic therapy in critically ill patients. At this point, the exact threshold for PCT to safely discontinue antibiotics remains unclear, and cost-effectiveness has not been studied adequately.

It should be mentioned that there are patients who require longer antibiotic therapy duration, and limiting the duration of therapy is not possible in all patients (e.g., endocarditis, device-associated infections, or abdominal infections with poor or incomplete source control). In these situations, the risks associated with removing the source of the infection should be weighed against the side effects of prolonged antibiotic exposure [6].

IMPLEMENTING A STRUCTURED ANTIBIOTIC STEWARDSHIP PROGRAM

From an implementation point of view, to successfully improve antimicrobial use at the ICU and to tackle AMR, it is important to define appropriate antimicrobial use and to measure current practice. Next, insight must be gained into the factors that influence appropriate antimicrobial prescription at the ICU and an improvement strategy or program should be developed based on these factors while applying social and behavioral change theories (see Chapter 4).

Antimicrobial prescription is a complex process influenced by many factors. The appropriateness of antimicrobial use in hospitals varies between physicians, hospitals, and countries due to differences in professional background, clinical experience, knowledge, attitudes, hospital antibiotic policies, professionals' collaboration and communication, care coordination and teamwork, care logistics, and differences in sociocultural and socioeconomic factors.

This renders changing hospital antimicrobial use into a challenge of formidable complexity. Given that many influencing factors play a part, the measures or strategies undertaken to improve antimicrobial use need to be equally diverse.

Even in a single ICU setting, using relatively simple methods, these challenges can be met. A well-structured group discussion focused at barriers and facilitators that influence appropriate antibiotic use can lead to surprising insights. Based on these insights and the supporting literature linking specific barriers to effective interventions, these can be selected and carried out.

It is clear that there is no one-size-fits-all approach possible here. Rather a more tailored approach is advocated, sometimes leading to multifaceted interventions. Also, it is of importance to pace the work: Rome was not built in one day. Plan-do-study-act (PDSA) cycles can be used to target one relevant aspect of antibiotic care at the time, preferably going for the "low-hanging fruit" first. If measurement of quality indicators on a regular basis is unattainable, performing a point-prevalence study in ICU once per year will provide a good impression of which areas of care are most in need for improvement [13,14].

MEASURING EFFECTIVENESS OF AN AMS IN ICU: QUALITY INDICATORS AND BUNDLE APPROACH; QUANTITY METRICS

It is important to define not only what "appropriate antimicrobial use in ICU patients" is but also how it can be validly and reliably measured (see Chapter 3). While quality indicators for hospital stewardship programs are well described, they may not all be so relevant for the ICU setting (e.g., antibiotic IV-oral switch therapy), or they may represent recommendations that are particularly relevant in an ICU setting (e.g., adequate performance of antibiotic concentration levels).

TABLE 2 A Five Day Bundle for Antibiotic Stewardship in ICU

1st	The clinical rationale for antibiotic start should be documented in the medical chart at the start of therapy
	Appropriate microbiological cultures according to local and/or international guidelines should be collected
	The choice of empirical antibiotic therapy should be performed according to local guidelines
2nd	Review of diagnosis based on newly acquired microbiological cultures
	De-escalation therapy (the narrowest spectrum as possible) according to available microbiological results
3rd–5th	Review of diagnosis based on newly acquired microbiological cultures
	De-escalation therapy (the narrowest spectrum as possible) according to available microbiological results
	Interruption of treatment should be considered according to local and/or international guidelines

As an example, a bundle of six quality indicators was developed to define and measure appropriate antimicrobial use in the ICU setting. European experts specified—in a RAND-modified Delphi procedure—that six professional performance interventions were crucial in antimicrobial use in ICU patients (Table 2).

Comparable sets of quality indicators, oriented at national or local settings and culture, have been developed [15].

To evaluate the effectiveness of an ASP, ideally, these qualitative data are reported together with quantitative data: antibiotic usage data and local resistance patterns to the most relevant causative microorganisms, specific for the ICU.

Regular (e.g., quarterly) feedback on the use of restricted (or rescue) antibiotics like carbapenems, glycopeptides, linezolid, and colistin, expressed in DDD/100 patient days, will add to awareness of intensive care physicians and will undoubtedly facilitate discussions at ICU patient meetings.

DECISION SUPPORT/USE OF MODERN TECHNOLOGY TO OPTIMIZE ANTIBIOTIC STEWARDSHIP PROGRAMS IN ICU

We live in an era where modern technology increasingly enters the medical space. Making use of the large amounts of data we collect in electronic medical records will increasingly be supported by analytic tools and artificial intelligence. In the complex space of appropriate antibiotic prescribing and preventing AMR, these techniques may prove to be of additional value (see Chapter 8).

Many electronic medical records currently offer the possibility to intelligibly display the factors that influence prescribing. Such a real-time overview can help antibiotic stewardship teams, and individual intensive care physicians make more appropriate clinical decisions. Increasingly, active interventions like (un)solicited decision support, automatic reminders, or prior authorization are integrated into these systems. Studies have shown that implementing an electronic decision support system improved guideline adherence, reduced inappropriate antibiotic use, and even reduced the development of AMR over time. Linking the antibiotic prescription to a (tentative) diagnosis greatly improves the value of the collected data. Also, automatically linking the entry of certain antibiotic prescriptions with additional required actions (e.g., TDM for vancomycin infusion) will improve the quality and safety of antibiotic use. Successful implementation of such systems requires intense collaboration with an IT specialist [16–18].

REFERENCES

[1] Infectious diseases and the future: policies for Europe. A non-technical summary of an EASAC report, European public health and innovation policy for infectious disease: the view from EASAC, [http://www.easac.eu/fileadmin/Reports/Infectious_Diseases/Easac_11_IDF.pdf].

[2] Bartlett JG. Antimicrobial stewardship for the community hospital: practical tools & techniques for implementation. Clin Infect Dis 2011;53(suppl. 1):S4–7.

[3] Kaki R, Elligsen M, Walker S, Simor A, Palmay L, Daneman N. Impact of antimicrobial stewardship in critical care: a systematic review. J Antimicrob Chemother 2011;66:1223–30.

[4] Singer M, Deutschman CS, Seymour CW, *et al.* The third international consensus definitions for sepsis and septic shock (sepsis-3). JAMA 2016;315:801–10.

[5] Dellinger RP, Levy MM, Rhodes A, *et al.* Surviving sepsis campaign: international guidelines for management of severe sepsis and septic shock. Intensive Care Med 2012;39:165–228.

[6] Barrett J, Edgeworth J, Wyncoll D. Shortening the course of antibiotic treatment in the intensive care unit. Expert Rev Anti Infect Ther 2015;1–9.

[7] Tsai D, Lipman J, Roberts JA. Pharmacokinetic/pharmacodynamic considerations for the optimization of antimicrobial delivery in the critically ill. Curr Opin Crit Care 2015;21:412–20.

[8] De Waele JJ, Lipman J, Carlier M, Roberts JA. Subtleties in practical application of prolonged infusion of β-lactam antibiotics. Int J Antimicrob Agents 2015;45:461–3.

[9] Abdul-Aziz MH, Lipman J, Akova M, *et al.* Is prolonged infusion of piperacillin/tazobactam and meropenem in critically ill patients associated with improved pharmacokinetic/pharmacodynamic and patient outcomes? An observation from the Defining Antibiotic Levels in Intensive care unit patients (DALI) cohort. J Antimicrob Chemother 2015;71:196–207.

[10] Leone M, Baumstarck K, Lefrant J. De-escalation versus continuation of empirical antimicrobial treatment in severe sepsis: a multicenter non-blinded randomized noninferiority trial. Intensive Care Med 2014;40(10):1399–408.

[11] Roberts JA, Abdul-Aziz MH, Lipman J, *et al.* Individualised antibiotic dosing for patients who are critically ill: challenges and potential solutions. Lancet Infect Dis 2014;14(6):498–509.

[12] Tabah A, Cotta M, Garnacho-Montero J, *et al.* A systematic review of the definitions, determinants, and clinical outcomes of antimicrobial de-escalation in the intensive care unit. Clin Infect Dis 2016;62(8):1009–17.

[13] Hulscher ME, Grol RP, van der Meer JW. Antibiotic prescribing in hospitals: a social and behavioural scientific approach. Lancet Infect Dis 2010;10:167–75.

[14] Rodrigues AT, Roque F, Falcão A, Figueiras A, Herdeiro MT. Understanding physician antibiotic prescribing behaviour: a systematic review of qualitative studies. Int J Antimicrob Agents 2013;41(3):203–12.

[15] De Angelis G, De Santis P, Di Muzio F, Palazzolo C, Brink-Huis A, Hulscher M, *et al.* Evidence-based recommendations to increase the appropriate usage of antibiotics in ICU patients: a 5-day bundle, In: 22nd European congress of clinical microbiology and infectious diseases, London 31/03–03/04/2012; 2012.

[16] Thursky KA, Buising KL, Bak N, *et al.* Reduction of broad-spectrum antibiotic use with computerized decision support in an intensive care unit. Int J Qual Health Care 2006;18:224–31.

[17] Yong MK, Buising KL, Cheng AC, Thursky KA. Improved susceptibility of Gram-negative bacteria in an intensive care unit following implementation of a computerized antibiotic decision support system. J Antimicrob Chemother 2010;65(5):1062–9.

[18] Nachtigall I, *et al.* Long-term effect of computer-assisted decision support for antibiotic treatment in critically ill patients: a prospective 'before/after' cohort study. BMJ Open 2014;4(12).

Chapter 17

Antimicrobial Stewardship in Hematology Patients

Murat Akova
Hacettepe University School of Medicine, Ankara, Turkey

INTRODUCTION

Among the several risk factors predisposing to infections in hematology patients, the most important one is neutropenia [1]. Patients with acute myeloid leukemia (AML) and those undergoing allogeneic hematopoietic stem cell transplantation (allo-HSCT) are at the highest risk of developing profound and persistent (lasting >10 days) neutropenia, which may abate inflammatory signs and symptoms of infections. Thus, fever may be the only sign of severe infection. Overall, only up to 50% of patients with febrile neutropenia will have a microbiologically and/or clinically documented infection. Febrile neutropenia is a medical emergency and would require the initiation of immediate empirical antimicrobial therapy targeting the most frequent infecting pathogens. The choice of empirical antibacterials should depend on the local epidemiology [1–3]. Although this empirical approach will be preventing early mortality, it may cause an ecological impact that leads to selection of highly resistant bacteria and pathogens like *Clostridium difficile* [4,5]. A rational antimicrobial stewardship (AMS) strategy in patients with febrile neutropenia should include prompt initiation of appropriate empirical antibacterials, which should be adapted upon results of cultures when available. The strategy should also address prophylactic antimicrobial use, duration of empirical and targeted antibiotic therapies, appropriate pharmacokinetics/pharmacodynamics (PK/PD) applications, and infection control measures in selected groups of hematology patients. Above all, close cooperation between primary physician(s) in charge (i.e., hemato-/oncologist) and the infectious diseases (ID) specialists, the clinical microbiologists, and the clinical pharmacists is strongly advised [6].

Antimicrobial Stewardship. http://dx.doi.org/10.1016/B978-0-12-810477-4.00017-9

EPIDEMIOLOGY OF INFECTING BACTERIA AND THEIR ANTIMICROBIAL SUSCEPTIBILITY IN PATIENTS WITH HEMATOLOGICAL MALIGNANCIES

Despite advances of intense microbiological interventions, the rate of positive blood cultures remains relatively low around 10%–25% in febrile neutropenia. This may increase up to 60% in allo-HSCT recipients. In addition, clinically documented infections can be diagnosed in up to 30% of neutropenic cancer patients [1,7]. However, a larger portion (10%–60%) of febrile neutropenic patients are treated for fever of unknown origin (FUO), eventually.

Epidemiology of bacterial infections varies between countries and institutions. In general, Gram-positive bacteria are more frequently isolated in bloodstream infections (BSIs) from patients with febrile neutropenia in whom coagulase-negative staphylococci (CNS) are the most frequent ones followed by viridans streptococci, enterococci, and *Staphylococcus aureus* [8]. Widespread use of indwelling catheters to which Gram-positives tend to adhere, introduction of aggressive chemotherapy regimens leading to severe mucosal damage, use of prophylactic fluoroquinolones, and early empirical broad-spectrum antimicrobial therapy targeting mainly Gram-negative bacteria have all been held responsible for their increased prevalence [9]. However, epidemiological data from different parts of the globe such as Southern and Eastern Europe and the Middle East indicate that Gram-negatives are more prominent among BSI isolates from cancer patients with febrile neutropenia [8].

Emerging antimicrobial resistance among isolates causing infections in patients with febrile neutropenia has become a significant challenge. So-called ESKAPE pathogens including *Enterococcus faecium*, *S. aureus*, *Klebsiella pneumoniae*, *Acinetobacter baumannii*, *Pseudomonas aeruginosa*, and *Enterobacter* spp. can be responsible of a significant portion of infections in these patients. In a recent analysis, ESKAPE pathogens have been found responsible for 34% of bacteremia among 1148 episodes of BSIs in cancer patients [10]. It has been shown consistently that the patients infected with resistant pathogens are less likely to receive an adequate empirical antibiotic therapy that may have deleterious consequences [11].

EARLY DIAGNOSTIC STRATEGIES

Every attempt should be made to identify presumptive bacterial etiology as early as possible during an febrile neutropenia attack. A more lenient practice may result in unnecessary and liberal use of broad-spectrum antibiotics for a prolonged period. Such exercise, in turn, causes dramatic increases in emerging resistance, and when coupled with poor application of infection control measures, resistant bacteria can easily disseminate to other patients and the environment [12].

Several new and rapid diagnostic tests have recently become available and may be extremely useful for early identification of the offending microorganisms. Among these are selective chromogenic agars, MALDI-TOF, and nucleic-acid-based methods including rapid polymerase chain reaction (PCR) tests and DNA hybridization. The details of these diagnostic methods are beyond the scope of this chapter, and the readers are referred to recent reviews for details [13,14].

The use of serum inflammatory markers such as C-reactive protein (CRP) and procalcitonin is neither sensitive nor specific for differentiation between bacterial infections and other types of infectious or noninfectious causes of inflammation [7,15].

ARE ALL PATIENTS WITH FEBRILE NEUTROPENIA EQUAL? RISK ASSESSMENT FOR A COMPLICATED INFECTION

Risk stratification has become a standard practice before initiating empirical antibacterial therapy for febrile neutropenia. Clinical scores such as MASCC score (Table 1) can be used for defining low-risk patients [16]. Many physicians and guidelines would include prolonged neutropenia (>7 days) along with the existence of several comorbidities (mucositis, pneumonia, diarrhea, perianal cellulitis, and new onset abdominal pain) and presence of septic shock at admission to describe a high-risk patient category with an increased risk of a severe, complicated infection [1]. This group of patients should be admitted to the hospital, and after careful evaluation of infectious foci,

TABLE 1 **Multinational Association for Supportive Care in Cancer (MASCC) Risk Score[a] [16]**

Clinical Parameters	Score
Burden of illness: no or mild symptoms	5
No hypotension	5
No chronic obstructive pulmonary disease	4
Solid tumor or no previous fungal infection	4
No dehydration	3
Outpatient status	3
Burden of illness: moderate symptoms	3
Patient's age <60 years	2

[a]*Maximum score is 26. Low-risk patient is defined when the score is >21. Scores for burden of illness are not cumulative.*

parenteral broad-spectrum antibiotics must be initiated empirically. Other patients without such risk factors or those with a MASCC score >21 whom can be classified as low-risk group should also be carefully evaluated but may be given oral antibiotics and followed up in an outpatient setting.

PREVENTION OF INFECTION AND PROPHYLAXIS

Single-room accommodation along with the contact precautions and basic infection control practices including strict hand hygiene methods are standard procedures for neutropenic patients [17]. The role of oral antibiotic (mostly fluoroquinolone) prophylaxis for preventing infection and related mortality in patients with neutropenia has been subjected to an intense debate. While individual studies showed no mortality benefit, the meta-analysis also produced conflicting results: only one recent meta-analysis found that prophylaxis may decrease mortality [18], while two others failed to do so [19,20]. Another meta-analysis indicated that combined intervention with protective isolation with air quality control, prophylactic antibiotics, and barrier isolation reduced all-cause mortality for around 40% in high-risk cancer patients and also reduced bacteremia, Gram-negative and Gram-positive infections, and candida infections [21]. The major criticism for those showing a survival benefit was that they included studies mostly done in the 1970s and 1980s. However, several guidelines proposed fluoroquinolone prophylaxis in high-risk patients with prolonged neutropenia unequivocally based on these data [7]. Although no controlled trails exist to prove, the major concern for fluoroquinolone prophylaxis is to select fluoroquinolone-resistant and multidrug-resistant (MDR) *E. coli* and other enteric Gram-negatives.

Evaluating the data between 2005 and 2014, the most recent European Conference on Infections in Leukemia (ECIL) guidelines have suggested that fluoroquinolone prophylaxis should be applied only to those patients with hematologic cancers and with prolonged neutropenia (>7 days) and without colonization with MDR Gram-negatives in a setting with <20% of fluoroquinolone-resistant *E. coli* prevalence (Mikulska M, *et al.* ECIL 6. Antibacterial prophylaxis: critical appraisal of previous ECIL guidelines. Manuscript in preparation). This approach currently seems the most rational practice in hematology patients until further data are available.

OPTIMIZING ANTIBIOTIC DELIVERY WITH PHARMACOKINETICS/PHARMACODYNAMICS (PK/PD) CONSIDERATIONS IN PATIENTS WITH FEBRILE NEUTROPENIA

High-risk cancer patients with sepsis display similar physiological characteristics of other critically ill patients with altered antimicrobial PK/PD. These include increased volume of distribution related with capillary leakage, hypoalbuminemia, and augmented renal clearance. Although few data are available for patients with febrile neutropenia, the general principles in severely ill ICU patients may apply for this situation as well [22].

ROLE OF INFECTIOUS DISEASES (ID) CONSULTATION AND COOPERATION WITH MICROBIOLOGY FOR THE MANAGEMENT OF PATIENTS WITH FEBRILE NEUTROPENIA

A multidisciplinary team approach and ward rounds with bedside discussions must be encouraged when managing the patients with febrile neutropenia. Clinical microbiologists and ID physicians along with the primary physician (i.e., hemato-/oncologist) of the patient should take the responsibility when managing cancer patients with fever [6]. Although specific studies in this population are lacking, in a recent systematic review, it has been clearly shown that ID specialist intervention is related with prudent antibiotic prescription and decreased consumption in acute care hospitals [23]. Microbiology laboratory should also be informed about the characteristics of cancer patients, and clinical microbiologist should be involved in the managing team because not only timely report of culture results is extremely important but also skin colonizers (e.g., CoNS and *Bacillus* spp.) can also cause significant infections and should be reported appropriately. Multidisciplinary protocols considering the local patterns, antimicrobial surveillance data, and advising about the management of patients with febrile neutropenia are extremely useful [6]. Several strategies of AMS have been applied with varying degrees in different settings [5,24]. Preauthorization before initiation of empirical therapy may delay effective therapy, which in turn causes deleterious results in such a medical emergency [5,24]. Therefore, in most centers, the core strategy is postprescription audit and feedback. However, in our center, a 1000-bed university hospital with a large oncology and transplant practice, we have successfully implemented the former strategy during the last three decades with 24/7 consultation service and every day bedside visits for each patient [25]. Admittedly, considering this practice could be very laborious and may not be suitable in every institution, formulary restriction coupled with prospective audit and feedback may be equally effective and cost limiting [24]. The bedside visits can be used as an opportunity to have in-depth discussion between the primary consultant(s) and the ID specialists and may lead to de-escalation or switch from intravenous to oral therapy or most importantly to discontinuation of the treatment. Antimicrobial cycling, another suggested element of AMS, has been reported from various centers as an adjunctive measure to other stewardship efforts; however, antimicrobial heterogeneity with all available antibiotics to be used in selected patients depending on individual characteristics is usually preferred [5].

EMPIRICAL ANTIMICROBIAL STRATEGIES

Since febrile neutropenia is a medical emergency, all patients with this syndrome should receive empirical antimicrobial therapy. However, in order to prevent unnecessary use of broad-spectrum antimicrobials for an undefined duration of period, a risk-based strategy is strongly advised. The main principle of the treatment is to provide enough coverage of antimicrobial spectrum for the first 72 h to keep patients alive after which the therapy can be modified according to the culture results becoming available. In general, empirical antimicrobial regimens can be classified in three different categories:

1. *Outpatient management with oral antibiotics and step-down therapy in low-risk patients* (Fig. 1): Low-risk patients (MASCC score >21, who can tolerate oral medication, with no comorbidities) who did not receive prophylactic fluoroquinolones and with a capacity to reach hospital if clinically deteriorated can be treated with amoxicillin-clavulanate (625 mg tid or 1000 mg bid) plus ciprofloxacin (500 mg bid) or alternatively moxifloxacin (400 mg od), orally as outpatients. A short period of observation in hospital or in the emergency room is usually required in order to assure that patients can tolerate oral antibiotics. Otherwise, low-risk patients may be admitted to the hospital for short term and given parenteral beta-lactam monotherapy chosen per local epidemiology. Examples for antibiotics are ceftazidime, cefepime, piperacillin-tazobactam, and cefoperazone-sulbactam. Carbapenems should be avoided. Once the fever is controlled and if the patient can swallow, a switch to oral therapy is provided and the patient can be discharged [1,2,7] (Fig. 1).
2. *Escalation therapy with parenteral antibiotics in high-risk patients* (Fig. 2): High-risk patients with an uncomplicated presentation, no known colonization or previous infection with MDR pathogens, and in those centers with low prevalence of resistant pathogens can initially be treated empirically with antipseudomonal beta-lactam monotherapy without covering ESBL/carbapenemase producers or production or MDR bacteria. If patient deteriorates or a resistant pathogen is cultured, the spectrum is

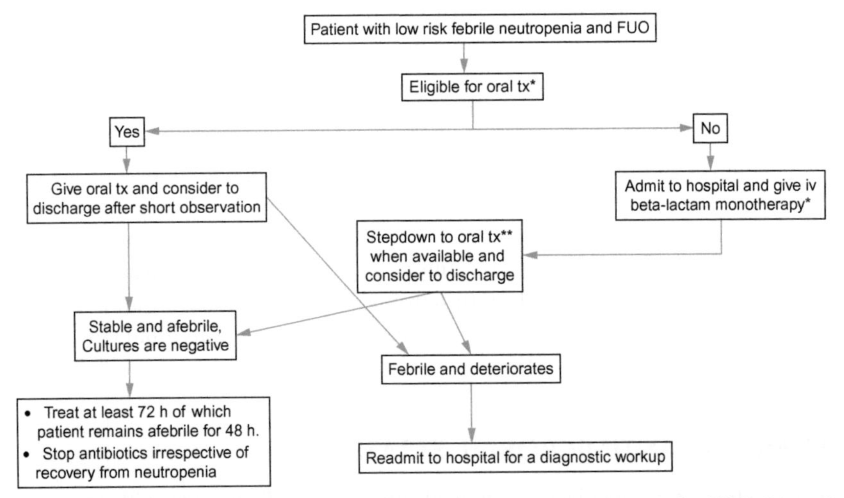

FIG. 1 Management algorithm for low-risk patients with febrile neutropenia. *FUO*, fever of unknown origin. *See text for antibiotics available for this indication. **See text for eligibility of step-down therapy.

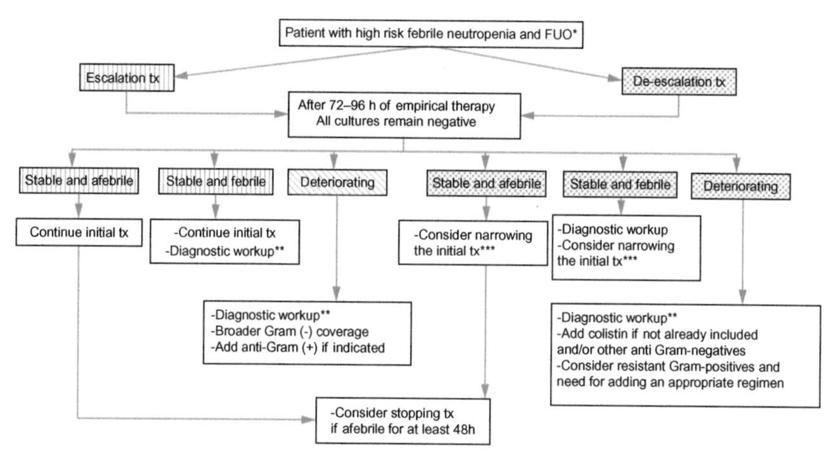

FIG. 2 Management algorithm for high-risk patients with febrile neutropenia. *FUO*, fever of unknown origin; tx, treatment. *See text for eligibility for either escalation or de-escalation therapy. **A diagnostic workup includes a detailed physical examination, blood and other cultures for bacteria and fungi, serial serological tests for fungi (e.g., serum galactomannan and beta-D-glucan), and possibly a CT scan of the lungs. ***Consider to discontinue aminoglycosides, quinolones, colistin, or glycopeptides if started initially in the combination and switch to a noncarbapenem antipseudomonal beta-lactam antibiotic.

escalated to a broader spectrum of activity. This approach prevents unnecessary use of broad-spectrum antibiotic(s) such as carbapenem or colistin upfront [26].

3. *De-escalation therapy in high-risk patients* (Fig. 2): Those high-risk patients in centers with high prevalence of MDR pathogens, who may have been colonized or previously infected with and those present with complicated clinical picture such as sepsis or septic shock, are the candidates for this approach [26]. The empirical therapy should cover Gram-negatives with ESBL, carbapenemase production, or MDR pattern and may also include agents for resistant Gram-positive pathogens if risk factors for these bacteria are present (e.g., previous colonization with MRSA, vancomycin-resistant enterococcus (VRE), and penicillin nonsusceptible *Streptococcus pneumonia*, catheter tunnel or exit site infection, and skin and soft tissue infection) [1,26]. The antibiotics that can be used for this category are carbapenem monotherapy and a combination of antipseudomonal beta-lactams including carbapenems and aminoglycosides or colistin. For Gram-positive coverage, vancomycin, teicoplanin, daptomycin, or linezolid can be used. This therapy is then de-escalated to narrower-spectrum antibiotic(s) once the culture results are available with nonresistant pathogens. Although this strategy has not been extensively tested in patients with febrile neutropenia, it is assumed that the early coverage of resistant bacteria may decrease mortality. However, since nonprudent use of this alternative may cause excessive antibacterial pressure and will turn itself a risk factor for increasing resistance, the eligible patients should be carefully selected.

Since no mortality differences have been shown in randomized trials, empirical inclusion of anti-Gram-positive antibiotics in the upfront therapy is strongly discouraged if no risk factors for these pathogens are present. On the other hand, if they are used in patients with defined risk factors, this coverage should immediately be stopped once blood and/or bronchoalveolar lavage cultures do not reveal any Gram-positive pathogens [7,26].

Removal of central venous catheters (CVCs) is required when a catheter-related infection is documented with MDR bacteria or candida. Other indications include hemodynamic instability, tunnel or port pocket infections, endocarditis, septic thrombosis, and if blood cultures are still positive after >72 h of appropriate therapy [1].

TARGETED THERAPY FOR DOCUMENTED INFECTIONS

In case of a documented infection, the antibiotic therapy should be adjusted according to the *in vitro* susceptibility data. As a general rule, aminoglycosides should not be used as monotherapy for bacteremia cases. Bacteremia caused by ESBL-producing *Enterobacteriaceae* is best treated with carbapenems, and less severe infections such as urinary tract infections can be managed with piperacillin-tazobactam. Infections due to carbapenemase-producing *K. pneumoniae* can be treated with colistin plus meropenem if the carbapenem MIC is <8 mcg/mL; otherwise, another effective antibiotic should be added to colistin [7]. Consequent bacteremia after intensive chemotherapy in carbapenemase-producing *K. pneumoniae*-colonized HSCT recipients has been reported up to 45%, which may result in mortality in 64.4% of these patients [27]. Similar association between colonization and consequent bacteremia has also been described with other bacteria including VRE, *P. aeruginosa*, *Stenotrophomonas maltophilia*, and ESBL-producing *Enterobacteriaceae*. Hence, empirical therapy in febrile neutropenic patients with known previous colonization should provide enough coverage for the colonizers until the results of blood cultures are available.

The outcome data for decolonization in these patients are lacking, but currently, the practice is discouraged due to the possibility of selecting resistance to the last resort antibiotics, which will be obviously the choice of treatment once the infection occurs [28]. The alternative agents and regimens for targeted therapy are reviewed in detail elsewhere [29].

HOW LONG IS LONG ENOUGH FOR EMPIRICAL AND TARGETED ANTIBACTERIAL THERAPY IN PATIENTS WITH FEBRILE NEUTROPENIA?

Since up to 50% of patients with febrile neutropenia may eventually be treated empirically for FUO, at least some of these patients will receive unnecessary broad-spectrum antibiotic therapy. On the other hand, a significant portion of these patients will benefit from the empirical antibiotics and defervesce immediately. However, it is not clear how long empirical therapy should be continued, and most importantly, if the therapy is ceased early, will that cause an increased mortality? The recent European guidelines by ECIL proposed that in those neutropenic patients with FUO who are stable and afebrile for the last 48 h and treated for at least 72 h with antibiotics, the therapy can safely be discontinued. These patients should be observed closely for 24–48 h, and if the fever recurs, empirical therapy should be reinstituted after obtaining a careful physical examination and blood cultures. In centers where fluoroquinolone prophylaxis is used, it may be considered to reinstate prophylaxis after cessation of empirical therapy. But routine secondary prophylaxis with fluoroquinolones is discouraged [26]. The strategy with early cessation of empirical therapy has not been tested in large randomized trials but obviously will decrease the unnecessary use of empirical antibiotics and consequently will have a positive impact on patient and environmental flora [30].

For targeted therapy, antibiotics should be continued for at least 7 days during which patient is afebrile at least 4 days and until the resolution of clinical signs and symptoms and microbiological cure is achieved. Similar to those patients receiving empirical therapy, the patients should be under close observation after stopping the treatment, and if the fever recurs, treatment should be restarted as described above [29].

CONCLUSIONS

Febrile neutropenia in hematology patients is a medical emergency. Early empirical antimicrobial treatment is lifesaving. However, every attempt should be provided promptly to identify the offending microorganisms. The initial empirical therapy should be modified according to the culture results; otherwise, the antibiotics should be discontinued at the earliest opportunity after defervescence in stable patients with FUO. The basic principles for an effective AMS strategy in these patients are summarized in Table 2.

CASE STUDIES-I

A 35-year-old male patient with autologous HSCT for Hodgkin's lymphoma develops fever (38.4°C) at the 7th day after transplantation in the hospital. He has been neutropenic during the last 10 days and his current absolute neutrophil count (ANC) is 100/mm [3]. He reports no cough, dysuria, or diarrhea. His physical examination reveals no other abnormality except fever. He has grade 1 mucositis but swallows with no difficulty. He has a CVC in place. The MASCC score is calculated as 23. The serum biochemistry, urinalysis, and chest X-ray are all unremarkable. He has not received any antibiotics for therapy or prophylaxis during the last month. Having taken two sets

TABLE 2 The 10 Basic Principles of Antimicrobial Stewardship in Hematology Patients With Febrile Neutropenia

1. *Local data availability*: Current epidemiology of bacteremia and other infections and the outcome data (e.g., infection-related mortality and length of hospital stay) should be readily available
2. *Early diagnosis*: Early microbiological diagnostic procedures should be implemented, and the results should immediately be communicated to the clinicians
3. *Prophylaxis*: Oral fluoroquinolone prophylaxis can be used in selective high-risk neutropenic patients (see text for eligibility). Surveillance for emergence of resistance should be maintained during prophylaxis. The prophylaxis should be discontinued once the patient recovers from neutropenia or becomes febrile and the empirical therapy is initiated
4. *Risk stratification*: The risk stratification and local epidemiology should prompt the type of empirical therapy (e.g., oral vs parenteral, inpatient vs outpatient, and escalation vs de-escalation)
5. *Colonization of resistant bacteria*: Previous colonization with MDR bacteria should guide the choice of initial empirical therapy
6. *De-escalation and switch therapy*: Once microbiology data become available, de-escalation to a narrower spectrum of antibiotics or switch to oral therapy should be provided
7. *Duration of therapy*: Empirical or targeted therapy should not be prolonged until recovery from neutropenia and may be discontinued earlier in stable patients (see text for suggested duration)
8. *Infection control*: Enteric decolonization is discouraged in MDR-colonized neutropenic patients. But basic infection control principles must meticulously be applied
9. *Team-based approach*: A team-based approach should be provided when managing patients with febrile neutropenia. The team members should include but not restricted to hemato-/oncologist, infectious diseases physician, clinical microbiologist, and a pharmacist, if available
10. *Guideline adherence*: Management protocols prepared according to the local epidemiologic data should be provided and guide the physicians in charge of these patients

of blood cultures, he is given a single dose of cefepime 2 g, and then the therapy is switched to amoxicillin-clavulanate 1000 mg bid and ciprofloxacin 500 mg bid, po. His general condition remains to be stable, and after 48 h of antibiotic therapy, he defervesces and the blood cultures return to be negative. At day 5 with therapy, while he is still neutropenic (ANC 240/mm [3]), but afebrile for the last 72 h, all antibiotics are discontinued and the patient is discharged home.

The learning objectives and main messages from this case: This is a low-risk febrile neutropenia patient who was treated with de-escalation therapy and with an early switch to oral therapy. A short duration of empirical treatment and inpatient care (i.e., 5 days) was sufficient. All these measures should assure a low resistance selection from patient's microbiome, no or fewer side effects of antibiotics, and obviously less cost for overall treatment.

CASE STUDIES-II

A 22-year-old female was consulted to the ID department for fever and neutropenia. She had been diagnosed with AML type-II, and having undergone complete remission, an allo-HSCT was performed 10 days ago. She is still severely neutropenic with no WBCs and receiving levofloxacin, fluconazole, and acyclovir, prophylactically. She had been given several courses of broad-spectrum antibiotics during remission-induction chemotherapy for AML. Surveillance stool cultures have been taken once a week while hospitalized and indicated that she is colonized with VRE. She has no respiratory symptoms of dysuria or diarrhea. Her physical exam indicated a blood pressure (BP) 90/50 mm Hg. Grade III oral mucositis is present. No other pathological clinical findings are detected. A Hickman catheter is in place. In serum biochemistry, apart from slightly elevated ALT and AMT, no abnormalities are detected. Chest X-ray and urinalysis are also normal. Blood cultures are taken from vein and the catheter. Empirical therapy with meropenem 2 g tid with prolonged infusion and linezolid 600 mg bid is given and levofloxacin is discontinued. Although she becomes stable after intensive fluid resuscitation, she is still febrile after 48 h, and her BP drops again to 90/45 mm Hg. She becomes somnolent. Two out of four blood cultures from both venous and catheter origin reveal *K. pneumoniae, in vitro* susceptibility pending. No Gram-positives were cultured. Colistin is added (9 mU loading and then 4.5 mU bid, iv) to meropenem and linezolid is discontinued. Catheter removal is suggested but declined since she is severely thrombocytopenic (10.000/mm [3]) and no other venous access seems possible at this time. With this treatment, she defervesces during the following 24 h with a normalizing BP 120/70 mmHg. The laboratory reports a *K. pneumoniae* resistant to meropenem (MIC 8 mcg/ml), ciprofloxacin, and amikacin, but sensitive to colistin, gentamicin, and tigecycline. Blood cultures taken after 72 h of antibiotic therapy grew no pathogens. The colistin plus meropenem combination is given a total of 7 days and then discontinued while the patient is under close observation. Quinolone prophylaxis is not reinstituted. Weekly stool surveillance cultures are maintained.

The learning objectives and main messages from this case: In contrast to the first one, this is a high-risk febrile neutropenia patient with several comorbidities. Due to previous colonization with VRE and quinolone prophylaxis, a very broad spectrum of empirical therapy was initiated upon detection of febrile neutropenia. Therapy was modified according to culture results, and most importantly, linezolid was discontinued depending on the lack of evidence for VRE (and other Gram-positive) bacteremia. Upon confirming the clearance of offending pathogens from the blood, (similar to the first case) early discontinuation of targeted therapy under close surveillance aimed to decrease further selection of resistant bacteria from the microbiome, to avoid side effects and for cost savings.

REFERENCES

[1] Freifeld AG, Bow EJ, Sepkowitz KA, Boeckh MJ, Ito JI, Mullen CA, *et al.* Clinical practice guideline for the use of antimicrobial agents in neutropenic patients with cancer: 2010 update by the infectious diseases society of America. Clin Infect Dis 2011;52(4):e56–93.

[2] Alp S, Akova M. Management of febrile neutropenia in the era of bacterial resistance. Ther Adv Infect Dis 2013;1(1):37–43.

[3] Kara O, Zarakolu P, Ascioglu S, Etgul S, Uz B, Buyukasik Y, *et al.* Epidemiology and emerging resistance in bacterial bloodstream infections in patients with hematologic malignancies. Int J Infect Dis 2015;47(10):686–93.

[4] Akova M, Paesmans M, Calandra T, Viscoli C, International Antimicrobial Therapy Group of the European Organization for Research and Treatment of Cancer . A European Organization for Research and Treatment of Cancer-International Antimicrobial Therapy Group Study of secondary infections in febrile, neutropenic patients with cancer. Clin Infect Dis 2005;40(2):239–45.

[5] Tverdek FP, Rolston KV, Chemaly RF. Antimicrobial stewardship in patients with cancer. Pharmacotherapy 2012;32(8):722–34.

[6] Gyssens IC, Kern WV, Livermore DM, ECIL-4, a joint venture of EBMT, EORTC, ICHS and ESGICH of ESCMID . The role of antibiotic stewardship in limiting antibacterial resistance among hematology patients. Haematologica 2013;98(12):1821–5.

[7] Gustinetti G, Mikulska M. Bloodstream infections in neutropenic cancer patients: a practical update. Virulence 2016;7(3):280–97.

[8] Mikulska M, Viscoli C, Orasch C, Livermore DM, Averbuch D, Cordonnier C, *et al.* Aetiology and resistance in bacteraemias among adult and paediatric haematology and cancer patients. J Infect 2014;68(4):321–31.

[9] Worth LJ, Slavin MA. Bloodstream infections in haematology: risks and new challenges for prevention. Blood Rev 2009;23(3):113–22.

[10] Bodro M, Gudiol C, Garcia-Vidal C, Tubau F, Contra A, Boix L, *et al.* Epidemiology, antibiotic therapy and outcomes of bacteremia caused by drug-resistant ESKAPE pathogens in cancer patients. Support Care Cancer 2014;22(3):603–10.

[11] Gudiol C, Tubau F, Calatayud L, Garcia-Vidal C, Cisnal M, Sanchez-Ortega I, *et al.* Bacteraemia due to multidrug-resistant Gram-negative bacilli in cancer patients: risk factors, antibiotic therapy and outcomes. J Antimicrob Chemother 2011;66(3):657–63.

[12] Akova M. Epidemiology of antimicrobial resistance in bloodstream infections. Virulence 2016;7(3):252–66.

[13] Kothari A, Morgan M, Haake DA. Emerging technologies for rapid identification of bloodstream pathogens. Clin Infect Dis 2014;59(2):272–8.

[14] Opota O, Croxatto A, Prod'hom G, Greub G. Blood culture-based diagnosis of bacteraemia: state of the art. Clin Microbiol Infect 2015;21(4):313–22.

[15] Ratzinger F, Haslacher H, Perkmann T, Schmetterer KG, Poeppl W, Mitteregger D, *et al.* Sepsis biomarkers in neutropaenic systemic inflammatory response syndrome patients on standard care wards. Eur J Clin Investig 2015;45(8):815–23.

[16] Klastersky J, Paesmans M, Rubenstein EB, Boyer M, Elting L, Feld R, *et al.* The Multinational Association for Supportive Care in Cancer risk index: a multinational scoring system for identifying low-risk febrile neutropenic cancer patients. J Clin Oncol 2000;18(16):3038–51.

[17] Engelhard D, Akova M, Boeckh MJ, Freifeld A, Sepkowitz K, Viscoli C, *et al.* Bacterial infection prevention after hematopoietic cell transplantation. Bone Marrow Transplant 2009;44(8):467–70.

[18] Gafter-Gvili A, Fraser A, Paul M, Vidal L, Lawrie TA, van de Wetering MD, *et al.* Antibiotic prophylaxis for bacterial infections in afebrile neutropenic patients following chemotherapy. Cochrane Database Syst Rev 2012;1:CD004386.

[19] Kimura S, Akahoshi Y, Nakano H, Ugai T, Wada H, Yamasaki R, *et al.* Antibiotic prophylaxis in hematopoietic stem cell transplantation. A meta-analysis of randomized controlled trials. J Infect 2014;69(1):13–25.

[20] Imran H, Tleyjeh IM, Arndt CA, Baddour LM, Erwin PJ, Tsigrelis C, *et al.* Fluoroquinolone prophylaxis in patients with neutropenia: a meta-analysis of randomized placebo-controlled trials. Eur J Clin Microbiol Infect Dis 2008;27(1):53–63.

[21] Schlesinger A, Paul M, Gafter-Gvili A, Rubinovitch B, Leibovici L. Infection-control interventions for cancer patients after chemotherapy: a systematic review and meta-analysis. Lancet Infect Dis 2009;9(2):97–107.

[22] Tsai D, Lipman J, Roberts JA. Pharmacokinetic/pharmacodynamic considerations for the optimization of antimicrobial delivery in the critically ill. Curr Opin Crit Care 2015;21(5):412–20.

[23] Pulcini C, Botelho-Nevers E, Dyar OJ, Harbarth S. The impact of infectious disease specialists on antibiotic prescribing in hospitals. Clin Microbiol Infect 2014;20(10): 963–72.

[24] Seo SK, Lo K, Abbo LM. Current state of antimicrobial stewardship at solid organ and hematopoietic Cell transplant centers in the United States. Infect Control Hosp Epidemiol 2016;37(10):1195–200.

[25] Guven GS, Uzun O, Cakir B, Akova M, Unal S. Infectious complications in patients with hematological malignancies consulted by the infectious diseases team: a retrospective cohort study (1997–2001). Support Care Cancer 2006;14(1):52–5.

[26] Averbuch D, Orasch C, Cordonnier C, Livermore DM, Mikulska M, Viscoli C, *et al.* European guidelines for empirical antibacterial therapy for febrile neutropenic patients in the era of growing resistance: summary of the 2011 4th European Conference on Infections in Leukemia. Haematologica 2013;98(12):1826–35.

[27] Girmenia C, Rossolini GM, Piciocchi A, Bertaina A, Pisapia G, Pastore D, *et al.* Infections by carbapenem-resistant *Klebsiella pneumoniae* in SCT recipients: a nationwide retrospective survey from Italy. Bone Marrow Transplant 2015;50(2):282–8.

[28] Girmenia C, Viscoli C, Piciocchi A, Cudillo L, Botti S, Errico A, *et al.* Management of carbapenem resistant *Klebsiella pneumoniae* infections in stem cell transplant recipients: an Italian multidisciplinary consensus statement. Haematologica 2015;100(9):e373–6.

[29] Averbuch D, Cordonnier C, Livermore DM, Mikulska M, Orasch C, Viscoli C, *et al.* Targeted therapy against multi-resistant bacteria in leukemic and hematopoietic stem cell transplant recipients: guidelines of the 4th European Conference on Infections in Leukemia (ECIL-4, 2011). Haematologica 2013;98(12):1836–47.

[30] Orasch C, Averbuch D, Mikulska M, Cordonnier C, Livermore DM, Gyssens IC, *et al.* Discontinuation of empirical antibiotic therapy in neutropenic leukaemia patients with fever of unknown origin is ethical. Clin Microbiol Infect 2015;21(3):e25–7.

Chapter 18

AMS in an Era of Multidrug-Resistant Bacteria

Pilar Retamar, Jesús Rodríguez-Baño, Mical Paul
and Khetam Hussein

18.1 INFECTIONS DUE TO ESBL-PRODUCING BACTERIA: HOW TO PROMOTE AMS?

Pilar Retamar and Jesús Rodríguez-Baño
Hospital Universitario Virgen Macarena, Sevilla, Spain
Universidad de Sevilla - Instituto de Biomedicina de Sevilla (IBiS), Sevilla, Spain

Introduction

Extended-spectrum beta-lactamases (ESBLs) are beta-lactamases that hydrolyze penicillins, cephalosporins, and aztreonam; do not hydrolyze cephamycins; and are inhibited by beta-lactamase inhibitors. Beyond that, ESBL-producing Enterobacteriaceae (ESBL-E) are frequently resistant to other antibiotics. ESBL-E have dramatically spread worldwide during the last two decades. The fact that carbapenems have been considered the preferred option for invasive infections due to ESBL-E has fueled the use of these drugs, both empirically and whenever an ESBL-E is isolated, which is contributing to the spread of carbapenem resistance.

Empirical Therapy for ESBL-E: When and How

Initial inappropriate antimicrobial therapy of severe infections leads to increased morbidity and mortality. This assertion needs to be considered from two perspectives: which patients should receive empirical treatment covering ESBL-E and, if ESBL-E is to be empirically covered, which drug to use. To decide on empirical treatment, AMS programs should include promotion of risk evaluation, including the severity of infection and the risk of ESBL-E, which depends on the local rates of ESBL-E and individual risk conditions,

Antimicrobial Stewardship. http://dx.doi.org/10.1016/B978-0-12-810477-4.00018-0

including previous colonization, recent antibiotic use, residence in a long-term care facility, recent hospitalization, and age >65 years [1]. Carbapenems are usually considered the drugs of choice targeting ESBLs, but there might be other options, including beta-lactam/beta-lactamase inhibitor combinations (BLBLI) [2,3], cephamycins [4], temocillin, colistin, fosfomycin, tigecycline, and aminoglycosides (Table 1). To date, there are no data from randomized trials. However, local susceptibility of ESBL-E using EUCAST break points to these antimicrobials should be reviewed as there are important regional variations. If a carbapenem is to be used, ertapenem may be considered as its spectrum is narrower than that of other carbapenems [5].

Targeted Therapy of ESBL-E

Rapid detection of ESBL-E is essential to allow sooner optimization of the treatment. Rapid methods are discussed elsewhere in this book (Chapter 6), but it is important to keep in mind that any rapid method can be irrelevant if not accompanied by an active stewardship intervention.

When choosing targeted therapy, the microbiological results, the source of infection, the severity of the patient, and the potential interactions and PK/PD features of each drug should be considered. In patients with infection who were empirically treated with a carbapenem, de-escalation to antibiotics with narrower spectrum (Table 1) must be promoted [6].

Helping to Avoid the Selection and Spread of ESBL-E

Exposure to antibiotics, particularly fluoroquinolones and cephalosporins, contributes to the selection of ESBL-E and facilitates transmission and acquisition. Therefore, promotion of prudent use of these antibiotics is nowadays a key activity of AMS programs. Because these antibiotics are frequently prescribed, interventions must be tailored to the local situation and availability of resources.

Implementing AMS Activities Related to ESBL-E

Continuous education and information on local ESBL-E rates, risk factors, choice of antibiotics, and interpretation of microbiology results are a must for all AMS programs as ESBL-E are present with higher or lower frequency in most hospitals. Specific recommendations not only about when and how to cover ESBL-E in empirical regimens but also about how to limit the use of cephalosporins, fluoroquinolones and carbapenems should be included in local protocols and prescription-assistance tools.

For improving targeted therapy, current evidence-based recommendations favor auditing of preselected patients as the most efficacious way to promote streamlining [7]. Such interventions should be based on a tight collaboration

TABLE 1 Treatment Options for ESBL and Carbapenem-Resistant Gram Negative Organisms

Antimicrobial Agent	Potential Spectrum of Coverage	Comments
Beta-lactam/beta-lactam inhibitors (AMC and TZP)	ESBL	Reasonable alternatives to carbapenems if active *in vitro* in many circumstances. High-dose recommended (TZP, 4.5 g every 6 h or every 8 h in prolonged infusion; AMC, 2.1 g every 8 h). Caution is needed for patients in septic shock, with difficult sources such as pneumonia and MIC >8 mg/L, until more data are available Oral AMC useful for cystitis
Cephamycins (cefoxitin, flomoxef, cefmetazole, and others)	ESBL	Might be useful if active *in vitro*, particularly for urinary-tract-related bacteremia due to ESBL-producing *E. coli*. Anecdotal failures with cefoxitin reported because of porin loss during treatment. High-dose recommended
Temocillin	ESBLs—TEM, SHV, CTX-M, and most OXA CRE—only KPC (not active against NDM, VIM, IMP, and OXA-48) and inherently inactive against *Acinetobacter* sp. and *P. aeruginosa*	Available only in a few countries. Comparative data with carbapenem for ESBL producers are not available. Noncomparative clinical data and animal models suggest reasonable efficacy. Suggested dosing 2 g every 8–12 h (prolonged or continuous possible)
Fluoroquinolones	Most ESBLs and CRGNBs are resistant	Available data suggest worse outcomes in comparison with carbapenems but probably due to resistance. There is no reason to suspect worse results if active *in vitro* (which is not frequent). Caution in case of isolates showing borderline MIC
Trimethoprim-sulfamethoxazole	ESBLs and CR *A. baumannii*, dependent on local susceptibilities	Resistance is frequent. Scarce experience with susceptible isolates but expected efficacy as in non-ESBL producers No data on clinical efficacy for CR *Acinetobacter* infections

Continued

TABLE 1 Treatment Options for ESBL and Carbapenem-Resistant Gram Negative Organisms—Cont'd

Antimicrobial Agent	Potential Spectrum of Coverage	Comments
Aminoglycosides	ESBLs and variably CRGNBs	Scarce experience with susceptible isolates but expected efficacy as in non-ESBL producers; therefore, potentially useful mainly for urinary tract infections
Colistin/polymyxin B	ESBK and CRE, both serine and metallo beta-lactamases	Not recommended for ESBLs. Comparative studies on colistin versus polymyxin B not available. Polymyxin given without adjustment to renal function and possibly less nephrotoxic than colistin
Tigecycline	ESBL and CRGNB depending on local susceptibilities. Not active against *P. aeruginosa*	Not recommended as first line because of worse outcomes than comparators in non-ESBL producers. May be an option in polymicrobial, nonsevere infections
Fosfomycin	ESBL and CRE	Useful for cystitis as an oral drug (trometamol formulation) and potentially for systemic infection (IV fosfomycin) in combination
Pivmecillinam	ESBL	Useful for cystitis as an oral drug
Nitrofurantoin	ESBL and variably CRE	Useful for cystitis as an oral drug
Ceftazidime-avibactam	ESBLs, KPC CREs, and *P. aeruginosa*	Recently FDA approved for abdominal and urinary tract infections. Proved as noninferior to best available therapy for ceftazidime-resistant Gram-negative bacteria [3]
Ceftolozane-tazobactam	ESBL and CR *P. aeruginosa*	Recently FDA approved for abdominal and urinary tract infections

TZP: piperacillin-tazobactam, AMC: amoxicillin-clavulanate, CRGNB: carbapenem-resistant gram negative bacilli, CRE: carbapenem-resistant Enterobacteriaceae.

of infectious disease specialists, microbiologists, and pharmacists and may target either patients treated with carbapenems in the form of postprescription auditing or patients with positive cultures for ESBL-E. Restriction by preutilization approval may have short-term positive effects but may not be well accepted and requires more resources. In addition, inclusion of de-escalation rules in local guidelines and educational activities based on case vignettes is recommended. Tips to implement these activities are summarized in Table 2.

TABLE 2 Tips for Antimicrobial Stewardship Activities in the Context of ESBL-Producing Enterobacteriaceae (ESBL-E) and Carbapenem-Resistant Gram-Negative (CRGNB) Infections

• Monitor your rate of ESBL-E and CRGNB and provide updated information to prescribers. If possible, provide data for community and nosocomial infections and specific data for high-risk wards (e.g., ICU)
• Work with your laboratory to define the antibiotic susceptibility panel used, when MIC data are needed and when ESBLs and CR have to be defined to the molecular level
• Implement infection control measures as appropriate
• Promote and check that microbiological samples are taken before starting antibiotics
• Include recommendations for empirical and targeted therapy for ESBL-E and CRGNBs in local guidelines according to local epidemiology and susceptibility data
• Train prescribers and provide tools to predict ESBL-E and CRGNBs in community and nosocomial infections to guide empirical antibiotic treatment. Key aspects to consider are local epidemiology, colonization status, prior antibiotic and healthcare exposure, source, and severity of the infection
• Guide and train prescribers to reserve carbapenems and polymyxins as empirical treatment for patients presenting with severe sepsis or septic shock and having at least one risk factor for ESBL-E and CRGNB, respectively. Consider ertapenem if *P. aeruginosa* is not a concern and carbapenem- or polymyxin-sparing options according to local epidemiology and susceptibility
• If possible, use rapid diagnostic testing for ESBLs and CREs and assure that the results are reported to the AMS and infection control teams for intervention
• Guide and train prescribers about the importance of source identification and drainage/device removal if appropriate
• Whenever an ESBL-E or CRGNB is isolated, provide expert advice if possible. Consider whether antibiotic treatment is necessary, de-escalation, carbapenem/polymyxin-sparing options, optimized PK/PD, and appropriate duration of treatment
• Implement interventions for auditing targeted therapy in invasive infections due to ESBL-E and CRGNBs
• Establish local indicators and standards for evaluating the appropriateness of the use of carbapenems and polymyxins and targeted therapy against ESBL-E CRGNBs

In outbreaks caused by ESBL-E, more intense activities aimed at avoiding as much as possible the selective pressure of some antibiotics (mainly fluoroquinolones and cephalosporins but might be others as well) should be implemented. Despite the fact that infection control measures are probably more important, AMS may be an important part of a bundle of activities. Beyond the previously recommended activities, restriction of the above antibiotics must be considered, particularly if their consumption was higher than desirable.

Practical Cases

Case 1: Empirical Therapy

A 71-year-old lady arrives to the emergency department with fever, flank pain, and dysuria. Her previous health problems include dyslipidemia and kidney stones. On clinical examination, her blood pressure is 110/56 mmHg, heart rate is 103 per minute, and temperature is 38.4°C. Blood tests show 14.360 leukocytes/ml (90% polymorphonuclear cells), creatinine 1.5 mg/dl, CRP 190 mg/dl, and lactate 1 mmol/L. A urine dipstick is strongly positive for leucocytes. Blood and urine cultures are taken. The attending physician decides to start therapy with imipenem. You are called as member of the AMS team to review the prescription. So you discuss with the prescriber the following aspects:

- In this situation, can Enterobacteriaceae be the causative pathogen? Yes, the most probably source of infection is the urinary tract, and Enterobacteriaceae are the most frequent pathogens isolated in urinary and intra-abdominal infections.
- Is the patient critically ill? No, the patient presented with nonsevere sepsis.
- Which is the risk for ESBL-E? A review of the primary care prescriptions disclosed that she had been prescribed levofloxacin 2 months ago because of "cystitis." Also, the patient is 71 years old. So, there is a low-to-moderate probability of ESBL-E in a patient without severe sepsis or shock.

The discussion with the prescriber may include the different options. If according to local susceptibility patterns the probability of ESBL is low, standard treatment with cefotaxime or ceftriaxone may be started; if the probability of ESBL is higher, the use of an aminoglycoside if local rates of resistance are low enough, is another option until the causative pathogen's susceptibility data are available. Other options may be included according to local susceptibility rates. A carbapenem is probably not needed in most areas in this situation.

Case 2: Targeted Therapy

The pharmacist informed the infectious diseases member of the AMS team that a patient in the ICU was being treated with meropenem 1 g every 8 h.

When reviewing the data, the patient was admitted after abdominal surgery. An ESBL-producing *K. pneumoniae* had been isolated 2 days before from blood and urine cultures; on that day, the patient had hypotension and elevated lactate. There was no evident source of the infection. The urinary catheter has been changed. The isolate was resistant to penicillins, beta-lactam inhibitors combinations, cephalosporins, and TMP-SMX but was susceptible to ciprofloxacin (MIC < 0.5 mg/L), carbapenems, amikacin, colistin, tigecycline, and fosfomycin. Contact precautions had been implemented. The patient is afebrile and stable after 2 days of treatment and is going to be transferred to a surgical ward for follow-up. Is there any opportunity of optimizing this treatment?

- The patient is now stable. De-escalation should be considered.
- The options include ciprofloxacin in this case. Despite the fact that many ESBL producers are resistant, in the case of a susceptible isolate and particularly when the urinary tract is the source, it should be considered as an option. This would allow to avoid further use of a carbapenem, and a switch to oral therapy as soon as the patient is able to use this route.
- Switch to oral fosfomycin when possible would be another option, but the experience available comes from patients with cystitis. There is less experience with fosfomycin in bacteremic patients.

18.2 AN OUTBREAK OF CARBAPENEM-RESISTANT OR POLYMYXIN-RESISTANT STRAINS: HOW CAN ANTIMICROBIAL STEWARDSHIP HELP?

Mical Paul and Khetam Hussein
Israel Institute of Technology, Haifa, Israel

Introduction

Carbapenem-resistant (CR) Gram-negative bacteria (GNB) and more so, polymyxin-resistant strains, pose a significant threat to patients, since effective treatment options are unavailable. The last resort antibiotics covering these bacteria, colistin and tigecycline, are probably less effective than beta-lactams, quinolones, and other modern antibiotics. Aminoglycosides, to which CR strains remain variably susceptible, are less effective than other antibiotics when treating infections other than urinary tract infections. The higher mortality observed among patients with CR strains compared with patients with ESBL-E infections or other patients with sepsis is multifactorial and largely due to patients' comorbidities, but an unmeasurable proportion of the higher mortality is due to inappropriate empirical antibiotic treatment, poor definitive treatment options, or intrinsic properties of these MDRs. Whatever the reasons, the mortality rates among patients with CR-GNB bacteremia, ventilator-associated pneumonia (VAP), and other infections are consistently around 40% in many observational studies, being the highest with VAP. Thus, preventing infections caused by such bacteria is an important priority of antimicrobial stewardship (AMS) programs and can save lives of patients.

The common CR strains include Enterobacteriaceae (CRE), *Acinetobacter baumannii* and *Pseudomonas aeruginosa*. CR resistance rates within these bacteria are variable worldwide, with foci of endemicity in Italy, Greece, Israel, India, China, and the United States, among centers publishing resistance rates (http:/ecdc.europa.eu/en/activities/surveillance/EARS-Net/Pages/index.aspx). Interestingly, incidence and prevalence of these bacteria has been described to change quite rapidly in response to infection control/AMS activities or as part of a natural fluctuation curve of outbreaks [8]. Polymyxin-resistant has been described as prevalent in Italy and Greece as of 2016, with up to 30% of *K. pneumoniae* isolates in certain hospitals resistant to colistin.

Prevention of Acquisition and Spread

Invasive infections caused by CRE frequently follow acquisition and colonization of the gut, but with *P. aeruginosa* and *A. baumannii*, colonization of vascular and other catheters or wounds is also important. Thus, our first priority is to prevent patients from acquiring these bacteria by infection control interventions. Early identification of colonized patients, contact isolation, cohorting if necessary in endemic setting, dedicated staff, dedicated equipment, revision of cleansing and disinfection procedures, monitoring and reporting, and if necessary more dramatic measures, such as closure of a ward, must be implemented to curtail cross infection.

Antimicrobial stewardship (AMS) is likely to contribute on top of an infection control program. Reducing antibiotic pressure works through two mechanisms. The first and probably most important is to reduce selective pressure. Antibiotics, especially the broad-spectrum antibiotics such as carbapenems, induce immediate changes in the composition of colonizing bacteria, whose largest reservoir is in the colon. Depleting the flora from bacteria susceptible to carbapenems provides a welcoming niche for CR bacteria. A second hypothetical occurrence is that the antibiotic administered to an individual patient induces a new resistant mutation. Carbapenems induce CRE colonization through both mechanisms. Studies have shown that carbapenem exposure is a strong risk factor for CRE, CR *P. aeruginosa* and CR *A. baumannii* acquisition. Conversely, reducing carbapenem consumption by preauthorization resulted in significant reduction in the incidence of CR *P. aeruginosa* and *A. baumannii* in a hospital endemic for these CRGNBs. Colistin used was the strongest risk factor for colistin-resistant carbapenemase-producing *K. pneumoniae* bacteremia in Italy, compared with patients without bacteremia or to patients with colistin-resistant carbapenemase-producing *K. pneumoniae* bacteremia [9].

AMS for CRGNBs

Tips for AMS are provided in Table 2. The spread of ESBL-E is an important driver for carbapenem overuse. The strategies proposed in Section 18.1 to reduce unnecessary carbapenem use constitute the first step in AMS for CRGNBs. Promoting de-escalation whenever an ESBL-E has been discarded, and providing individualized, reasonable alternatives to carbapenems for ESBL-E when appropriate is an important activity.

Empirical polymyxin treatment should be reserved for patients at highest risk for CRGNB infection to avoid unnecessary exposure of patients to this last resort antibiotic. The most likely candidate groups are patients colonized by CR Enterobacteriaceae or other GNBs developing severe sepsis/septic shock, especially if neutropenic. Screening for CRE colonization can assist in judicious empirical antibiotic prescription, in addition to its being important tool for patient isolation. More cautiously, polymyxins might have to be used empirically in an outbreak setting, for example, in a unit where most patients are colonized or when frequent new acquisitions have been recently identified, again reserving such empirical treatment to patients with severe sepsis or septic shock.

The microbiology laboratory is an essential partner in AMS programs (see Chapter 9). Working in collaboration with the microbiology lab should ensure optimal specimen collection (e.g., minimization of contamination and time documentation), immediate transport to the lab, lab processes to speed up the time to result and document the resistance patterns relevant to the epidemiological setting, efficient communication of results to the relevant healthcare workers, the AMS program, and the infection control team. Optimal use of the limited antibiotic armamentarium available to treat CRE infections currently requires that resistance mechanisms be defined to the molecular level. Few new antibiotics targeting CR or polymyxin-resistant GNBs have been developed and approved and none using novel mechanisms of action (Table 1) [10]. Both recently approved beta-lactam/beta-lactamase antibiotics and those advanced in the pipeline (meropenem-vaborbactam, aztreonam-avibactam, and imipenem/cilastatin-relebactam) are active only against KPC-type carbapenemases and not against the metallo-beta-lactamases (NDM, VIM, and IMP types). Thus, knowledge of the local molecular mechanisms of carbapenem resistance is necessary before considering the use of these new agents. The AMS team needs to decide whether and which new antibiotics should be introduced to the hospital considering the local epidemiology and working in collaboration with the microbiology laboratory.

A debate is currently ongoing as to whether combination therapy is beneficial in the treatment of CRGNB infections. The combination most frequently proposed is that of a polymyxin and a carbapenem, with or without an addition antibiotic, such as tigecycline. The rationale behind combination therapy is the high mortality of these infections, *in vitro* demonstration of synergistic interactions between and *in vitro* prevention of polymyxin resistance development [11]. The long-term ecological implications of carbapenem-based combination therapy policy for treatment of CRGNB infections in endemic locations for these bacteria are difficult to predict. Carbapenem treatment of patients infected by CRGNB might promote extended colonization and increased excretion of these bacteria to the environment fueling the epidemic [12]. Such a policy implies that carbapenems will also be used empirically for patients suspected of CRGNB infections, thus exposing many who are not colonized to a very significant risk factor for CRE acquisition. Observational studies show uniformly lower mortality with antibiotic combinations compared with polymyxin. However, there are many limitations to these studies inherent to the observational design that was described elsewhere [13]. Clearly, polymyxins are less effective antibiotics than some of the more modern antibiotics, and outcomes reported for patients with CRGNB infections treated with polymyxin monotherapy are dismal. However, we do not know at this time whether combination therapy offers an advantage, while we are certain that carbapenems constitute a risk factor for acquisition and persistent carriage of CRE. As a policy of combination therapy in endemic settings poses hazards to an AMS program, this practice must be based on evidence from randomized controlled trials.

Decolonization is an approach to be considered in the AMS strategy against CRGNB, especially for CREs whose reservoir in the gut might be amenable to oral decolonization. Many randomized controlled trials have assessed selective digestive decontamination or selective oral decontamination among critically ill patients, showing survival benefit to decontamination [14] and those examining resistance development as an outcome and have not found more resistant strains in the decontamination arm. Most trials have used a polymyxin in the decontamination strategy. However, many of the studies were conducted in settings with low ESBL rates and almost no CRE, and patients were not colonized with MDR bacteria at the start of therapy. Observational studies and two RCTs show short-term effects of decolonization of ESBL or CRE carriers during or immediately after, but no long-term effect or no longer follow-up [15]. Further studies are ongoing and will be important to examine the ecological safety of this appealing strategy.

Finally, avoiding underdosing might delay resistance development. Dosing of colistin has been recently revised, mainly in Europe (EMA/643444/2014), following the completion of modern PK/PD studies [16–18].

General Considerations in AMS for ESBL-E and CRGNB

Besides infection control activities that are crucial for limiting the spread of ESBL-E and CRE, antimicrobial stewardship (AMS) programs are a core element to help in controlling the spread of these organisms, improving outcomes and avoiding further resistance selection.

Firstly, we should ask ourselves who does not need antibiotics. As in all positive culture results, the first consideration is to discern whether the patient is infected or just colonized. Common examples of misinterpretations are positive urine cultures in patients with asymptomatic bacteriuria or superficial positive cultures from chronic ulcers; these patients should not be treated with antibiotics. A recent study challenged the need to treat asymptomatic bacteriuria in pregnancy, considering the 2.4% incidence of pyelonephritis among untreated women as acceptable [19], as did a study on asymptomatic bacteriuria among renal transplant recipients after 2 months posttransplantation [20]. When these patients are colonized by ESBL or CRE, indeed no treatment might be the safest choice given the need for intravenous and potentially toxic treatment. A final group of patients is that at the end of life among which antibiotic treatment is very frequent and increases with approaching death. Implementing these interventions mandates a very active AMS program. Efforts, personnel, and time are needed for education of hospital staff, supervision, monitoring of behavior, and feedback.

Observational studies showed a significant and large benefit to appropriate empirical antibiotic treatment initiated within the first 24–48 h of infections [21], and mortality has been shown to increase with each hour of covering antibiotic delay among critically ill patients. However, we should remember that all patients in these studies had microbiologically documented infections and severe infections, mostly bacteremia or VAP, while we are implementing the knowledge empirically, on all patients with suspected infections of whom, only some will have an infection, and less than 30% will have a microbiologically documented infection. Thus, a clear strategy must be in place to review empirical antibiotics and consider stopping unnecessary treatment.

De-escalation is associated with lower mortality in observational studies; however, confounding by indication is unavoidable in these studies [22]. Two RCTs to date have shown no harm in de-escalation, and although these studies did not attempt to prove beneficial ecological effects, we believe that switching to narrower-spectrum antibiotics is beneficial to selection pressure and induction of resistance, especially when the empirical antibiotics are last resort agents such as carbapenems and polymyxins.

An even more important intervention is to stop antibiotic altogether as soon as safe. High-quality evidence proves a 5–7-day treatment course as safe for pyelonephritis, 7–8-day treatment duration for ventilator-associated pneumonia (although patients with multidrug-resistant (MDR) bacteria were usually excluded from these trials), a 4-day course for complicated intra-abdominal infections following source control and with somewhat fewer data, and 5–7 days of treatment for Gram-negative bacteremia. Further trials are examining treatment durations for other types of infections and will be a welcomed information source for AMS in the future. But even more importantly, AMS programs should address stopping unnecessary empirical antibiotics.

In summary, AMS has gained more responsibility and more opportunities of affecting patient safety and outcomes than ever, in the era of ESBLs and CREs. High-quality research is needed to direct optimal management strategies.

REFERENCES

[1] Rodriguez-Bano J, Cisneros JM, Cobos-Trigueros N, Fresco G, Navarro-San Francisco C, Gudiol C, *et al.* Diagnosis and antimicrobial treatment of invasive infections due to multidrug-resistant Enterobacteriaceae. Guidelines of the Spanish Society of Infectious Diseases and Clinical Microbiology. Enferm Infec Microbiol Clin 2015;33:337.e1–337.e21.

[2] Gutierrez-Gutierrez B, Perez-Galera S, Salamanca E, de Cueto M, Calbo E, Almirante B, *et al.* A multinational, preregistered cohort study of beta-lactam/beta-lactamase inhibitor combinations for treatment of bloodstream infections due to extended-spectrum-beta-lactamase-producing enterobacteriaceae. Antimicrob Agents Chemother 2016;60:4159–69.

[3] Carmeli Y, Armstrong J, Laud PJ, Newell P, Stone G, Wardman A, *et al.* Ceftazidime-avibactam or best available therapy in patients with ceftazidime-resistant Enterobacteriaceae and Pseudomonas aeruginosa complicated urinary tract infections or complicated intra-abdominal infections (REPRISE): a randomised, pathogen-directed, phase 3 study. Lancet Infect Dis 2016;16:661–73.

[4] Matsumura Y, Yamamoto M, Nagao M, Komori T, Fujita N, Hayashi A, *et al.* Multicenter retrospective study of cefmetazole and flomoxef for treatment of extended-spectrum-beta-lactamase-producing Escherichia coli bacteremia. Antimicrob Agents Chemoth 2015;59:5107–13.

[5] Gutierrez-Gutierrez B, Bonomo RA, Carmeli Y, Paterson DL, Almirante B, Martinez-Martinez L, *et al.* Ertapenem for the treatment of bloodstream infections due to ESBL-producing Enterobacteriaceae: a multinational pre-registered cohort study. J Antimicrob Chemother 2016;71:1672–80.

[6] Lew KY, Ng TM, Tan M, Tan SH, Lew EL, Ling LM, *et al.* Safety and clinical outcomes of carbapenem de-escalation as part of an antimicrobial stewardship programme in an ESBL-endemic setting. J Antimicrob Chemother 2015;70:1219–25.

[7] Barlam TF, Cosgrove SE, Abbo LM, MacDougall C, Schuetz AN, Septimus EJ, *et al.* Executive Summary: Implementing an Antibiotic Stewardship Program: Guidelines by the Infectious Diseases Society of America and the Society for Healthcare Epidemiology of America. Clinical Infectious Diseases: An Official Publication of the Infectious Diseases Society of America 2016;62:1197–202.

[8] Schwaber MJ, Carmeli Y. An ongoing national intervention to contain the spread of carbapenem-resistant enterobacteriaceae. Clin Infect Dis 2014;58:697–703.

[9] Giacobbe DR, Del Bono V, Trecarichi EM, De Rosa FG, Giannella M, Bassetti M, *et al.* Risk factors for bloodstream infections due to colistin-resistant KPC-producing Klebsiella pneumoniae: results from a multicenter case-control-control study. Clin Microbiol Infect 2015;21:1106 e1101–1108.

[10] Deak D, Outterson K, Powers JH, Kesselheim AS. Progress in the fight against multidrug-resistant bacteria? A review of U.S. food and drug administration-approved antibiotics. Ann Internal Med 2010–2015;2016.

[11] Zusman O, Avni T, Leibovici L, Adler A, Friberg L, Stergiopoulou T, *et al.* Systematic review and meta-analysis of in vitro synergy of polymyxins and carbapenems. Antimicrob Agents Chemother 2013;57:5104–11.

[12] Giannella M, Trecarichi EM, De Rosa FG, Del Bono V, Bassetti M, Lewis RE, *et al.* Risk factors for carbapenem-resistant Klebsiella pneumoniae bloodstream infection among rectal carriers: a prospective observational multicentre study. Clin Microbiol Infect 2014;20:1357–62.

[13] Paul M, Carmeli Y, Durante-Mangoni E, Mouton JW, Tacconelli E, Theuretzbacher U, *et al.* Combination therapy for carbapenem-resistant Gram-negative bacteria. J Antimicrob Chemother 2014;69:2305–9.

[14] Price R, MacLennan G, Glen J. Selective digestive or oropharyngeal decontamination and topical oropharyngeal chlorhexidine for prevention of death in general intensive care: systematic review and network meta-analysis. BMJ (Clin Res Ed) 2014;348:g2197.

[15] Bar-Yoseph H, Hussein K, Braun E, Paul M. Natural history and decolonization strategies for ESBL/carbapenem-resistant Enterobacteriaceae carriage: systematic review and meta-analysis. J Antimicrob Chemother 2016;71:2729–39.

[16] Garonzik SM, Li J, Thamlikitkul V, Paterson DL, Shoham S, Jacob J, *et al.* Population pharmacokinetics of colistin methanesulfonate and formed colistin in critically ill patients from a multicenter study provide dosing suggestions for various categories of patients. Antimicrob Agents Chemother 2011;55:3284–94.

[17] Gregoire N, Mimoz O, Megarbane B, Comets E, Chatelier D, Lasocki S, *et al.* New colistin population pharmacokinetic data in critically ill patients suggesting an alternative loading dose rationale. Antimicrob Agents Chemother 2014;58:7324–30.

[18] Plachouras D, Karvanen M, Friberg LE, Papadomichelakis E, Antoniadou A, Tsangaris I, *et al.* Population pharmacokinetic analysis of colistin methanesulfonate and colistin after intravenous administration in critically ill patients with infections caused by gram-negative bacteria. Antimicrob Agents Chemother 2009;53:3430–6.

[19] Kazemier BM, Koningstein FN, Schneeberger C, Ott A, Bossuyt PM, de Miranda E, *et al.* Maternal and neonatal consequences of treated and untreated asymptomatic bacteriuria in pregnancy: a prospective cohort study with an embedded randomised controlled trial. Lancet Infect Dis 2015;15:1324–33.

[20] Origuen J, Lopez-Medrano F, Fernandez-Ruiz M, Polanco N, Gutierrez E, Gonzalez E, *et al.* Should asymptomatic bacteriuria be systematically treated in kidney transplant recipients? Results from a randomized controlled trial. Am J Transpl 2016. Epub.

[21] Paul M, Shani V, Muchtar E, Kariv G, Robenshtok E, Leibovici L. Systematic review and meta-analysis of the efficacy of appropriate empiric antibiotic therapy for sepsis. Antimicrob Agents Chemother 2010;54:4851–63.

[22] Paul M, Dickstein Y, Raz-Pasteur A. Antibiotic de-escalation for bloodstream infections and pneumonia: systematic review and meta-analysis. Clin Microbiol Infect 2016;22:960–7.

Section D

AMS Experiences Around the World

Chapter 19.1

AMS Initiatives and Policies: The International Picture

Bojana Beović* and Sara E. Cosgrove**
**Faculty of Medicine, University of Ljubljana, Ljubljana, Slovenia*
***Johns Hopkins University School of Medicine, Baltimore, MD, United States*

Antimicrobial resistance (AMR) was recognized as an international threat by the European authorities in the last decade of the 20th century. In 1998, the Danish government organized the conference "The Microbial Threat" leading to the Copenhagen recommendations, which drew an outline for the fight against antimicrobial resistance for the next two decades. These recommendations focused on antimicrobial resistance and consumption surveillance, prudent use of antimicrobials, development of new drugs, and research; they were followed by the establishment of the two major surveillance systems, the European Antimicrobial Resistance Surveillance System (EARSS) and the European Surveillance of Antimicrobial Consumption (ESAC). Further political interest in AMR resulted in the European Council recommendations in 2001 and several European Council conclusions. The 2008 conclusions introduced the European Antibiotic Awareness Day (EAAD) on November 18th each year. The EAAD is an ECDC-led initiative providing platform and tools for raising awareness in professionals and the general public. In 2009, the Transatlantic Task Force on AMR (TATFAR) was established by the highest level politicians and health authorities from the European Union (EU) and the United States of America. In the same year, EU Council conclusions on innovative incentives for effective antibacterials were adopted during Swedish presidency. Council conclusions from 2012 forwarded the importance of a one-health approach. The emphasis was given on all aspects of fight against AMR in human and veterinarian sectors, which should mirror the human sector in the surveillance and the principles of prudent use of antimicrobials. In 2011, the European commission adopted the action plan against the raising threats of antimicrobial resistance reinforcing the one-health approach and collaboration of all stakeholders. The progress report of the action plan summarizing all the activities was published in 2015 [1–3].

Antimicrobial Stewardship. http://dx.doi.org/10.1016/B978-0-12-810477-4.00035-0

The Dutch presidency of EU resulted in the conclusions on the next steps under a one-health approach to combat antimicrobial resistance, which were adopted by the Council of the EU in June 2016. The conclusions call upon the member states to have in place a national action plan against antimicrobial resistance, based on the one-health approach, before mid-2017. Together with the member states, the European Commission should develop comprehensive EU action plan based on one-health approach and EU guidelines on prudent use of antimicrobials in humans [4].

WHO published the global strategy for containment of antimicrobial resistance in 2001. The strategy included education of patients, prescribers, and drug dispensers; improvement of drug prescribing with use of guidelines and supervision; active management of antimicrobial drug prescribing in hospitals; improved microbiology diagnostics; and regulation of pharmaceutical industry marketing activities. The strategy also included the control of the use of antimicrobials in food-producing animals, which should follow the same principles as the antimicrobial use in humans. Several World Health Assemblies called for the implementation of the strategy in the following years. AMR was the focus of the World Health Day in 2011. On this occasion, WHO published a policy package that includes national plans, surveillance including improved laboratory performance, access to good-quality drugs, prudent prescribing in humans and animals, infection control, and drug development. In 2014, WHO published a report on AMR situation including estimation of its health and economic burden. Globally, significant resistance was found in microorganisms causing most common infections. The report was followed by the Global action plan on antimicrobial resistance, which includes five objectives: awareness, knowledge, infection prevention, prudent use of antimicrobials, and investments related to antimicrobial resistance. Since 2015, Antibiotic Awareness Week has been a WHO-led initiative raising awareness of AMR [5].

Other initiatives in AMS with an international scope (the list is nonexhaustive) are the following:

- The Chennai declaration, which started in India in 2012 (www.chennaideclaration.com)
- The Alliance for the Prudent Use of Antibiotics, which was founded in 1981 (http://emerald.tufts.edu/med/apua/)
- The World Alliance Against Resistance to Antibiotics, which includes predominately French-speaking individuals and organizations (http://www.ac2bmr.fr/index.php/fr/)

Several other initiatives have been started on the national level, but their activities have an international range (ReAct in Sweden and Antibiotic Action in the United Kingdom); the others have added the combat against antimicrobial resistance to their already existing activities (Center for Disease Dynamics, Economics, and Policy, CDDEP) [6].

In this section, we will present several country case reports, describing AMS programs in different settings; we hope that it will be a source of inspiration for you!

REFERENCES

[1] http://ecdc.europa.eu/en/healthtopics/antimicrobial_resistance/Pages/index.aspx. Accessed 25 April 2016.

[2] http://eur-lex.europa.eu/legal-content/EN/TXT/?uri=celex%3A52012XG0718%2801%29. Accessed 26 April 2016.

[3] Ganter B, Stelling J, editors. Expert consultation on antimicrobial resistance. Copenhagen: WHO Regional office for Europe; 2011.

[4] http://www.consilium.europa.eu/en/press/press-releases/2016/06/28-euco-conclusions/. Accessed 17 September 2016.

[5] http://www.who.int/mediacentre/factsheets/antibiotic-resistance/en/. Accessed 26 April 2016.

[6] Carlet J, Pulcini C, Piddock LJV. Antibiotic resistance: a geopolitical issue. Clin Microbiol Infect 2014;2014(20):949–53.

Chapter 19.2

Missions and Objectives of EUCIC and ESGAP

Evelina Tacconelli* and Bojana Beović,‡**
**Medical University Hospital Tübingen, Tübingen, Germany*
***Faculty of Medicine, University of Ljubljana, Ljubljana, Slovenia*
‡University Medical Centre Ljubljana, Ljubljana, Slovenia

European Society of Clinical Microbiology and Infectious Diseases (ESCMID) is the Europe's leading society in clinical microbiology and infectious diseases with more than 30,000 members and affiliated societies from all European countries and all continents. Antimicrobial resistance and related activities such as infection control and antimicrobial stewardship belong to the most important ESCMID research and professional topics.

In order to further contribute to the fight against the increasing spread of healthcare-associated infections due to antimicrobial-resistant bacteria, ESCMID founded the European Committee on Infection Control (EUCIC) in March 2014. Major objectives of EUCIC are to take a leading role in creating a gold standard for infection control in hospitals, long-term and rehabilitation facilities, and community in all European countries in order to reduce morbidity and mortality related to healthcare-associated infections. The group includes a steering committee, three boards (advisory, stakeholders, and education), and national committees representing more than 85% of European countries. Numerous experts in the field of infection prevention and control are involved with the aim to contribute to the harmonization of the infection prevention and control measures across Europe, by developing new educational and training tools, guidelines, and increasing political awareness of the pivotal role of infection control to guarantee patients safety and security. This target is achieved through the strict cooperation of EUCIC and ESCMID study groups with focus on infection control and with other major European stakeholders and societies (e.g., European Center for Disease Prevention and Control, World Health Organization, British Society for Antimicrobial Chemotherapy, Healthcare Infection Society, International Federation of Infection Control, European Network to Promote Infection Prevention for Patient Safety, and European Committee on Antimicrobial Susceptibility Testing).

Antimicrobial Stewardship. http://dx.doi.org/10.1016/B978-0-12-810477-4.00036-2

Major fields for EUCIC action include to define limitations of implementing infection control approaches as suggested from national and international stakeholders, to develop locally calibrated mitigation actions, and to develop new educational tools (e.g., e-learning) and promote continuously updated infection control training across all European countries. Dedicated working group focuses on educational activities (including a new ESCMID Certificate in Infection Control), guidance and consensus documents with "real-life" applicability, and research projects (e.g., prevention of surgical site infections, selective susceptibility reporting, mapping, and limitations of surveillance systems in Europe, among others). In the field of antimicrobial stewardship, EUCIC actively collaborates with ESCMID Study Group for Antibiotic Policies (ESGAP) organizing basic and advanced education and in projects aimed to push forward the integration between antimicrobial stewardship and infection control procedures at regional level according to local epidemiology.

ESGAP is a common forum for scientists and health-care professionals active in antimicrobial stewardship. The aim of ESGAP is to improve the knowledge and practice in antimicrobial prescribing and stewardship and thus optimize the antimicrobial therapy for individual patients and diminish the public threat of antimicrobial resistance.

The cornerstone of ESGAP's educational activity is the organization of postgraduate educational courses on various aspects of antimicrobial stewardship either alone or in collaboration with other ESCMID study groups. In addition, ESGAP organizes many workshops in different parts of Europe coorganized by national professional societies or embedded in various regional and international conferences and educational events, which help bringing the knowledge on antibiotic stewardship to various parts of the Europe and beyond.

Research activities have been focused to various problems in antimicrobial stewardship. More recent research projects include the global survey on antimicrobial stewardship activities with ISC, colistin dosing, antibiotic treatment practices for endocarditis, and various aspects of education of health-care professionals including knowledge, attitudes and practices of antibiotic prescribing and resistance in medical students and young doctors, and antimicrobial stewardship in the European medical schools' curricula. The problem of forgotten antibiotics was addressed in 2011 and again in 2015. The surveys on knowledge, attitudes, and practices of veterinarian students, the legal issues in antimicrobial stewardship, the duration of antimicrobial treatment, and the development of antimicrobial prescribing and stewardship competencies are among the most important ongoing projects.

The ESGAP's open virtual learning community (OVLC) is a web-based platform that provides information on all important achievements in the field of antimicrobial stewardship such as guidelines, initiatives, and interesting papers and invites all professionals interested in antimicrobial stewardship for discussion. ESGAP newsletter is another form of the contact with membership bringing short notices on most relevant ESGAP activities.

The future perspectives of ESGAP include the continuation and intensification of all educational and research activities together with broadening international collaboration within ESCMID and also beyond by the joint activities with ECDC, WHO, and ISC.

Information on EUCIC, including news, activities, reports, and documents (statutes, pamphlets, etc.), are available at the EUCIC website (https://www.escmid.org/eucic/).

Information on ESGAP activities is available at https://www.escmid.org/research_projects/study_groups/antibiotic_policies/ and http://esgap.escmid.org/.

Chapter 19.3

Antimicrobial Stewardship in Latin America

Gabriel Levy-Hara*, Pilar Ramón-Pardo**, José L. Castro**, Cristhian Hernández-Gómez‡, Luis Bavestrello§ and María V. Villegas‡
*Hospital Carlos G Durand, Buenos Aires, Argentina
**Pan American Health Organization/World Health Organization, Washington, DC, United States
‡Centro Internacional de Entrenamiento e Investigaciones Médicas (CIDEIM), Cali, Colombia
§Clínica Reñaca, Viña del Mar, Chile

Antimicrobial stewardship programs (ASPs) are relatively new in Latin American countries, compared with other parts of the world—like Australia, the United States, and Europe. During the last years, some hospitals have begun to create ASP in different countries—in most cases as individual projects—as there are no official policies of national governments currently supporting or regulating these interventions.

What do we know about Antimicrobial stewardship programs in Latin American countries?

Two regional surveys were performed during the last 4 years aiming to investigate the current state of ASP in hospitals.

The first international survey that included Latin American countries was conducted jointly between ESCMID study group for antibiotic policies (ESGAP) and the antimicrobial stewardship (AMS) working group of the International Society of Chemotherapy (ISC) and published by Howard *et al.* [1]. This was done between March and September 2012 and included overall 660 eligible responses (67 countries from 6 continents).

One hundred and three inputs came from South and Central America. Most responders were from Argentina (39), Peru (18), Brazil (9), Venezuela (9), Chile (8), Colombia (6), and Uruguay (5). More than half were teaching tertiary care hospitals. Standards for AMS in hospitals varied from country to country; overall 46% already had an ASP, compared with 56% of the rest of the world, and much less than Europe and the United States (66% and 67%,

Antimicrobial Stewardship. http://dx.doi.org/10.1016/B978-0-12-810477-4.00037-4

respectively). The median duration for ASP was 3 years overall. Hospitals from Brazil (67%), Chile (88%), and Colombia (83%) had more ASPs in place, although figures are too small for comparison.

The main objectives of ASP were comparable between all countries: to reduce or stabilize resistance (87%), to reduce the amount of prescribing antibiotic (53%), and to improve clinical outcomes (49%). In this regard, main objectives were similar to the rest of the world.

Dedicated weekly hours of AMS team members were different than other regions. For example, Latin American country hospitals reported a mean of 9 h of a pharmacist with experience in antimicrobials or infectious diseases (ID) (world mean, 18 h), 12 h of ID physician (world mean, 10 h), and 7 h of a clinical microbiologist (world mean, 9 h). Interestingly and similarly to lower income regions, nurses had a critical role: 14 h/week dedicated to the ASP, compared with a mean of 6 h in the rest of the globe.

Fifty-six percent of hospitals had a published AMS policy, with around 50% having treatment guidelines and 70% surgical prophylaxis guidelines. Most had some sort of microbiology advice (by phone or during ward rounds) for all or some areas of the hospital. In turn, just around 40% had preauthorized pharmacy-driven dose optimization (e.g., automated adjustment according to renal function). Inflammatory markers (e.g., C-reactive protein and procalcitonin) for initiating or discontinuing antimicrobials treatment were used in less than half of the cases.

Antimicrobial ward rounds performed jointly with the medical staff of the corresponding unit were conducted on a regular basis by 67% of centers, figures similar to Europe (70%) and Oceania (61%). In Latin American countries, the frequency varied from country to country; in the ICUs, a daily round was the commonest.

Overall, 64% of Latin American country hospitals informed a restricted or reserve antimicrobial list or formulary, figures lower than the worldwide mean (81%). Carbapenems and cephalosporins were the classes more frequently restricted.

Communication regarding AMS (e.g., guidelines, antimicrobials consumption, and learning points from incidents) was mainly driven through staff meetings (more than across intranet or booklets).

Main barriers—common for all Latin American countries—to deliver a functional and effective ASP were the lack of personnel or funding, opposition from prescribers, lack of information technology support and/or ability to get data, and other higher-priority initiatives. These obstacles were similar to the rest of the globe.

The second survey was exclusively targeted to Latin American countries (Muñoz *et al.* [2]) and addressed complementary issues. It was conducted between October 2014 and April 2015 among representatives participating in an AMS training program. Participating countries were Argentina, Peru, Brazil, Chile, Colombia, Mexico, Bolivia, Panama, Ecuador, and Dominican Republic.

Twenty-seven respondents from 10 hospitals completed the survey. Teaching hospitals were 18/27 (66.7%). Half of the hospitals reported no information technology tools or training support for their ASPs. Pharmacists were only present in the 37% of the teams. Sixty-seven percent had recommendations based on local susceptibility for treatment of common clinical conditions.

Antibiotic time-out audits after 48 h were performed in just 37%, and preauthorization of specific antimicrobials was required in 74%—mainly, carbapenems, colistin, vancomycin, and linezolid. Automatic alerts for redundant therapy and automatic stop for antibiotic orders were performed in just 7% and 37% of hospitals, respectively. Only seven (25.9%) laboratories agreed to perform and report confirmatory testing for multidrug-resistant organisms. Fifteen (55.6%) monitored antibiotic consumption using defined daily dose/ATC system, and just 11 (40.7%) reported to deliver some kind of education on AMS.

A long way ahead...

Both surveys have inherent limitations, as representative concerns, a number of responders in different surveys, self-exclusion bias of those who have not ASP, and unverifiable responses. However, they provide some evidence on the current state of ASP in Latin American countries. The implementation of these programs is somehow delayed compared with other regions. Obstacles have shown to be similar all around the world, although, from a practical point of view, issues should be addressed in a specific way in Latin America.

In Latin American countries, there is an urgent need for improving the awareness of governments about the huge morbidity related to the misuse of antibiotics, the benefits of ASP, and the need of support for implementing ASPs in hospitals. Simultaneously, it would be essential to directly reach lawmakers to sanction the respective laws concerning the creation of ASP, in a context of a National Action Plan on antimicrobial resistance. This would grant sustainability to the interventions in spite of political changes.

The ASP could be progressively installed, beginning with those bigger and more complex institutions—that usually have greater antimicrobials consumption and misuse. Then, a nationwide policy should be implemented to change the current situation, where AMS is carried on just based on the consciousness and desires of key actors (IDs, microbiologists, pharmacists, intensivists, clinicians, and pediatricians). Hospital authorities might stimulate the creation of these programs and support those already in place. This support must be real, not just declarative. The potential reduction of antimicrobials use, progression of antimicrobials resistance, lengths of stay, and adverse events should undoubtedly justify the initial investments.

Human and material resources will facilitate the development and sustainment of all ASP components (i.e., control of antimicrobials utilization, audits, education, communication, monitoring resistance, and antimicrobials consumption). Also, the progressive incorporation of rapid diagnosis techniques would contribute to significantly decrease the misuse of antimicrobials.

A strong AMS team with a continuous education program in stewardship could overcome many of the challenges. For example, even the opposition from prescribers—one of the most frequently reported barriers, it is not so hard to overcome when AMS teams are able to perform antimicrobial joint rounds with local staff.

The World Health Organization (WHO) and its regional office for the Americas, the Pan-American Health Organization, have approved action plans for the containment of antimicrobial resistance [3,4]. The design of the National Action Plans—in coherence with the WHO Global Action Plan proposal—has already begun in Latin American countries.

As part of the actions, being AMS's one of the most effective interventions to contain antimicrobial resistance, PAHO is engaged in supporting the organization of AMS teams and programs throughout Latin America and the Caribbean. Since 2014, some workshops have been developed in different countries (e.g., Paraguay and Barbados) with the intention of starting pilot projects and disseminating the experience to other countries [3]. For countries' antibiotic sale estimation, data from IMS are being used. Also, local and national mechanisms to estimate sales—and whenever possible, consumption—are under discussion and evaluation.

In summary, these are some good news for Latin American countries. Hence, we hope that in the near future we will count with more data and more active ASP in place in our region.

REFERENCES

[1] Howard P, Pulcini C, Levy Hara G, West RM, Gould IM, Harbarth S, *et al.* An international cross-sectional survey of antimicrobial stewardship programmes in hospitals. J Antimicrob Chemother 2015;70(4):1245–55.

[2] Muñoz JS, Motoa G, Escandón-Vargas K, Bavestrello L, Quirós R, Hernandez C, *et al.* Current antimicrobial stewardship practices in Latin America: where are we? Open Forum Infect Dis Fall 2015;2(Suppl 1):S92.

[3] WHO 68.7. 2015. Global action plan on antimicrobial resistance. Access: http://apps.who.int/iris/bitstream/10665/193736/1/9789241509763_eng.pdf?ua=1.

[4] Pan American Health Organization/World Health Organization. Plan of Action on Antimicrobial Resistance [Internet]. 54th Directing Council of PAHO, 67th meeting of the WHO Regional Committee for the Americas; 2015 Sep 28–Oct 2, Washington (DC), United States. Washington (DC): PAHO; 2015.

Chapter 19.4

Antimicrobial Stewardship in Australia

Kirsty Buising and Karin Thursky
University of Melbourne, Melbourne, VIC, Australia

Since 1978, Australia has had national antibiotic prescribing guidelines, called THERAPEUTIC GUIDELINES: ANTIBIOTIC. These are authored by independent Australian expert groups and as such have been well accepted and widely endorsed. In the community, the pharmaceutical benefits scheme (PBS) governs reimbursement for drug costs for a specified list of antimicrobial drugs and indications, thus acting as a formulary with restriction. This scheme has successfully limited access to some antimicrobial drugs such as fluoroquinolone antibiotics in the community by requiring a phone approval before a prescription will be reimbursed.

Surveillance of hospital antibiotic consumption has occurred since 2004 through the National Antimicrobial Utilization Surveillance Program. This produces reports on volumes of antimicrobial drug consumption, adjusted for bed occupancy, and allows comparisons between hospitals.

Restriction of access to broad-spectrum antimicrobial agents in hospitals has a long history of acceptance by clinicians in Australia. In the past, phone approval programs predominated in which an infectious disease (ID) specialist had to approve the use of restricted drugs for individual patients. Since 2001, computerized approval systems, often with some decision support, have been implemented, with most functioning in the absence of electronic prescribing. These systems enabled triaging of patients for individual prescriber feedback from ID physicians (pre- or postprescription) and detailed auditing of indications for antimicrobial use. Audits of the appropriateness of antibiotic prescribing in hospitals, relative to the national guidelines and taking microbiology results and patient attributes into account, have occurred since 2010 through the National Antimicrobial Prescribing Survey. Participation rates are high in hospitals of all types.

Antimicrobial Stewardship. http://dx.doi.org/10.1016/B978-0-12-810477-4.00038-6

In 2011, the Australian Commission on Safety and Quality in Health Care (ACSQHC) published recommendations for antimicrobial stewardship (AMS) in Australian hospitals. Since 2013, all hospitals have been required to have an AMS program in place to be accredited according to National Safety and Quality in Health Care (NSQHC) standards. These programs must have appropriate governance structures, access to national guidelines must be provided to staff, auditing of drug use and antibiotic-resistant pathogens must occur, and action to promote improvement needs to be taken. It is also recommended that laboratories provide annual facility specific antibiograms.

Most tertiary hospitals now have dedicated resources for AMS—usually an ID physician and clinical pharmacist, though at some sites nursing staff and infection control practitioners are also part of the AMS team. Antimicrobial stewardship is part of undergraduate and postgraduate education for all healthcare staff. Most hospital AMS programs provide some form of education on antibiotic use as part of continuing professional development for nurses, doctors, and pharmacists. Public hospitals generally restrict the access of pharmaceutical company representatives to influence the learning of doctors in training.

In the community, the NPS MedicineWise (formally the National Prescribing Service) is a Commonwealth government-funded service that conducts prescriber education for general practitioners. They provide feedback on antibiotic consumption against peers using pharmaceutical benefits scheme data and recently began feedback on individual practice patterns relative to other practices by indication. They also conduct traditional media and social media campaigns for the public.

In 2015, the National Centre for Antimicrobial Stewardship was established. This group oversees auditing of antimicrobial use in hospitals and residential aged-care settings, provides education to build capacity in staff from rural hospitals, and develops resources and tools for prescribers.

In 2013, the antimicrobial use and resistance in Australia (AURA) project was established as a Commonwealth government-funded national initiative overseen by the ACSQHC. The AURA project functions at a high level to coordinate data from the community and hospital sectors on antibiotic usage and resistance. An important initiative has been the launch of the clinical care standards in antimicrobial stewardship, which define expectations of appropriate antimicrobial prescribing for consumers and healthcare workers. The first national antimicrobial resistance strategy was published in 2015, and it describes objectives of ensuring effective AMS practices in human health and animal care.

Major challenges have included the following:

- Developing models of care to bring expertise to rural/regional/remote sites
- Gathering meaningful data on prescribing behavior in the community
- Gathering consistent laboratory data on bacterial resistance patterns

Major priority areas

The next key focus will primarily be the community/general practice setting:

- Better use of data to identify inappropriate antimicrobial use
- Attention to antibiotic overuse in residential aged-care settings, particularly prophylactic antibiotics, overtreatment of asymptomatic bacteriuria or colonized skin ulcers
- Focus on antibiotic overuse in upper respiratory tract infections

REFERENCES

[1] Therapeutic Guidelines—https://www.tg.org.au/.

[2] NAUSP—http://www.sahealth.sa.gov.au/wps/wcm/connect/public+content/sa+health+internet/clinical+resources/clinical+programs/antimicrobial+stewardship/national+antimicrobial+utilisation+surveillance+program+nausp.

[3] NCAS https://ncascre.wordpress.com/who-are-we/.

[4] NAPS—https://www.naps.org.au/Default.aspx.

[5] Commissions recommendations for AMS in hospitals book—http://www.safetyandquality.gov.au/wp-content/uploads/2011/01/Antimicrobial-stewardship-in-Australian-Hospitals-2011.pdf.

[6] NPS MedicineWise—http://www.nps.org.au/.

[7] NSQHC standards—http://www.safetyandquality.gov.au/wp-content/uploads/2011/09/NSQHS-Standards-Sept-2012.pdf.

[8] First National strategy—http://health.gov.au/internet/main/publishing.nsf/Content/1803C433C71415CACA257C8400121B1F/$File/amr-strategy-2015-2019.pdf.

[9] Clinical care standard—http://www.safetyandquality.gov.au/our-work/clinical-care-standards/antimicrobial-stewardship-clinical-care-standard/.

[10] AURA—http://www.safetyandquality.gov.au/national-priorities/amr-and-au-surveillance-project/.

[11] Atlas of variation—http://www.safetyandquality.gov.au/atlas/.

Chapter 19.5

Antimicrobial Stewardship in Austria

Agnes Wechsler-Fördös
Krankenanstalt Rudolfstiftung, Vienna, Austria

The health-care system is characterized by a high density of easily accessible health-care facilities. In 2011, 273 hospitals with about 64,000 beds were available for inpatient care. By the end of 2011, about 19,500 physicians (general practitioners and specialists) were self-employed working in their own practices [1].

In Austria, health care is based on a social insurance model that guarantees equitable access to high-quality health services—irrespective of age, sex, origin, social status, or income. The social insurance system is based on the principles of compulsory insurance, solidarity, and self-governance and is primarily funded through insurance contributions. Insurance is usually linked to gainful employment and thus begins automatically; some groups (self-employed and voluntarily insured people) have to apply for insurance, however. Usually half of the contributions are paid by the employer and half by the employee. The total contribution rate is 7.65% of the contribution base in the majority of cases. Social insurance is the most important source of health-care funding, contributing around EUR 13.8 billion in 2011, which corresponds to about 45% of current health expenditure [1].

In Austria, the first antibiotics list was implemented in Rudolfstiftung hospital in Vienna in 1992. After the first year, costs for antibiotics had decreased by 4.68% (€ 88,000); until 2002, costs decreased by 35%. Concomitantly with the decreased prescription of third-generation cephalosporins, resistance rates in Gram-negative bacteria also decreased [2].

In 1998, the "Guidelines to Further Develop and Define Antibiotic Use in Hospitals" were published, initiated, and sponsored by the Ministry of Health [3]. Until 2005, the Ministry of Health funded antibiotic stewardship (ABS) counseling and workshops to establish local ABS teams in 71 hospitals.

Antimicrobial Stewardship. http://dx.doi.org/10.1016/B978-0-12-810477-4.00039-8

Since then, regular ABS courses are organized by the Austrian Society of Antimicrobial Chemotherapy [4]. In Upper Austria, annual workshops with updates for all hospitals in this region are offered [4]. In the "ABS International" project funded by the EU, the Austrian ABS concept as a model for good practice was presented to nine cooperating countries [5].

Compared with many other European countries, resistance rates in Austria are lower [6]. Data for antibiotic consumption in primary health care are available since 1998. In 2014, the level of antimicrobial use was 17.4 prescriptions per 10,000 inhabitants with a slightly decreasing trend since 1998 [6] but biased as only antibiotics paid by health insurance funds were recorded.

In 2014, a hospital-based antibiotic consumption surveillance system was developed by the Austrian Agency for Health and Food Safety (AGES) in cooperation with the German Robert-Koch Institute and the Charité Hospital in Berlin. Consumption data are available for 5 Austrian hospitals back to 2014 and can be benchmarked with 75 participating hospitals in Germany [7].

Especially in hospitals that participated in ABS trainings, there is a high motivation to carry on stewardship activities with the help of ABS teams. To enable meaningful changes, either a legal basis or other incentives should be created for a countrywide compulsory implementation of ABS teams with sufficient time and funding.

To provide a solid basis for judicious medication, knowledge about antibiotics has to be strengthened, starting in medical school: Only 21% of the 495 Austrian participants in the ESGAP Prepare Study stated that they had received enough training; 40% specifically wanted more education about prudent use. Regular updates on antibiotic stewardship should also be provided for junior and senior doctors (unpublished data).

National therapy guidelines for the most important bacterial infections should be endorsed to provide basis for adherence measurements and benchmarking.

Targets concerning changes in antibiotic consumption should be defined according to resistance rates.

In order to secure also cooperation of patients, there is a need for information of the public concerning the problem of antibiotic resistance and its implications for the public health.

REFERENCES

[1] Austrian Federal Ministry of Health. The Austrian Health Care System. Key Facts. Updated version August 2013. ISBN 978-3-902611-38-3.

[2] Ludwig Boltzmann-Institut für Medizin- und Gesundheitssoziologie. Wiener WHO-Modellprojekt Gesundheit und Krankenhaus. Wie man im Krankenhaus die Hygiene-Organisation verbessern kann. Wien, Dezember 1995.

[3] Allerberger F, Gareis R, Janata O, Krause R, Meusburger S, Mittermayer H, Rotter-le Beau M, Watschinger R, Wechsler-Fördös A. ABS antibiotics strategies. Guidelines to Further Develop and Define Antibiotic Use in Hospitals. Vienna, 2nd revised ed., November 2002.
[4] http://www.oegach.at/asp-seminare/. Accessed 15 Feb 2017.
[5] Allerberger F, Lechner A, Wechsler-Fördös A, Gareis R. Optimization of antibiotic use in hospitals—antimicrobial stewardship and the EU project ABS international. Chemotherapy 2008;54:260–7. http://dx.doi.org/10.1159/000149716.
[6] Bundesministerium für Gesundheit. Resistenzbericht Österreich AURES 2014: Antibiotikaresistenz und Verbrauch antimikrobieller Substanzen in Österreich. Vienna, November 2015. http://www.bmg.gv.at/home/Schwerpunkte/Krankheiten/Antibiotikaresistenz/.
[7] Nationales Referenzzentrum für Surveillance von nosokomialen Infektionen - Antibiotic Consumption Surveillance http://www.nrz-hygiene.de/surveillance/antibiotikaverbrauch/. Accessed 1 August 2016.

Chapter 19.6

Antimicrobial Stewardship in Belgium

Patrick Lacor* and Peter Messiaen**
**UZ Brussel, Brussel, Belgium*
***Jessaziekenhuis, Hasselt, Belgium*

Belgium has a publicly funded health-care system organized both on a federal level (compulsory health-care insurance and financing of hospitals) and on the level of the federated communities (preventive care and health promotion). Self-employed general practitioners constitute a strong and broad base for primary health care. The vast majority of secondary and tertiary care hospitals are public, university, or semiprivate institutions publicly funded. Patients are free to choose their own medical professionals. The health-care provider has a large degree of therapeutic freedom and is paid on a fee-for-service base. Overall, the health-care system scores reasonably good in terms of accessibility [1].

Almost 99% of the population is covered by mandatory health insurance funded mainly by taxes; costs are paid up front and for a large proportion reimbursed. The patient's own contribution can run up till 17.7% of overall costs [2]. Antibiotics are nearly 100% reimbursed as "essential medicines." Antibiotic use is mostly contained in the standard fee for hospital admission.

In 1999, the multidisciplinary Belgian Antibiotic Policy Coordination Committee (BAPCOC) was created by royal decree [3] with the objective to reduce development and spread of antibiotic resistance. The committee reports on the use of antibiotics and the evolution of antibiotic resistance in Belgium sets up campaigns to promote rational use in the community and to improve hand hygiene compliance in hospitals and publishes guidelines for antibiotic use in primary and hospital care. "Antibiotic management teams" were introduced in all acute care hospitals, with well-defined responsibilities such as monitoring the use of antibiotics and the evolution of

Antimicrobial Stewardship. http://dx.doi.org/10.1016/B978-0-12-810477-4.00040-4

microbial resistance in the hospital and issuing local practice guidelines on antibiotic treatment and prophylaxis [4,5].

The public campaigns launched in the 2000–01 winter decreased outpatient antibiotic use by 36% over 7 years when expressed as PID, the number of packages/1000 inhabitants/day [6]. This longitudinal trend in PID differed from the estimates using DID, the defined daily dose (DDD)/1000 inhabitants/day. Data from the European Surveillance of Antimicrobial Consumption (ESAC) show that the consumption of antibiotics for systemic use in the Belgian community in 2014 amounted to 28.2 in DID, which is above the population-weighted EU/EEA mean consumption [7] but to 2.41 in PID, which is below the mean. The question was raised whether the number of packages might not be a better proxy of antibiotic prescribing than the number of DDDs [6,8]. In the hospital sector, Belgium ranked rather favorably in 2013, staying below the mean consumption (in DDD).

Future research should focus on the tools most reliable to evaluate the use of antibiotics, on other factors predicting antibiotic resistance [9] and on how to make interventions most effective. Efforts should be continued to improve the quality of prescription, avoiding over- and underprescription [10]. This requires a strong political commitment and a joint engagement of the many involved players, integrated in the BAPCOC 2014–19 strategic plan [11].

REFERENCES

[1] OECD. Health at a glance 2015: OECD indicators, Paris: OECD Publishing; 2015. http://dx.doi.org/10.1787/health-glance-2015-en.

[2] Gerkens S, Merkur S. Belgium: health system review. Health Syst Trans 2010;12(5):1–266.

[3] Goossens H, Coenen S, Costers M, De Corte S, De Sutter A, Gordts B, *et al.* Achievements of the Belgian Antibiotic Policy Coordination Committee (BAPCOC). Eurosurveillance 2008;13(46):1–4.

[4] Van Gastel E, Costers M, Peetermans WE, Struelens MJ. Nationwide implementation of antibiotic management teams in Belgian hospitals: a self-reporting survey. J Antimicrob Chemother 2010;65:576–80.

[5] Lambert ML, Bruyndonckx R, Goossens H, Hens N, Aerts M, Catry B, *et al.* The Belgian policy of funding antimicrobial stewardship in hospitals and trends of selected quality indicators for antimicrobial use, 1991-2010: a longitudinal study. BMJ Open 2015;5:e006916. http://dx.doi.org/10.1136/bmjopen-2014-006916.

[6] Coenen S, Gielen B, Blommaert A, Beutels Ph, Hens N, Goossens H. Appropriate international measures for outpatient antibiotic prescribing and consumption: recommendations from a national data comparison of different measures. J Antimicrob Chemother 2014;69:529–36.

[7] http://ecdc.europa.eu/en/activities/surveillance/ESAC-Net/Pages/index.aspx.

[8] Bruyndonckx R, Hens N, Aerts M, Goossens H, Molenberghs G, Coenen S. Measuring trends of outpatient antibiotic use in Europe: jointly modelling longitudinal data in defined daily doses and packages. J Antimicrob Chemother 2014;69:1981–6.

[9] Blommaert A, Marais C, Hens L, Coenen S, Muller A, Goossens Beutels P. Determinants of between-country differences in ambulatory antibiotic use and antibiotic resistance in Europe: a longitudinal observational study. J Antimicrob Chemother 2014;69:535–47.

[10] Coenen S, Costers M, De Corte S, De Sutter A, Goossens H. The first European Antibiotic Awareness Day after a decade of improving outpatient antibiotic use in Belgium. Acta Clin Belg 2008;63:296–300.

[11] http://consultativebodies.health.belgium.be/sites/default/files/documents/policy_paper_bapcoc_executive_summary_2014-2019_english.pdf.

Chapter 19.7

Antimicrobial Stewardship in Bulgaria

Emma Keuleyan* and Tomislav Kostyanev**
**Medical Institute of the Ministry of Interior, Sofia, Bulgaria*
***University of Antwerp, Antwerp, Belgium*

Republic of Bulgaria's GDP (44.162 million EUR, 2015) and the share for health-care costs (7.6% of GDP) are lower compared with the EU average rate, that is, 10.1% [1]. The estimation of public health expenditure is around 4% of GDP (2008–14), compared with an average of 7.3% for the EU countries.

National Health Insurance Fund covers basic level of health-care services, which are reimbursed. National Health Insurance Fund is the only compulsory health-care insurance pillar. A substantial share of health-care costs needs to be funded directly by the patients (~40%) [2], and in general, the health-care system in Bulgaria is underfunded. The physician/population ratio in Bulgaria (3.76/1000) is among the first 20 countries in the world.

A lot of measures have been undertaken by microbiologists and the state authorities to contain antimicrobial resistance (AMR) and to regulate prescription of antibiotics (only by physicians) and dispense from pharmacies (upon prescription), thus contributing to conserve antibiotics for future generations [3]. In 1999, an expert multidisciplinary committee within the Ministry of Health was created to work on rational antibiotic policy (AP). The first national program for AP was approved in 2001 and included (i) quality control in clinical microbiology laboratories and antimicrobial resistance surveillance, (ii) rational antibiotic treatment and prophylaxis, (iii) infection control, and (iv) education. A reference laboratory for quality control in clinical microbiology laboratories and antimicrobial resistance surveillance was created within the National Center for Infectious and Parasitic Diseases. National antibiotic consumption was registered by the Bulgarian Drug Agency (BDA). Hospital committees developed guidelines for AP and infection control.

Antimicrobial Stewardship. http://dx.doi.org/10.1016/B978-0-12-810477-4.00041-6

The program was updated in 2004 [4]. National antimicrobial resistance surveillance program BulSTAR (www.bam-bg.net) provided data on etiology of infections and AMR of pathogens. International cooperation with the European Society of Clinical Microbiology and Infectious Diseases, the Alliance for the Prudent Use of Antibiotics, the European Society of Chemotherapy/Infectious Diseases (ESCID), the International Society of Chemotherapy (ISC), European Center for Disease Prevention and Control (ECDC), and others played important role [5]. In 2005, the Bulgarian societies, ESCMID Study Group for Antibiotic Policies (ESGAP), and the Alliance for the Prudent Use of Antibiotics organized "rational antibiotic policies" conference in Sofia (www.APUA-Bulgaria – www.tuft.edu). AMR and antibiotic stewardship (ABS) were leading topics in many congresses in Bulgaria, and the European Antibiotic Awareness Day has been celebrated since 2008. Several medical standards were adopted by the Ministry of Health. In 2015, the Ministry of Health started issuing pharmacotherapeutic guidelines. BDA strengthened the control on "over-the-counter" (OTC) sale of antibiotics. Work on improving usage of antibiotics was performed in the veterinary sector, and the national program for control of AMR for zoonotic pathogens was approved in 2012. Education of society by experts was supported by mass media. On 1 June 2016, the parliamentary commission on health-care, the patient organizations, and experts organized a round table discussion "rational antibiotic use—a common mission of physicians, patients, and institutions," which decided to support the introduction of National Action Plan on antibiotic stewardship.

Many regulations have been adopted by the state authorities in Bulgaria concerning the improvement of responsible antibiotic use in the country, including regulations banning the sale of antibiotics without prescription. There are still some pharmacies (<10%) where patients can buy antibiotics OTC due to the lack or low level of control. Some GPs still prescribe antibiotics without microbiology testing.

Main challenges to implement hospital AP are economic. Wide usage of cheap antibiotics (e.g., generic ceftriaxone ~0.8 DDD/1000 inhabitants/day) [6] is responsible for the emergence and spread of AMR strains, such as ESBL-producing *Enterobacteriaceae*, which further contributes to the emergence of carbapenem-resistant Gram-negative bacteria [7]. Another problem is the lack of patient isolation to prevent spread of AMR.

Rational AP should receive much more attention from the official authorities including appropriate regulative, human and financial resources. An updated antibiotic stewardship program is needed to guarantee:

- appropriate antibiotic use,
- the introduction of rapid methods for identification of pathogens and their resistant mechanisms,
- the screening of patients at risk for carriage of multidrug-resistant microorganisms and appropriate facilities for their isolation.

Bulgarian scientists and physicians should contribute to the international efforts to contain AMR and guarantee antibiotic stewardship.

REFERENCES

[1] The World Bank. http://data.worldbank.org/ (accessed 15 February 2016).

[2] Ministry of Finance of Republic of Bulgaria, Report on Healthcare System Management and Funding (in Bulgarian). www.minfin.bg/document/2891:1.

[3] Antibiotic resistance: synthesis of recommendations by expert policy groups. WHO/CDS/CSR/DRS/2001.10, pp. 140–146.

[4] Ministry of Health of Republic of Bulgaria. http://www.mh.government.bg/bg/novini/aktualno/szdava-se-natsionalna-sistema-za-nadzor-na-antibio/ (in Bulgarian) (accessed 15 February 2016).

[5] Keuleyan E. IM Gould. Key issues in developing antibiotic policies: from an institutional level to Europe wide. ESGAP. Clin Microbiol Infect 2001;7(Suppl 6):16–21.

[6] Todorova B, Velinov T, Kantardjiev T. Extended-spectrum beta-lactamase surveillance in Bulgaria in view of antibiotic consumption. Probl Infect Parasit Dis 2014;42(1):14–8.

[7] ECDC Antimicrobial surveillance report 2014. http://ecdc.europa.eu/en/publications/Publications/antimicrobial-resistance-europe-2014.pdf.

Chapter 19.8

Antimicrobial Stewardship in Croatia

Arjana Tambić Andrašević* and Vera Vlahović-Palčevski**
**University Hospital for Infectious Diseases, Zagreb, Croatia*
***University Hospital Rijeka, Rijeka, Croatia*

INTRODUCTION

In Croatia, there is nearly 100% health-care coverage predominately provided through public insurance. The number of doctors and nurses per capita is below the European average, and the number of doctor consultations and average length of stay is higher than the European average [1]. This indicates personnel shortage and patient overcrowding in hospitals that promote defensive medicine and poor compliance with infection prevention precautions. Antibiotics are reimbursed in full or in part and are prescription-only medicines.

ACTIONS TAKEN

In line with the national strategy, Interdisciplinary Section for Antimicrobial Resistance Control (ISKRA) coordinates various activities in the field of antimicrobials and resistance. It is a joint platform for the three ministries and brings together experts in the field of human and veterinary medicine, agriculture, science, and education, which proves to be important in rising awareness on rational antibiotic use in all sectors. Antibiotic consumption and resistance surveillance network are well developed with >90% of health-care facilities coverage. National surveillance data are used when developing national guidelines in education and public campaigns. Since 2008, Croatia follows European Center for Disease Prevention and Control initiative for public campaigns, which are held at regional and national level with special emphasis on education of preschool children and their parents. Since 2008, a slight decrease in ambulatory antibiotic consumption is observed at the national

Antimicrobial Stewardship. http://dx.doi.org/10.1016/B978-0-12-810477-4.00042-8

level. This may be attributed to implementation of not only ISKRA guidelines and public campaigns but also general financial restrictions in health care. ISKRA collaborates closely with the national committee for infection control, which seems to be the key strategy in restricting the initial spread of *Klebsiella pneumoniae* carbapenemase (KPC). Although hand hygiene campaign is very active since 2009, a decrease of MRSA is probably more related to the general trend observed in many European countries [2].

MAIN CHALLENGES

In spite of the many activities at regional and national level, Croatia has a huge problem with multidrug-resistant bacteria, principally carbapenem-resistant *Acinetobacter baumannii*, carbapenem-resistant *Pseudomonas aeruginosa*, and MRSA [2]. There are microbiology facilities in all country regions with regular continuous educational programs on resistance mechanisms and their detection at the laboratory level. However, financial restrictions might lead to underuse of rapid tests for timely detection of MDR carriers. Also, interventions following detection of MDR are often not adequate. In some areas, timely reporting of microbiology findings and consulting role of a microbiologist are obstructed by the fact that a laboratory is not within the hospital and does not provide service 7 days a week. Greater recognition of antimicrobial stewardship programs by hospital managements and deeper involvement of infectious disease physicians, clinical microbiologists, clinical pharmacologists, and clinical pharmacists are needed.

ACTIONS NEEDED

1. As >90% of antibiotics are dispensed in primary care settings, it is very important to develop national guidelines for antimicrobial drug use in primary care. Regional guidelines that exist are well accepted. At the moment, national guidelines are available for only few indications, and it is planned to develop more in the near future.
2. Implementation of guidelines is a further challenge. Engagement from medical schools is needed in order to accept the guidelines as a tool in responsible antibiotic prescribing. There are a lot of initiatives on introducing learning packages on antibiotics and infection for junior and senior school children (e-Bug project), and there are serious efforts to harmonize curricula for specialist doctors' training throughout Europe (UEMS activities), but programs for undergraduate medical students are not widely discussed.
3. In hospitals, antimicrobial stewardship (AMS) teams should be introduced, which will require substantial restructuring and education of our CM and ID doctors, clinical pharmacologists, and clinical pharmacists. Traditionally, in many Eastern and Central European countries, microbiology

laboratories are seated at public health institutes and not hospitals, while ID physicians are clustered in specialized hospitals, so many patients in intensive care units and other high-risk wards are left without the consulting assistance of full AMS team, if any.

REFERENCES

[1] OECD. Health at a glance: Europe 2014. OECD Publishing; 2014.

[2] European Centre for Disease Prevention and Control. Antimicrobial resistance surveillance in Europe 2014. Annual Report of the European Antimicrobial Resistance Surveillance Network (EARS-Net). Stockholm: ECDC, 2015

Chapter 19.9

Antimicrobial Stewardship in England

Philip Howard
Leeds Teaching Hospitals NHS Trust, Leeds, United Kingdom

England is the largest of four countries in the United Kingdom with a population of 55 million. More than 83% of health care is delivered by the state through the National Health Service (NHS). Public funding accounts for 98% of health-care costs. Care is delivered free at the point of care with patients registered with a single family doctor (general practitioner). Patients can be referred or directly attend any state hospital for care. The United Kingdom has average antibiotic consumption rate among European countries as 20.8 DDD/1000 inhabitants in 2014 (ECDC).

The "path of least resistance" [1] first described the causes and solutions for AMR in the United Kingdom in 1998. In 2000, the first AMR strategy and action plan was published, and a Standing Advisory Committee on AMR (SACAR) to the government was formed a year later.

Investment in AMS took place in 2003 with the hospital pharmacy initiative: an investment of £12 million over 3 years to establish an antimicrobial pharmacist in every hospital to develop antimicrobial guidelines [2]. In 2008, the Health and Social Care Act was introduced and included a code of practice for healthcare-acquired infections. This mandated that all state hospitals have an ongoing antimicrobial stewardship program that includes guidelines that were audited, revised, and updated [3]. These focused on achieving the national targets for a reduction in *Clostridium difficile* infection [4] and MRSA bacteremia. Both were successful, and both achieved over 75% reductions over a 6-year period [5] mainly through setting targets for each hospital and levying large financial penalties for not achieving objectives. For the CDI reduction, it was primarily based on avoiding the use of cephalosporins and fluoroquinolones.

Antimicrobial Stewardship. http://dx.doi.org/10.1016/B978-0-12-810477-4.00043-X

In 2011, antimicrobial stewardship guidelines for hospitals, Start smart then focus [6], and primary care, TARGET—Target antimicrobial guidelines, education, and tools[1], were introduced.

The 2011 Chief Medical Officer's annual report volume 2 that was published in March 2013 on "infections and the rise of antimicrobial resistance" [7] was the prequel of the UK five-year antimicrobial resistance strategy (2013–18) [8] This was a One Health strategy produced by the Department of Health (human health) and Department for Environment, Food, and Rural Affairs (animal health and agriculture). It covered the four countries within the United Kingdom and had three strategic aims:

1. Improve the knowledge and understanding of AMR
2. Conserve and steward the effectiveness of existing treatments
3. Stimulate the development of new antibiotics, diagnostics, and novel therapies

There were seven key areas for action:

1. Optimizing prescribing practice
2. Improving infection prevention and control
3. Improving professional education, training, and public engagement
4. Better access to and the use of surveillance data
5. Improving the evidence base through research
6. Developing new drugs, diagnostics, vaccines, and treatments
7. Strengthening the United Kingdom and international collaboration

In 2015, the code of practice for IPC in the Health and Social Care Act was updated to include AMS [9]. The National Institute for Clinical Excellence (NICE) published the guidance of antimicrobial stewardship—processes and systems for effective antimicrobial medicine use [10]—and undertook the development of AMS: behavioral interventions in public and professionals [11]. Antimicrobial prescribing and stewardship competencies have been developed for prescribers [12]. Introductory-level learning tool for all health-care staff and carers was developed for AMR.

---

1. Royal_College_of_General_Practitioners. TARGET Antibiotics Toolkit. TARGET stands for: Treat Antibiotics Responsibly, Guidance, Education, Tools.

The TARGET antibiotics toolkit aims to help influence prescribers' and patients' personal attitudes, social norms, and perceived barriers to optimal antibiotic prescribing. It includes a range of resources that can each be used to support prescribers' and patients' responsible antibiotic use, helping to fulfill CPD and revalidation requirements.

Who is it for and how can it be used?

The TARGET antibiotics toolkit is designed to be used by the whole primary care team within the GP practice or out-of-hours setting. These resources can be used flexibly, either as standalone materials or as part of an integrated package. We do recommend that all resources are used if this is feasible. Available from http://www.rcgp.org.uk/clinical-and-research/target-antibiotics-toolkit.aspx.

In 2015, two incentive schemes were introduced to decrease antibacterial prescribing in primary care and improve the management of sepsis in hospitals in England. The former quality premium reduced primary care prescribing by 7% as items per population and broad-spectrum antibiotics by 10% [13]. Large improvements were also seen for sepsis management. Both were rolled over to push for further improvements in 2016.

In 2016, quality improvement incentives were set to reduce antibiotic prescribing to below 2013 levels for total antibacterials (J01), carbapenems, and piperacillin-tazobactam for hospitals based on DDD per admission. These were selected because of increasing consumption and resistance in these antibiotics. There is also a requirement for a documented day 3 (48–72 h) review and submission of antibiotic consumption data for benchmarking. The data on antibiotic consumption and resistance for both hospitals and primary care are publicly available on PHE AMR Fingertips portal http://fingertips.phe.org.uk/profile/amr-local-indicators. It has been agreed to roll the sepsis and AMR incentives forward to 2017–19 as a serious infection CQUIN. Targets have also been set for antimicrobial consumption in food animals. Since 2014, there has been a public and health-care engagement campaign called Antibiotic Guardian [14]. There was a pilot AMR television campaign over the winter of 2016–17 in North West England to guide a national media program in the following year.

REFERENCES

[1] Antimicrobial resistance subgroup SMAC-. The path of least resistance. In Health Do, (ed): Department of Health; 1998.

[2] Wickens HJ, Jacklin A. Impact of the Hospital Pharmacy Initiative for promoting prudent use of antibiotics in hospitals in England. J Antimicrob Chemother 2006;58(6):1230–7. PubMed PMID: 17030518.

[3] Department of Health. The Health and Social Care Act 2008 Code of Practice on the prevention and control of infections and related guidance. In: Health Do, (ed), 2010.

[4] England PH. Clostridium difficile: what it is, how to prevent, how to treat. London 2009.

[5] Department of Health. Annual Epidemiological Commentary: Mandatory MRSA, MSSA and E. coli bacteraemia and C. difficile infection data, 2013/14. In: England PH, (ed). London 2014.

[6] Department of Health. Antimicrobial stewardship: Start smart—then focus. 2011.

[7] Davies SC, Fowler T, Watson J, Livermore DM, Walker D. Annual Report of the Chief Medical Officer: infection and the rise of antimicrobial resistance. Lancet 2013;381(9878):1606–9. PubMed PMID: 23489756.

[8] UK Five Year Antimicrobial Resistance Strategy 2013 to 2018; 2013.

[9] Department of Health. The Health and Social Care Act 2008: code of practice on the prevention and control of infections and related guidance 2015. Available from: https://www.gov.uk/government/publications/the-health-and-social-care-act-2008-code-of-practice-on-the-prevention-and-control-of-infections-and-related-guidance.

[10] NICE antimicrobial stewardship: right drug, dose, and time? Lancet. 2015;386(9995):717. PubMed PMID: 26333955.

[11] Team NPHaSCG. Antimicrobial stewardship—changing risk-related behaviours in the general population 2016. Available from: https://www.nice.org.uk/guidance/indevelopment/gid-phg89.

[12] Public Health England. Antimicrobial prescribing and stewardship competencies. 2013.

[13] ESPAUR. UK 5 year antimicrobial resistance (AMR) strategy 2013 to 2018: annual progress report 2016. Dept Health; 2016.

[14] Ashiru-Oredope D, Hopkins S. Antimicrobial resistance: moving from professional engagement to public action. J Antimicrob Chemother 2015;70(11):2927–30. PubMed PMID: 26377862.

Chapter 19.10

Antimicrobial Stewardship in France

Céline Pulcini* and Guillaume Béraud**
**Université de Lorraine and Nancy University Hospital, Nancy, France*
***CHU de Poitiers, Poitiers, France*

France is a European high-income country with 66 million inhabitants. Everyone benefits from compulsory health insurance financed by social contributions and taxes based on revenues. Employees from the private sector also benefit from a complementary health insurance provided by employers, whereas self-employed people, the civil servants, the unemployed, and the retirees can subscribe to a complementary health insurance (most are not for profit) on a voluntary basis. Deprived people get free complementary health insurance. The copayments are the rule except for people suffering from chronic illnesses. About 75% of health-care expenditures are paid by compulsory health insurance and roughly 15% by complementary health insurances. The out-of-pocket payments represent less than 10% of health-care expenditures. The majority (75%) of hospital beds belong to the public sector or to the not-for-profit sector. The public hospitals and the not-for-profit hospitals share the same public health duties. The for-profit hospitals are mostly specialized in surgery. Concerning the outpatient sector, most of the doctors are self-employed and are paid on a fee-for-service basis. Since 2005, some gatekeeping has been introduced; patients must register to a general practitioner (GP) of their choice. They still may access freely to specialists or to hospitals, but they are economically incited to request advice from their GP first.

It is illegal to sell antibiotics over the counter, and this practice is reported to be quite rare in France. Only medical doctors and dentists (and midwives for very specific indications) can prescribe antibiotics. Contrary to other European countries, clinical microbiologists are mostly laboratory-based and rarely give clinical advice at the bedside, and so do hospital pharmacists.

A National Antibiotic Plan led by the French Ministry of Health has been implemented since 2001 (Table 1), and a dedicated task force was set up in

Antimicrobial Stewardship. http://dx.doi.org/10.1016/B978-0-12-810477-4.00044-1

TABLE 1 The Main Actions of the French Ministry of Health Antibiotic Plan (Running Since 2001, http://www.plan-antibiotiques.sante.gouv.fr)

- National policy, with regional coordination of actions
- National monitoring of antibiotic use and resistance
- Antibiotic stewardship teaching recently part of the mandatory curriculum of medical students
- National information campaigns targeting the general public
- Information documents on antibiotic stewardship made freely available to long-term care facilities and day care centers
- Educational pack on antibiotic stewardship (e-Bug, http://www.e-bug.eu) made freely available to schools for children and teenagers
- In hospitals, structure, process, and outcome indicators assessing the antibiotic stewardship program (ICATB2 composite indicator, http://www.sante.gouv.fr/IMG/pdf/Instruction_no66_TdBIN.pdf) are part of the certification process of hospitals (as an example, an expert on antibiotics must be named in each hospital)
- In the outpatient setting:
  - Regional antibiotic stewardship networks with expert advice available
  - Pay-for-performance indicators on antibiotic prescribing for general practitioners (GPs) and pediatricians
  - Rapid antigen diagnostic tests for streptococcal tonsillitis made freely available to GPs, ENT (ear-nose-and-throat) specialists and pediatricians
  - Regular visits from the National Health Insurance delegates to GPs, to discuss their antibiotic prescriptions and their practices and promote rapid diagnostic tests

2015 to suggest priority actions (http://social-sante.gouv.fr/IMG/pdf/rapport_carlet_anglais.pdf). All these actions led to a 26.5% decrease of outpatient antibiotic use from 2002 to 2007, but prescriptions are on the rise again over the last years; the impact in hospitals was more modest.

The national campaigns targeting the public were successful initially, less so lately. In hospitals, the antibiotic stewardship program certification indicator has really been a trigger for change. In the outpatient setting, France was one of the few countries that implemented regional networks coordinating the antibiotic stewardship program, with expert advice available for all prescribers. Moreover, microbiology services and monitoring of antibiotic use and resistance are of excellent quality nationwide, and national policies with regional coordination are usual practice in France.

However, antimicrobial stewardship (AMS) has not really been a priority on the political agenda so far, and lack of funding makes it difficult to fully implement the measures recommended by the national antibiotic plan (e.g., AMS teams in hospitals).

France lacks a well-organized system of assessment of medical practices and continuing medical education (CME). There is also a paucity of information technology (IT) resources and computerized decision support systems (CDSS), and data on diagnoses linked to every antibiotic prescription are very rare, making the use of internationally validated quality indicators currently impossible.

French doctors are attached to their autonomy of prescription, and non-compliance with rules is culturally accepted in France; uncertainty avoidance is also the rule rather than the exception (precautionary principle), leading to unnecessary prescriptions. Overprescription of drugs is a national problem in France, not limited to antibiotics, and the patients' expectations for getting a prescription at each consultation are very high.

The main targets of the national actions have so far been GPs and hospital doctors, but other health-care professionals, such as nurses, community pharmacists, and outpatient doctors (other than GPs), must also be involved.

Finally, some organizational specificities could also be a challenge to improve antibiotic use; AMS teams rarely work closely together with infection control teams in hospitals, although they pursue the same aim of reducing bacterial resistance and usually use the same behaviour change strategies; the fee-for-service system in the outpatient setting has also been identified as a potential risk factor for antibiotic overprescribing in some studies.

The top three actions that would be most needed in France, in our opinion, could be

1. to fund AMS teams in hospitals and regional networks in the outpatient setting [1];
2. to make AMS part of the certification/pay-for-performance system in all settings (hospitals, long-term care facilities, and all prescribers working in the outpatient setting), with improved quality indicators and IT systems allowing to measure them automatically;
3. to make undergraduate training and CME on AMS mandatory for all health-care professionals and to develop freely available CDSS based on national guidelines (including shortened durations of treatment).

REFERENCE

[1] Le Coz P, Carlet J, Roblot F, Pulcini C. Human resources needed to perform antimicrobial stewardship teams' activities in French hospitals. Med Mal Infect 2016;46:200–6.

Chapter 19.11

Antimicrobial Stewardship in Germany

Katja de With* and Winfried V. Kern**
**University Hospital Carl-Gustav-Carus at the TU Dresden, Germany*
***University Hospital and Medical Center, and Faculty of Medicine, Albert-Ludwigs-University Freiburg, Germany*

BACKGROUND

The German health-care system ensures free health care via health insurance funds. The ratio of acute-care hospital beds per 100,000 population is relatively high (>500) compared with many other countries, and there are many small-size hospitals (>1000 with <200 beds vs. <100 with >800 beds) [1]. Only one-third of Germany's general hospitals are public hospitals. Antibiotic use density appears relatively low both in the outpatient and inpatient setting, but this needs to be interpreted cautiously in view of the large number of available hospital inpatient beds. Regional variation in antibiotic use and in antibiotic resistance appears substantial. Overall, cephalosporin use has remained unusually high in many settings [2–5].

The rate of in-hospital microbiology laboratories is low (<25%). Most (small- and medium-sized) hospitals use private (outsourced/external) diagnostic laboratory services. The ratios (per million population) and also the in-hospital workforce (numbers) of ID specialists, medical microbiologists, and infection-control physicians are low.

KEY DEVELOPMENTS AND PROGRAMMES ASSOCIATED WITH IMPROVEMENTS IN RESPONSIBLE ANTIBIOTIC USE

National Action Plan and New Legislations

The Federal Government released the "German Antimicrobial Resistance Strategy" (DART), a policy document for the period 2008–13 [6], with multiple recommendations including intensified surveillance and postgraduate training in the infection field. This was followed by "DART 2020" (released

Antimicrobial Stewardship. http://dx.doi.org/10.1016/B978-0-12-810477-4.00045-3

in 2015 [7]) that summarizes further policy steps in particular regarding "one-health" aspects. Important DART components have been the establishment of a national antimicrobial resistance council (Kommission ART) and the provision of (limited) applied research funds.

Since few years, hospital antibiotic consumption data must now be monitored continuously and evaluated in relation to antimicrobial resistance data as part of internal quality management processes [8,9]. There have been new regulations for refinancing postgraduate training and new hospital positions for medical microbiologists and infection control personnel (starting 2011), extramural training in hospital antibiotic stewardship (see below) for pharmacists and physicians, and subspecialty training of ID physicians (starting 2016) upon budget agreements with health insurance representatives [10].

More Transparency and Better Coverage of Consumption and Resistance Data

Data on antibiotic consumption and resistance in human and veterinary medicine are summarized and critically evaluated in national (biannual) surveillance reports (GERMAP) since 2008 [11]. Data sources (human medicine) for GERMAP are a variety of ongoing projects (www.ars.rki.de, www.p-e-g.org/econtext/resistenzdaten, www.avs.rki.de, www.antiinfektiva-surveillance.de, www.versorgungsatlas.de, and www.wido.de/arzneiverordnungs-rep.html).

Hospital Antibiotic Stewardship (ABS) Practice Guideline

An evidence-based German-Austrian guideline with recommendations for implementation of hospital ABS programs was developed under the leadership of the German Infectious Diseases Society (DGI) in 2013 [12] (English language version in Ref. [13]). It specifies essential requirements (personnel & infrastructure) plus core and supplemental ABS strategies and tools. A key recommendation has been the minimum academic personnel staffing (at least 1 full-time equivalent [FTE] per 500 inpatient beds) for a program to be successful. Following the guideline publication and with increasing availability of trained experts in the field, the number of multidisciplinary ABS teams in hospitals has increased considerably [14,15].

Hospital Antibiotic Stewardship (ABS) Expert Training and Certification

An intensive training program in hospital ABS (established at Freiburg University Hospital under the auspices of DGI and the Association of Hospital Pharmacists (ADKA)) includes four weekly structured courses (with limited attendance to allow maximal interaction) plus execution and mandatory presentation of a pertinent project or study in the attendee's working environment and final certification as a trained "ABS expert." The program was financially supported in part by the Federal Ministry of Health during 2010–13 and has been continued since then as a nonprofit fee-for-training project on behalf of DGI and DGI's infection academy. The idea was that these trained ABS experts will take over parts of the responsibility and tasks of ID specialists in the field of ABS in smaller hospitals [16,17].

The ABS experts (certified so far, 500) have set up a national ABS expert network, which convenes on a yearly basis for exchange of experience, continuous education, and as forum for cooperative quality improvement projects. The response has been overwhelming with a waiting list of >1000 physicians and pharmacists. The current 2014–17 program is likely to be extended for another two years before it will transform into an ID physician-focused training course while continuing to offer the exchange and CME platform for the ABS expert network.

MAIN CHALLENGES

A critical challenge will be the increasing reduction in inpatient beds resulting in more admissions per hospital and a higher patient turnover that, in turn, will increase not only antibiotic consumption density in hospitals but also antibiotic prescribing in the community. More expert personnel on-site including those performing timely bedside consultations will be needed in hospitals.

The challenges in outpatient medicine are the reduction of oral cephalosporins and the critical assessment of the reasons for the regional disparities in antibiotic consumption. The sanctions used in ambulatory medicine have primarily addressed economic targets (overspending), and concepts to incentivize/sanction prescribing quality (instead of quantity, and independence of budget constraints) are poorly developed.

A third challenge is to increase regional representativeness of antibiotic resistance surveillance for ambulatory medicine, and hospital antibiotic use data representativeness - with inclusion of more small-size acute-care hospitals that are underrepresented in current (voluntary) surveillance systems.

WHAT IS NEEDED?

Among the top three actions most helpful to enhance responsible antibiotic use are the better recognition and the provision of a higher number of ID physicians (1000 by 2020) with official responsibility in antibiotic stewardship and infection management supported by hospital leadership. The availability of dedicated expert personnel on-site and other infrastructure as recommended by the German ABS guideline needs to be publicly reportable. Germany's ABS expert training initiative is a welcome interim solution that, however, cannot compensate for the lack of ID (and ID/infection control) doctors on the long term.

Other priorities are to (i) increase awareness among physicians and the general community that outpatient prescribing eventually needs monitoring and to (ii) develop and apply appropriate instruments and quality indicator assessments including regional analyses, specialty-focused analyses, anonymous individual physician/practice analyses, and reporting.

A third and urgent action (not Germany-specific and likely requiring international efforts and coordination) is the midterm thorough clinical evaluation of (advanced) biomarker and point-of-care diagnostic-test-based algorithms in outpatient and hospital antibiotic drug prescribing.

REFERENCES

[1] OECD. List of variables in OECD health statistics 2016, 2016. http://www.oecd.org/els/health-systems/List-of-variables-OECD-Health-Statistics-2016.pdf (accessed 18 September 2016)

[2] Augustin J, Mangiapane S, Kern WV. A regional analysis of outpatient antibiotic prescribing in Germany in 2010. Eur J Public Health 2015;25:397–9.

[3] Bätzing-Feigenbaum J, Schulz M, Schulz M, Hering R, Kern WV. Outpatient antibiotic prescription. Dtsch Arztebl Int 2016;113:454–9.

[4] Kern WV, Fellhauer M, Hug M, Hoppe-Tichy T, Först G, Steib-Bauert M, *et al.* Antibiotika-Anwendung 2012/13 in 109 deutschen Akutkrankenhäusern. Dtsch Med Wochenschr 2015;140:e237–46.

[5] Kern WV, Först G, Steib-Bauert M, Fellhauer M, de With K. Patterns of recent antibiotic use in German acute care hospitals—an analysis focussed on 128 non-university hospitals. Abstracts of the European Congress of Clinical Microbiology and Infectious Diseases, Amsterdam, 2016, abstract #P1292.

[6] DART Deutsche Resistenzstrategie. Bundesministerium für Gesundheit, 2011. http://www.bmg.bund.de/fileadmin/dateien/Publikationen/Gesundheit/Broschueren/Deutsche_Antibiotika_Resistenzstrategie_DART_110331.pdf (accessed 18 September 2016).

[7] DART 2020 Antibiotika-Resistenzen bekämpfen zum Wohl von Mensch und Tier. Bundesministerium für Gesundheit, 2015. http://www.bmg.bund.de/fileadmin/dateien/Publikationen/Ministerium/Broschueren/BMG_DART_2020_Bericht_dt.pdf (accessed 9 June 2016).

[8] Infektionsschutzgesetz vom 20. Juli 2000 (BGBl. I S. 1045), das zuletzt durch Artikel 6a des Gesetzes vom 10. Dezember 2015 (BGBl. I S. 2229) geändert worden ist. https://www.gesetze-im-internet.de/ifsg/BJNR104510000.html (accessed 18 September 2016).

[9] Schweickert B, Kern WV, de With K, Meyer E, Berner R, Kresken M, Fellhauer M, Abele-Horn M, Eckmanns T. Surveillance of antibiotic consumption: clarification of the "definition of data on the nature and extent of antibiotic consumption in hospitals according to § 23 paragraph 4 sentence 2 of the IfSG." Bundesgesundheitsbl 2013;56:903–12.

[10] Krankenhausentgeltgesetz vom 23. April 2002 (BGBl. I S. 1412, 1422), das durch Artikel 4 des Gesetzes vom 10. Dezember 2015 (BGBl. I S. 2229) geändert worden ist. https://www.gesetze-im-internet.de/khentgg/BJNR142200002.html (accessed 18 September 2016).

[11] GERMAP2015. http://media.econtext.de/v1/stream/16-424/0d8597f0c9715208faa712caaf9d70d9/1475671620/16/424.econtext. (accessed 10 February 2017).

[12] S3-Leitlinie—Strategien zur Sicherung rationaler Antibiotika-Anwendung im Krankenhaus. AWMF-Nr. 092-001, 2013. http://www.awmf.org/uploads/tx_szleitlinien/092-001l_S3_Antibiotika_Anwendung_im_Krankenhaus_2013-12.pdf (accessed 11 February 2014).

[13] de With K, Allerberger F, Amann S, Apfalter P, Brodt H-R, Eckmanns T, *et al*. Strategies to enhance rational use of antibiotics in hospital: a guideline by the German Society for Infectious Diseases. Infection 2016;44:395–439.

[14] Christoph A, Ehm C, de With K. Auswirkungen der ABS-Fortbildungsinitiative auf die ABS-Strukturqualität teilnehmender Krankenhäuser. Z Evid Fortbild Qual Gesundhwes 2015;109:521–7.

[15] Combating antimicrobial resistance—examples of best-practices of the G7 countries. BMG Federal Ministry of Health, 2015. http://www.bmg.bund.de/fileadmin/dateien/Downloads/G/G7-Ges.Minister_2015/Best-Practices-Broschuere_G7.pdf (accessed 18 September 2016).

[16] Kern WV, de With K. Rational antibiotic prescribing—challenges and successes. Bundesgesundheitsbl 2012;55:1418–26.

[17] Kern WV, Fätkenheuer G, Tacconelli E, Ullmann A. Klinische Infektiologie in Deutschland und Europa. Z Evid Fortbild Qual Gesundhwes 2015;109:493–9.

FURTHER READING

[1] Fätkenheuer G, Kern WV, Salzberger B. An urgent call for infectious diseases specialists. Infection 2016;44:269–70.

Chapter 19.12

Antimicrobial Stewardship in Greece

Antonis Valachis and Diamantis P. Kofteridis
University Hospital of Heraklion, Crete, Greece

Greece has one of the highest rates of antimicrobial consumption in outpatients and antimicrobial resistance among European countries [1,2]. In the community, the antimicrobial consumption in 2014 was 34.1 (defined daily doses (DDDs) per 1000 inhabitants per day), which is 1.6 times higher than the median EU/EEA consumption [1], whereas the consumption of specific antimicrobial groups used for infections due to multidrug-resistant bacteria (carbapenems and polymyxins) was 2.5–6 times higher than the median EU/EEA consumption [1]. Antimicrobial resistance contributes to higher mortality and is associated with high healthcare-related costs due to the morbidity and the need for prolonged hospitalization [3,4].

Considering the economic crisis in Greece that has resulted in decreased resources of public funding to the health care system [5] and the increasing problem of antimicrobial resistance in the country, it is reasonable to argue that antimicrobial stewardship (AMS) programs would be of high priority for policy makers as an effort to improve clinical outcomes and reduce healthcare-related costs [6–8].

This is, however, not the case. The other side of the same coin of economic crisis is that the limited public funding to the health-care system is a drawback for successful organization and implementation of infection control and AMS programs.

Apart from the economic dimension of the problem, other contributing factors to the high antimicrobial resistance in Greece are the high consumption of antimicrobials (inappropriate prescription by physicians and dispensing of antimicrobials without prescription) [9–11] and the population mobility (immigration and tourism) due to the country's geographic location.

Antimicrobial Stewardship. http://dx.doi.org/10.1016/B978-0-12-810477-4.00046-5

EFFORTS AND CHALLENGES IN THE GREEK CONTEXT

The key elements for successful AMS programs have been described [6,12]. An important aspect is the presence of clinicians that recognize the necessity of such programs and serve as leaders on the design and implementation of AMS programs. The Hellenic Center for Disease Control and Prevention (HCDCP; http://www.keelpno.gr) is moving forward to this direction by implementing guidelines for infection treatment, establishing the Greek System for the Surveillance of Antimicrobial Resistance, which is a national network for continuous monitoring of bacterial antimicrobial resistance in the Greek hospitals, and organizing action plans for specific endemic multidrug-resistant bacteria (Procrustes action plan for carbapenem-resistant Gram-negative pathogens).

Another important aspect is the presence of legislation that would support and help AMS practices. In Greece, such legislation has been implemented since January 2014 [13], although organization of AMS teams and implementation of core and supplemental AMS strategies have not yet been applied.

A third aspect is the need for financial support to implement AMS strategies. Despite the fact that AMS programs are cost-saving and their funding can be drawn through savings in both antimicrobial expenditures and indirect costs [6–8], the Greek health-care system has been suffered from several inefficiencies before the financial crisis [14], and there is a need for financial support to design, organize, and implement AMS programs.

WHAT IS THE NEXT STEP?

Taking into account the challenges and certain circumstances in Greece, the most important step forward to AMS efforts is to convince policy makers that this is crucial.

The monitoring and outcome of current initiatives by the HCDCP are the starting points for the discussion with the policy makers to prove in practice that AMS programs are beneficial for the public health. The application of legislation and the public financial support will hopefully be the outcome of such discussions, and the AMS initiatives will have the opportunity to be expanded.

REFERENCES

[1] European Centre for Disease Prevention and Control. Summary of the latest data on antibiotic consumption in the European Union. http://ecdc.europa.eu/en/eaad/Documents/antibiotics-consumption-EU-data-2014.pdf [accessed 10 April 2016].

[2] European Centre for Disease Prevention and Control. Summary of the latest data on antibiotic resistance in the European Union. http://ecdc.europa.eu/en/publications/Documents/antibiotic-resistance-in-EU-summary.pdf [accessed 10 April 2016].

[3] Cosgrove SE. The relationship between antimicrobial resistance and patient outcomes: mortality, length of hospital stay, and health care costs. Clin Infect Dis 2006;42(Suppl 2):S82–9.
[4] deKraker ME, Davey PG, Grundmann H, BURDEN Study Group. Mortality and hospital stay associated with resistant *Staphylococcus aureus* and *Escherichia coli* bacteremia: estimating the burden of antibiotic resistance in Europe. PLoS Med 2011;8:e1001104.
[5] Simou E, Koutsogeorgou E. Effects of the economic crisis on health and healthcare in Greece in the literature from 2009 to 2013: a systematic review. Health Policy 2014;115:111–9.
[6] Dellit TH, Owens RC, McGowan Jr JE, Gerding DN, Weinstein RA, Burke JP, Infectious Diseases Society of America, Society for Healthcare Epidemiology of America, *et al.* Infectious Diseases Society of America and the Society for Healthcare Epidemiology of America guidelines for developing an institutional program to enhance antimicrobial stewardship. Clin Infect Dis 2007;44:159–77.
[7] Standiford HC, Chan S, Tripoli M, Weekes E, Forrest GN. Antimicrobial stewardship at a large tertiary care academic medical center: cost analysis before, during, and after a 7-year program. Infect Control Hosp Epidemiol 2012;33:338–45.
[8] Malani AN, Richards PG, Kapila S, Otto MH, Czerwinski J, Singal B. Clinical and economic outcomes from a community hospital's antimicrobial stewardship program. Am J Infect Control 2013;41:145–8.
[9] Kourlaba G, Kourkouni E, Spyridis N, Gerber JS, Kopsidas J, Mougkou K, *et al.* Antibiotic prescribing and expenditures in outpatient paediatrics in Greece, 2010-13. J Antimicrob Chemother 2015;70:2405–8.
[10] Plachouras D, Kavatha D, Antoniadou A, Giannitsioti E, Poulakou G, Kanellakopoulou K, *et al.* Dispensing of antibiotics without prescription in Greece, 2008: another link in the antibiotic resistance chain. Euro Surveill 2010;15:19488. pii.
[11] Contopoulos-Ioannidis DG, Koliofoti ID, Koutroumpa IC, Giannakakis IA, Ioannidis JP. Pathways for inappropriate dispensing of antibiotics for rhinosinusitis: a randomized trial. Clin Infect Dis 2001;33:76–82.
[12] Pollack LA, Srinivasan A. Core elements of hospital antibiotic stewardshipprograms from the Centers for Disease Control and Prevention. Clin Infect Dis 2014;59(Suppl 3):S97–S100.
[13] Government Gazette (Efimeris tis Kyverniseos) Issue 388, 2014. http://www.eeel.gr/articlefiles/nomothesia/fek_388_2013_peri_enl.pdf [accessed 10 April 2016].
[14] Economou C. Greece: health system review. Health Syst Trans 2010;12:1–180.

Chapter 19.13

Antimicrobial Stewardship in India

Abdul Ghafur
Chennai Declaration, Chennai, India

Tackling antimicrobial resistance (AMR) in the current Indian scenario is a huge and complicated task with more than a billion population, 75,000 hospitals, one million doctors, half a million pharmacies, and inadequate infection control facilities in hospitals, socioeconomic disparity, and sanitation issues in the community.

However, there may be some light at the end of this rather challenging culture. Initiatives such as "Chennai Declaration" have stimulated extensive discussion on AMR issue among medical community and policy makers in India. Coverage of antibiotic resistance in the mass media and medical journals has made clinicians very concerned about the potential negative outcome of their patients, leading to a significant positive change in their attitude toward infection control practices. Hospital managements also started recognizing the importance of having an infection control team and were more cooperative in listening to the recommendations of their infection control nurses and doctors. When it comes to prescribing antibiotics prudently, there appears to be a lot less progress.

Indian Health Ministry published a document on national policy for containment of AMR, in 2011 [1]. Due to various political and administrative reasons, the implementation was delayed. This was the key stimulus for the development of "The Chennai Declaration" [2].

The Indian medical community was waiting to actively engage with an initiative to tackle the threat of antibiotic resistance. This initiative formed as "a roadmap—to tackle the challenge of antimicrobial resistance." In 2012, Chennai was the first joint meeting of medical societies in India on the subject with participation of all major stakeholders [2].

India needs "an implementable antibiotic policy" and not "a perfect policy" [2]. Authors of the Chennai Declaration are well versed with the background Indian scenario and the practicalities and impediments of

Antimicrobial Stewardship. http://dx.doi.org/10.1016/B978-0-12-810477-4.00047-7

implementing a national antibiotic policy. Recommending at the onset, a complete and strict antibiotic policy in a country where there is currently no functioning antibiotic policy at all was not considered to be the most appropriate or immediately viable option. Rather, introduction of step-by-step regulation of antibiotic usage, concentrating on higher-end antibiotics first and then slowly extending the list to second- and first-line antibiotics, appeared to be a more viable option [3]. "The Declaration" has made clear recommendations and plan of action to be initiated by all stakeholders. "The Chennai Declaration" is not a policy by itself, but a call for implementation of the national policy. Shortly after publication of the Chennai Declaration and multiple meetings in Health Ministry to discuss recommendations of Chennai Declaration in 2013, the Drug Controller General of India published a modified rule (H1 rule) to rationalize over-the-counter (OTC) sales of antibiotics in the country [4]. In 2014, Chennai Declaration team published a "five-year plan" to help the implementation of Chennai Declaration recommendations and the national policy [5].

In 2016, Indian Health Ministry released a national antibiotic guideline with clear-cut and simple recommendations on appropriate anti-infectives to treat common community- and hospital-acquired infections [3]. The Honorable Health Minister in a high-level meeting by Indian Health Ministry and WHO released the document. India signed the "WHO AMR Global Action Plan" in 2015 and is in the process of preparing a national action plan in concordance with the WHO plan.

In India, health is predominantly the responsibility of various states (29 states and 7 union territories), and so implementation of the national policy is also primarily the responsibility of states. Kerala, an Indian state famous for her high human development indexes and very high political awareness, is probably one of the ideal states to organize a mass AMR campaign and implement various recommendations of the policy in the country. Kerala government is now planning to implement "Kerala AMR action plan" based on the recommendations of Chennai Declaration. Prime Minister of India has initiated a very ambitious program "Clean India Campaign" (Swachh Bharat Abhiyan) to improve sanitation scenario in the country. Gujarat, one of the Indian states, has made significant progress in implementing the sanitation initiative.

Antimicrobial resistance is a global problem. Tackling this serious menace needs global efforts and collaboration along with serious and sincere local efforts. The success of Chennai Declaration in creating a serious attitude change among Indian medical communities and authorities and inspiring various policy changes is due to the persistent and perseverant efforts by the Chennai Declaration team [6,7]. Collaboration with various international organizations like WHO, UK Antibiotic Action, GARP, ReAct, and WAAAR has significantly helped Chennai Declaration in achieving this success.

Think global and act local!

REFERENCES

[1] Directorate General of Health Services, Ministry of Health and Family Welfare: National policy for containment of antimicrobial resistance, India, 2011.

[2] Ghafur A, Mathai D, Muruganathan A, *et al.* "The Chennai Declaration" Recommendations of "A roadmap- to tackle the challenge of antimicrobial resistance"—a joint meeting of medical societies of India. Indian J Cancer 2013;50:71–3.

[3] http://www.ncdc.gov.in/writereaddata/linkimages/AMR_guideline7001495889.pdf.

[4] Department of Health and Family Welfare. The Gazette of India. 441, part 2, August 30, 2013. http://www.cdsco.nic.in/588E30thAug2013.pdf.

[5] Team C. "Chennai Declaration": 5-year plan to tackle the challenge of anti-microbial resistance. Indian J Med Microbiol 2014;32:221–8.

[6] Ghafur A. 'The Chennai Declaration' and the attitude change in India. Future Microbiol 2015;10:321–3.

[7] Ghafur Abdul. Perseverance, persistence, and the Chennai declaration. Lancet Infect Dis 2013;13:1007–8.

Chapter 19.14

Antimicrobial Stewardship in Israel

Mitchell J. Schwaber
National Center for Infection Control, Ministry of Health, Tel Aviv, Israel

The Israeli health-care system is under the comprehensive regulatory oversight of the Ministry of Health (MOH), which functions as the state institutional licensing authority. Inpatient services are provided by hospitals that are owned by the state, by a health-care maintenance organization (HMO), or privately. There is universal health-care coverage under the aegis of four HMOs.

The MOH body charged with confronting antimicrobial resistance in health care is the National Institute for Infection Control and Antibiotic Resistance (NIICAR), through its subsidiary the National Center for Infection Control. In September 2012, the MOH published Director General's Circular 16/12, *A National Program for Judicious Use of Antibiotics* [1].

While certain local stewardship activities have been in place since before the circular was issued [2], this document, whose implementation is mandatory in all inpatient and outpatient health-care institutions nationwide, including long-term care hospitals, requires each institutional chief executive officer (CEO) to enact a program of antimicrobial stewardship. The circular places overall responsibility for antibiotic use in each institution on the CEO, who is directed to appoint a program chief—a specialist in infectious diseases with a background in stewardship—and a stewardship committee. The stewardship team should have access to all useful data, data processing services, and ancillary support required to implement the program.

The stewardship program at each institution must include the following elements: appropriate diagnostic testing prior to beginning treatment; empirical therapy based on local microbiological and epidemiological trends; timely reporting of microbiological results; definitive treatment based on these results, including drug-bug tailoring and discontinuation of unnecessary or overly broad-spectrum agents; and addressing duration of therapy, dosing, and appropriate routes of treatment.

Antimicrobial Stewardship. http://dx.doi.org/10.1016/B978-0-12-810477-4.00048-9

Required stewardship activities are geared to preventing antibiotic overuse and underuse, inappropriate use, and delay in appropriate therapy. These include surveillance of antibiotic use, indications, and outcomes; institutional treatment guidelines based on local patterns; educational activities among institutional staff along with oversight of and incentives for compliance; and restricted use of specific agents. Guidance and consultation regarding judicious use of antibiotics must be provided to professional staff and the institutional administration.

The stewardship committee is intended to function as a subcommittee of the institutional committee on infection control and antibiotic resistance. It is to be multidisciplinary, with representation from all relevant units, including at least the institutional CEO or senior deputy who functions as the committee chair, the institutional stewardship program chief, at least two senior clinicians representing the main clinical fields, a clinical pharmacologist, the chief of infection control, the chief of infectious diseases, and the chief of the clinical microbiology laboratory. Each HMO is required to maintain its own committee.

The committee is tasked with meeting at least biannually and issuing recommendations to the CEO regarding periodic objectives for judicious use of antibiotics, mechanisms of oversight of antibiotic use, and interventions to foster judicious use of antibiotics.

All institutions must report annually to the MOH regarding antibiotic use. The report includes the records of antibiotics distributed by the institutional pharmacy to each ward or clinical unit, divided according to the Anatomical Therapeutic Chemical (ATC) classification system. Annual reports have been submitted to the National Center for Infection Control for each year beginning with 2012. The National Center for Infection Control in turn distributes an annual report to health-care institutions and the MOH administration regarding use as measured in DDD/100 patients/day in different inpatient ward types and DDD/1000 patients insured per day for each HMO. In addition, the National Center for Infection Control distributes annually to each acute care hospital an institutional antibiogram based on blood cultures yielding the common healthcare-associated bacterial pathogens.

The National Center for Infection Control collaborates with the Veterinary Services of the Ministry of Agriculture in joint efforts to enhance oversight of use of antimicrobial agents and antimicrobial growth promoters in agriculture.

The main challenges encountered in our work so far have included convincing the MOH administration of the need for national mandatory guidelines for stewardship, gaining the acceptance of infectious disease clinicians of the authority of the MOH to regulate stewardship activities at the national level, securing the compliance of institutional CEOs in appointing and supporting the required stewardship bodies, and attaining across-the-board capabilities of clinical pharmacies to generate the required usage data.

Looking ahead, the top priorities in leading responsible antibiotic use in Israel are (1) training appropriate medical personnel in adequate numbers in the principles of stewardship; (2) developing and enhancing computerized support systems for surveillance of antibiotic use, including by indication; and (3) augmenting coordinated surveillance and oversight of antibiotic use in agriculture. While the tactical approach needed to attain these goals will of course be influenced by nuances of the Israeli political and health-care systems, the strategic objective of achieving judicious antibiotic use in all relevant spheres throughout the nation is generalizable to all countries in the developed world.

REFERENCES

[1] http://www.health.gov.il/hozer/mk16_2012.pdf, last accessed 24 August 2016.

[2] Raveh-Brawer D, Wiener-Well Y, Lachish T, *et al.* Effect of a computer application on appropriate use and control of broad spectrum antibiotics (article in Hebrew, English abstract). Harefuah 2015;154:166–70. 212.

Chapter 19.15

Antimicrobial Stewardship in Italy

Leonardo Pagani* and Pierluigi Viale**
**Bolzano Central Hospital, Bolzano, Italy*
***University of Bologna, Bologna, Italy*

The Italian national health system is composed by structures and resources that are aimed to guarantee the universal access to health services, according to Article 32 of the Italian Constitutional Chart. The three grounding principles of such an article are universality, equity, and equality of rights.

Every three years, the Ministry of Health outlines the National Health Plan, which focuses on identifying the essential assistance levels: provisions of services assured countrywide free of charge or with a shared fee, thanks to the incomes that stem from the taxation system.

The national prevention plan 2014–18 ranks the circulation of MDR strains and the infectious risk associated to health-care organizations as a national priority; it states also the core principles for such control and prevention.

Laws and ministerial documents contain ground rules, but the running action is conveyed to the single regions. Italy is divided into 19 administrative regions and 2 autonomous provinces. Each area is in charge of carrying out the National Health Plan and their essential assistance levels, customizing both the policy and the interventions according to the local organization models and the funding budget. This situation has produced heterogeneity of approaches and of activation level: The majority of Italian regions have indeed agreed to implement the measures of infection control and antibiotic stewardship; however, their factual implementation has been and still is inconsistent. National scientific societies play definitely a major role in supporting such programs, but a plan for AMS policy reaching all the different situations is still far from being affected.

Therefore, it is conceivable to summarize few interventions and programs of specific regions traditionally more involved in infectious risk management

Antimicrobial Stewardship. http://dx.doi.org/10.1016/B978-0-12-810477-4.00049-0

associated to health-care organizations, rather than to draft a national, too heterogeneous scenario.

THE REGION EMILIA ROMAGNA

Since 2005, the region Emilia-Romagna (ER) performs a systematic surveillance of any clinically relevant microbiological isolate from all laboratories, providing a yearly report with updated resistance trends. These data are completed by data on antimicrobial consumption by drug classes. An annual report edited by the Emilia-Romagna Infectious Risk Area of Health and Welfare Agency is then fed back to all medical directions.

In 2013, the regional council enacted a resolution to rearrange the committees for infection control and prevention in every public institution. A strategic core was created that would be bound to coordinate the activities of two operative cells, the first one for the control of health-care-associated infections and the second one for the antimicrobial stewardship (AMS). This reorganization gave to the single center autonomy of planning interventions to resolve peculiar concerns.

A quantitative indicator represented by the annual consumption of antimicrobials in each hospital has also been computed and adjusted by the index of case-mix complexity, with interinstitutional benchmarking.

THE REGION FRIULI VENEZIA GIULIA

Since 2011, the activities of AMS have been included in the regional program of clinical risk management; such a regional plan is centrally coordinated and deems the figure of a risk manager and a referring professional for each specific program (e.g., antimicrobial resistance and safer drug usage, among others) in each health institution.

Yearly, the specific goals for health organizations are defined on a regional level; "antimicrobial therapy experts" were identified and trained; multidisciplinary panels of experts draw up updated recommendations for the treatment of common and/or complex infections and surgical antimicrobial prophylaxis. The implementation of a regional microbiological alert system enabled a standardized report with therapeutic indications; a regional restrictive list of antimicrobials has been defined, and antimicrobials are subjected to conditional prescription; reports about antimicrobial consumption and main resistant microorganisms for each health agency are sent back and spread; information tools about specific risks according to different health-care levels have been provided; both health-care professionals and citizens have been involved in several specific educational events.

THE PROVINCE OF BOLZANO

The autonomous province of Bolzano (South Tyrol) is actively engaged in carrying out a network on antibiotic stewardship among the territorial hospitals under the leadership of the "Unit for Hospital Antimicrobial Chemotherapy," based in the Bolzano Central Hospital. Established in 2007 as an interdepartmental unit encompassing ID specialists, clinical microbiologists, and pharmacists, its activity has relentlessly expanded, providing an ever larger collaboration through regular meetings on antimicrobial resistance trends and shared strategies on antibiotic usage and monitoring. Isolation of MDR-Enterobacteriaceae, VRE, or MDR-*P. aeruginosa* is indeed yet sporadic in the whole area. One point of concern still remains the reluctance of changing historical behaviors to some extent and the need for more effective educational programs.

Chapter 19.16

Antimicrobial Stewardship in Japan

Keigo Shibayama
National Institute of Infectious Diseases, Tokyo, Japan

Efforts against AMR in Japan include research and development of antimicrobial agents and related technologies, improvement of infection control in health-care facilities, and ensuring the appropriate use of antibiotics in livestock. Japan has developed a number of antibiotics, such as meropenem and colistin, which are still effective against many types of multidrug-resistant bacteria. Alert to the increase of nosocomial infections, Ministry of Health, Labor, and Welfare (MHLW) introduced premiums in medical fee to promote nosocomial infection control of health-care facilities. In 2006, Medical Care Act was revised, and all medical institutions were required to formulate nosocomial infection control guidelines, to organize an infection control committee, and to hold training for all staffs. Control of AMR in livestock includes the appropriate use of antibiotics according to the guidelines and related legislation.

In order to enhance the activities against AMR, the government of Japan developed a national action plan on AMR in 2016 [1]. The objectives of the national plan consist of six areas, public awareness and education, surveillance and monitoring, infection prevention and control, appropriate use of antimicrobials, research and development, and international cooperation. Here, I will introduce surveillance of AMR of Japan, which is one of our successful experiences providing fundamental and basic knowledge and evidence required for the other objectives.

In Japan, MHLW organizes Japan Nosocomial Infections Surveillance (JANIS) as a national AMR surveillance [2]. Participation is on a voluntary basis, and approximately 1800 hospitals are participating in JANIS in 2016. JANIS analyzes the data of all bacterial isolates tested at clinical laboratories of the participating hospitals. JANIS publishes robust nationwide antimicrobial resistance data based on the collection of approximately four million cultured bacterial isolates every year. According to the report of 2014,

Antimicrobial Stewardship. http://dx.doi.org/10.1016/B978-0-12-810477-4.00050-7

carbapenem resistance rates were generally below 1% among Enterobacteriaceae species. Imipenem resistance rate of *Acinetobacter* spp. was 3.6%. Vancomycin resistance rate of *Enterococcus faecium* was below 1%. The resistance rates were considerably lower compared with many other developed and developing countries. In contrast, erythromycin resistance rate of *Streptococcus pneumoniae* was 86.6%, which is quite higher than that of many developed countries. JANIS publishes information of antimicrobial resistance of twenty clinically important bacterial species.

Another key point of Japanese national action plan is that international cooperation was included in the objectives to support achievement of the global action plan on antimicrobial resistance. We plan to introduce JANIS system into several Asian countries. Japan will actively take part in the promotion of international cooperation to contribute to global health security.

REFERENCES

[1] National Action Plan on Antimicrobial Resistance (AMR), The Government of Japan. http://www.who.int/drugresistance/action-plans/library/en/.

[2] Japan Nosocomial Infections Surveillance. http://www.nih-janis.jp/english/index.asp.

Chapter 19.17

Antibiotic Stewardship in the Netherlands

Inge C. Gyssens*, Jan Prins and Jeroen Schouten‡**
**Radboud University Medical Center, Nijmegen, The Netherlands*
***Academic Medical Center, Amsterdam, The Netherlands*
‡*Canisius Wilhelmina Hospital, Nijmegen*

The Dutch Working Party on Antibiotic Policy was founded in 1996 as an initiative of the Society for Infectious Diseases, the Dutch Society for Medical Microbiology, and the Dutch Association of Hospital Pharmacists. Its primary goal was to contribute to the containment of antimicrobial resistance and the expanding costs incurred for the use of antibiotics. It coordinates the national surveillance of antibiotic resistance, in collaboration with the National Institute for Public Health and the Environment (RIVM); coordinates the surveillance of the in- and outpatient use of antibiotics; and runs a guideline development program. Information about consumption of antimicrobial agents and antimicrobial resistance among medically important bacteria is presented annually in NethMap (www.swab.nl). In 2006, SWAB introduced our electronic national antibiotic guide "SWAB-ID" for the antibiotic treatment and prophylaxis of common infectious diseases in hospitals. Every hospital in the Netherlands has been offered the opportunity to adapt this version to local circumstances and resources and distribute it through an independent Web site. At present, approximately 60% of Dutch hospitals use a local, customized version of SWAB-ID. It has made a significant contribution to a more comprehensive, guideline-compliant, and evidence-based antibiotic policy in the Dutch hospitals.

After a hospital outbreak of KPC, the SWAB was published in 2012, at the request of Dutch Healthcare Inspectorate (IGZ), a vision document answering the question what measures are required to contain the spread of antimicrobial resistance at the national level. In this white paper, the SWAB stressed the need to establish antimicrobial stewardship teams (A-teams) in every Dutch hospital, responsible for the implementation of a local antimicrobial stewardship program. In response to the recommendation by SWAB, Healthcare

Antimicrobial Stewardship. http://dx.doi.org/10.1016/B978-0-12-810477-4.00051-9

Inspectorate, and the Ministry of Health, A-teams have been established in the hospitals in the Netherlands.

Practical support is provided by the "Antimicrobial Stewardship Practice Guide for the Netherlands," which was developed by SWAB in 2014 (available at www.ateams.nl). In 2015, the SWAB has started to develop the *antimicrobial stewardship monitor* program, to measure the progress and effects of the implementation of the national antimicrobial stewardship program. These data will provide insight into the process of implementation of the antimicrobial stewardship program in the Netherlands and its effects.

As of 2015, the Dutch Healthcare Inspectorate performs audits to ascertain that each hospital indeed runs a stewardship program. The main challenge is the lack of financing. Each hospital has to run a stewardship program, but no centralized funds have been allocated, which urges negotiation with the board of directors of each hospital to ensure sufficient funding. This is at present a major barrier. To facilitate this process, SWAB has performed a national Delphi procedure to assess what funds are minimally required to run the program.

At present, the Dutch Ministry of Health has installed a number of working groups to strengthen this system of stewardship in the Netherlands, in particular to cover also general practice and nursing homes. For general practice, up-to-date guidelines are available, and documenting the indication for each antibiotic prescription appears a feasible option. As virtually all GP's use an electronic patient record, these data can be retrieved easily. Individualized prescription behavior can be evaluated for every GP, benchmarked, and fed back to the GP and those with unusual or excessive prescribing habits addressed. For nursing homes, the situation is more complicated. Up-to-date guidelines for pneumonia and urinary tract infections are not available, and retrieval of the total antibiotic consumption for each nursing home is difficult, as several prescribing and drug delivery systems are in use. In addition, the prevalence of antibiotic resistance is not very well monitored. The lack of guidelines, registration of antibiotic consumption, and point prevalence measurements of resistance are currently being addressed.

REFERENCES

[1] www.swab.nl, site of the Dutch Working Party on Antibiotic Policy (SWAB).

[2] www.ateams.nl, Dutch Antimicrobial Stewardship Practice Guide.

[3] Schuts EC, Hulscher ME, Mouton JW, Verduin CM, Stuart JW, Overdiek HW, *et al.* Current evidence on hospital antimicrobial stewardship objectives: a systematic review and meta-analysis. Lancet Infect Dis 2016;16:847–56.

[4] Schuts EC, van den Bosch CM, Gyssens IC, Kullberg BJ, Leverstein-van Hall MA, Natsch S, *et al.* Adoption of a national antimicrobial guide (SWAB-ID) in the Netherlands. Eur J Clin Pharmacol 2016;72:249–52.

Chapter 19.18

Antimicrobial Stewardship in Scotland

Dilip Nathwani* and Jacqueline Sneddon**
*Ninewells Hospital and Medical School, Dundee, United Kingdom
**Scottish Antimicrobial Prescribing Group, Glasgow, United Kingdom

Within the United Kingdom, Scotland manages its own National Health Service (NHS) and has a unique national antimicrobial stewardship program. The Scottish Management of Antimicrobial Resistance Action Plan (ScotMARAP) published in 2008 provided a framework for delivery of stewardship via a new national leadership group, the Scottish Antimicrobial Prescribing Group (SAPG) [1] and regional antimicrobial management teams (AMTs) in each of the 14 health boards. SAPG provides leadership for stewardship across NHS hospitals and community settings including long-term care facilities both public and privately owned. ScotMARAP was updated in 2014 [2] to reflect progress made and to align with the UK 5-year AMR strategy. SAPG receives government funding, and AMTs receive additional central funding for one antimicrobial pharmacist to support local delivery of stewardship, which is complemented by any locally agreed funding.

An initial focus for SAPG was addressing high rates of *Clostridium difficile* infection (CDI) through national implementation of antimicrobial guidelines, which restricted antibiotics associated with a high risk of CDI ("4C" antibiotics) across all settings. This contributed to a >80% reduction in CDI rates, which has been maintained. In 2009, Scottish Government introduced a target for reduction of CDI supported by prescribing indicators developed by SAPG for hospital and primary care. We have further developed our hospital prescribing indicators to encompass review of therapy. Following successful reduction of 4C antibiotics in primary care, a target was introduced in 2014 to reduce the total use of antibacterials using a "best-in-class" approach. This initiative supported by an interactive education program has brought early success. Further initiatives in primary care include a feasibility study of using C-reactive protein point-of-care testing in respiratory infections

Antimicrobial Stewardship. http://dx.doi.org/10.1016/B978-0-12-810477-4.00052-0

and providing quarterly personalized feedback on antibiotic use benchmarked at local and national levels.

Hospital interventions initially focused on compliance with local empirical treatment policy and surgical prophylaxis with resultant reduction in the use of "4C" antibiotics, particularly quinolones and cephalosporins. Piperacillin-tazobactam and carbapenem use is increasing annually, so our current focus, using specific national guidance implemented by local AMTs, is ensuring review and de-escalation of IV therapy and using alternative Gram-negative treatments. A recent national bespoke point prevalence survey showed encouraging results for controlling carbapenem use, but further work is required to rationalize piperacillin-tazobactam use and to consistently ensure 72-hour review of such therapy.

SAPG has developed a range of resources to support clinicians including algorithms for management of key infections and online dosage calculators for gentamicin and vancomycin.

Surveillance of antimicrobial use and resistance has been a key element of SAPG work, and this has evolved to provide an innovative informatics resource [3] to enable linkage of key routinely available clinical datasets to inform future risk-based use of antimicrobials.

Development of education resources [4] to meet the training needs of health-care professionals across hospital and community settings has been a key success. Recently, an interactive workbook was launched to support the evolving contribution of nurses and midwives to stewardship. Several SAPG members have also contributed to the recent Massive Open Online Course on Antimicrobial Stewardship [5] that provides a global free learning opportunity and has been utilized by >10,000 learners to date.

We are fortunate in Scotland to have adequate financial resources that support an established infrastructure for stewardship and surveillance. Key challenges include maintaining and further developing clinician engagement to ensure that stewardship is everyone's business and measuring the positive and negative impact of our interventions. Across the UK hospital, prescribing continues to increase for a variety of reasons such as hospital activity, patient demographics, and the successful implementation of the sepsis campaign. In relation to the latter, there is a clear need to balance the need for prompt treatment with reliable review and de-escalation.

For sustainability and future development, Scotland requires the current level of funding for stewardship and informatics to be maintained. Building of additional stewardship capacity in hospitals, by engaging the broader clinical team and expertise, is also critical to ensure optimal antimicrobial use. In primary care, we require to engage with all prescribers to address the variation in quantity and quality of antibiotics prescribed through providing innovative tools to support clinical decision-making.

REFERENCES

[1] Scottish Antimicrobial Prescribing Group (SAPG). http://www.scottishmedicines.org.uk/SAPG.

[2] Scottish Management of Antimicrobial Resistance Action Plan 2014–18 (ScotMARAP2). http://www.gov.scot/Publications/2014/07/9192.

[3] Infection Intelligence Platform. http://www.isdscotland.org/Health-Topics/Health-and-Social-Community-Care/Infection-Intelligence-Platform/.

[4] Antimicrobial Resistance and Stewardship. Educational resources to support antimicrobial prescribing, resistance and stewardship within NHSScotland. http://www.nes.scot.nhs.uk/education-and-training/by-theme-initiative/healthcare-associated-infections/educational-programmes/antimicrobial-resistance-and-stewardship.aspx.

[5] Antimicrobial Stewardship: Managing Antibiotic Resistance. https://www.futurelearn.com/courses/antimicrobial-stewardship.

Chapter 19.19

Antimicrobial Stewardship in Slovenia

Bojana Beović* and Milan Čižman**
*University Medical Centre Ljubljana, Ljubljana, Slovenia
**Faculty of Medicine, University of Ljubljana, Ljubljana, Slovenia

Slovenia is geographically a central European country with some Mediterranean influence and an Eastern-type exsocialistic country by its recent history. According to census in 2015, Slovenia has 2.1 million inhabitants. The health-care system is a combination of Bismarck and Beveridge model with one national insurance company. Over 99% of population has compulsory health insurance. GDP per capita was 31,627 USD in 2015 (80% of OECD average). In 2013, the total health expenditure represented 8.7% of GDP (OECD 8.9%), slightly more than two-thirds from public sector. The private expenditure is almost evenly split into voluntary insurance and out of pocket payment [1].

A prescription is needed for every antibiotic purchase, and in human medicine, antibiotics may only be prescribed by physicians. Slovenia has monitored outpatient antibiotic consumption since 1974. Hospital consumption has been measured periodically every 5 years since 1985. In the year 2000, Slovenia joined to EARS and in 2001 to ESAC projects [2]. The national system for surveillance of antimicrobial use in all hospitals and outpatients has been developed on the basis of the ESAC contribution. Slovenia has contributed to European Surveillance of Veterinary Antimicrobial Consumption (ESVAC) since 2010 [3]. The intersectorial coordinating mechanism was established in 2005. In 2011, the rules on mandatory surveillance of antimicrobial use in hospitals, antibiotic stewardship teams and programs, education of team members, and auditing were adopted by the Ministry of Health [4,5]. Various educational activities are available during continuous medical education. Restriction for prescriptions of fluoroquinolones, co-amoxiclav, oral third-generation cephalosporins, and macrolides in outpatients has been in place since 2000, 2005, and 2009, respectively.

Antimicrobial Stewardship. http://dx.doi.org/10.1016/B978-0-12-810477-4.00053-2

The outpatient antibiotic use is among the lowest top six among the 30 EU/EEA countries contributing to ESAC-NET, and it decreased from the maximum 19.76 in 1999 to 14.21 defined daily doses (DDDs) per 1000 inhabitants a day (DID) in 2014. A significant decrease of the use of coamoxiclav and fluoroquinolones was observed after the restriction; the decrease of macrolide and oral third-generation cephalosporin use was not statistically significant [6].

Hospital use remained stable (1.6 DID) and is between the middle and lower third of the 23 EU/EEA EARS-net contributing countries [2].

In spite of relatively low antibiotic consumption in Slovenian hospitals, the resistance rates especially in Gram-negative bacteria are high. The structure of antibiotic use is unfavorable with high use of beta-lactams/beta-lactam inhibitors and carbapenems, which are used more often than in EU/EAA countries contributing to ESAC-NET on average. Antimicrobial stewardship programs are in place formally; their activity is variable. There is no stable financing of antimicrobial stewardship programs. Another problem is infection control in hospitals with few isolation rooms. Outpatient antibiotic use reached a plateau at approximately 14.5 DID 10 years ago; the highest rates remain in young children.

In hospitals, a step further from the cumulative antibiotic use, monitoring should be made toward patient-based assessment of quality of prescribing. Computerized drug prescribing in Slovenian hospitals is expected in the next few years. Outpatient drug prescribing is already computerized; next year, the diagnosis will be a mandatory entry in electronic drug prescriptions. Intensified undergraduate and postgraduate education in antimicrobial prescribing and stewardship should be achieved in collaboration with all education providers. National intersectorial coordinating mechanism is currently working on the new national antimicrobial stewardship strategy, which will include the educational interventions, monitoring of the quality of prescribing, and adequate staffing for hospitals. The strategy will also include the activities in animal sector. Stable financing is expected after the adoption of the strategy on the national level.

REFERENCES

[1] Anon. Health policies and data. http://www.oecd.org, 2003 [accessed 1 May 2016].

[2] Anon. Antimicrobial resistance interactive database (EARS-Net). http://ecdc.europa.eu/en/healthtopics/antimicrobial_resistance/database/Pages/database.aspx#sthash.drWGBBIG.dpufhttp://ecdc.europa.eu/en/healthtopics/antimicrobial_resistance/Pages/index.aspx, 2016 [accessed 1 May 2016].

[3] Anon. European Surveillance of Veterinary Antimicrobial Consumption (ESVAC). http://www.ema.europa.eu/ema/index.jsp?curl=pages/regulation/document_listing/document_listing_000302.jsp, 2016 [accessed 1 May 2016].

[4] Anon. Pravilnik o strokovnem nadzoru izvajanja programa preprečevanja in obvladovanja bolnišničnih okužb. Uradni list 2011;10:1022–4.

[5] Anon. Pravilnik o dopolnitvah Pravilnika o pogojih za pripravo in izvajanje programa preprečevanja in obvladovanja bolnišničnih okužb. Uradni list 2011;10:2024–5.
[6] Fuerst J, Čižman M, Mrak J, Kos D, Campbell S, Coenen S, *et al*. The influence of a sustained multifaceted approach to improve antibiotic prescribing in Slovenia during the past decade: findings and implications. Expert Rev Anti Infect Ther 2015;13:279–89.

Chapter 19.20

Antimicrobial Stewardship in South Africa

Marc Mendelson
University of Cape Town, Groote Schuur Hospital, Cape Town, South Africa

Human immunodeficiency virus (HIV) and tuberculosis dominate South Africa's health landscape. Overall, infection including malaria and neglected tropical diseases such as schistosomiasis, constitutes its major burden of disease. Approximately 84% of health care is delivered in the public sector, controlled by the nine provincial departments of health, with coordination from the National Department. Guidance on antimicrobial prescribing is provided by the National Essential Medicines list and structured treatment guidelines. Private health care is delivered by three major hospital groups (Netcare, Mediclinic, and Life Healthcare) and a number of smaller groups. Infectious disease specialists are a scarce resource in South Africa, the discipline only having been recognized as a subspecialty of medicine in 2005. Indeed, human resources in the field of infection across disciplines are a major challenge in our resource-limited health system.

South Africa's national action plan for antimicrobial resistance (AMR) was developed from a bottom-up approach rapidly transitioning to national governance. A national situational analysis in 2011, catalyzed by the Global Antimicrobial Resistance Partnership (GARP), stimulated the formation of the South African Antibiotic Stewardship Programme (SAASP), a multidisciplinary, One Health, public-private partnership of stakeholders from across the infection disciplines. SAASP's advocacy to the National Minister and Director-General of Health led to the rapid adoption of AMR as a major health issue in South Africa. Both distinguished leaders have demonstrated the importance of leadership in AMR coming from the very top. In 2014, a landmark Ministerial Summit was held where South Africa's AMR National Strategy Framework 2014–24 was adopted and subsequently published. This has been rapidly followed by the publication of South Africa's

Antimicrobial Stewardship. http://dx.doi.org/10.1016/B978-0-12-810477-4.00054-4

2014–19 Implementation Plan 2, which embeds AMR into the government's annual performance plan. A blueprint for antibiotic stewardship (ABS) at national, provincial, district, and institutional levels will be published in early 2017. One Health governance at the national level is through the intersectoral Ministerial Advisory Committee on AMR and a national secretariat embedded within Rational Medicines Use Directorate. South Africa has a well-developed national surveillance system for reporting antibiotic resistance in humans, which is coordinated by the National Institute of Communicable Diseases. SAASP has led ABS interventions in hospitals; a conventional multidisciplinary team-based approach in the public sector 3 and a pharmacist-led intervention 4 in a private group of 47 hospitals. Nurse-led models of stewardship for antiretroviral therapy are well established in South Africa, and we aim to develop a similar ABS model for antibiotic therapy.

Surveillance of antibiotic resistance in animal health and of antibiotic use in both animals and humans represents major current challenges to South Africa's AMR strategy framework. Furthermore, human resources for infection prevention and ABS are lacking. Two national training centers for ABS have been funded and have begun addressing the skills shortage in seven of the country's nine provinces. Prescriber-pharmacist-manager teams from major hospitals in each province attend a practical, hands-on, 5-day train-the-trainer program. Mentorship and follow-up visits to the hospitals occur to enable experience to be shared and programs to be developed. Community ABS programs are currently being planned with relevant stakeholders.

The top three priority areas for allocation of resources and actions currently needed are to address the paucity of resistance surveillance in livestock and companion animals and developing surveillance of use. These challenges are shared throughout the Southern African region and indeed in most low- and middle-income countries globally. Much needs to be done to address behavior change in prescribers and to address infection prevention within hospitals to prevent spread of resistant bacteria. Lastly, sustained public awareness campaigns within South Africa and the region are needed to aid advocacy for control of antibiotic use and basic infection prevention practice. Funding for AMR-focused research is starting to be addressed through the South African Medical Research Council and its bilateral programs with partner countries such as the United Kingdom.

REFERENCES

[1] National Dept of Health. Antimicrobial Resistance National Strategy Framework 2014–2024.
[2] National Dept of Health. Implementation Plan for the Antimicrobial Resistance Strategy Framework 2014–2019.

[3] Boyles TH, *et al.* Antibiotic stewardship ward rounds and a dedicated prescription chart reduce antibiotic consumption and pharmacy costs without affecting inpatient mortality or re-admission rates. PLoS One 2013;8(12). e79747. http://dx.doi.org/10.1371/journal.pone.0079747.

[4] Brink AJ, Messina AP, Feldman C, Richards GA, Becker PJ, Goff DA, *et al.* Netcare Antimicrobial Stewardship Study Alliance. Antimicrobial stewardship across 47 South African hospitals: an implementation study. Lancet Infect Dis. 2016; 16(9):1017–25. http://dx.doi.org/10.1016/S1473-3099(16)30012-3.

Chapter 19.21

Antimicrobial Stewardship in Argentina

Gabriel Levy-Hara
Hospital Carlos G Durand, Buenos Aires, Argentina

There are three health-care delivery systems in Argentina, public, private, and social security. Public hospitals are funded by ministries of health with varying resources significantly between richest and poorer cities or provinces. Social security institutions also have different levels of funding, depending upon their size. Financial situation of private institutions is also variable.

Policy making for improvement of the antibiotic use at national level started in 2008. The pivotal community-based intervention has been the creation of "Remediar" program, consisting in about 8000 nationwide primary care centers that provide free medical assistance and medicines for all. Physicians are stimulated to participate in training interventions. Rational use of antibiotics is extensively and well addressed along the educational material. However, analyses of ATB prescribed according to their indication still show around 40% of inappropriate use.

Public national campaigns targeting the community have never been performed. This is a shortfall compared with initiatives facing other public health problems—such as seasonal or mosquito-transmitted diseases—where nationwide dissemination of the problem was done through massive multimedia campaigns.

Many training interventions for physicians, pharmacists, and microbiologists have been performed. Scientific societies, Medical and Pharmacist Colleges, and the Pan-American Health Organization (PAHO) are continuously actively involved in educational activities.

On the other hand, there are no current policies regarding the implementation of antimicrobial stewardship programs (ASPs) in hospitals.

In the context of an international survey [1] performed in 2012, 39 responders were from Argentina, 13 reported an ASP already implemented, and 11 were under planning. Most ASPs were rather new, and most have published AMS policies and treatment guidelines.

Antimicrobial Stewardship. http://dx.doi.org/10.1016/B978-0-12-810477-4.00055-6

The Buenos Aires City's Infectious Diseases Network of Public Hospitals has been working during the last 10 years in implementing at least some degree of ASP in public hospitals. Human resources dedicated to ASP are very limited: there are generally around three infectious disease specialists (ID) per hospital of 250–300 beds, a few clinical pharmacists involved only in some hospitals, and a scarce of information technology specialists. Most ASPs include ward rounds in main units (e.g., ICUs and internal medicine) jointly with infectious disease specialists (ID), local prescribers such as staff and residents, and a clinical pharmacist, when available. Preauthorization strategies are not generally used, due to both restrained resources and consented position privileging educational interventions over restrictive ones. When present, preauthorization is generally required for expensive new antifungal drugs (e.g., lipidic formulations of amphotericin, voriconazole, and echinocandins).

Overall, the use of some drugs such as carbapenems, fosfomycin, tigecycline, colistin, vancomycin, and linezolid is somehow audited or controlled. However, successes in reducing their use are frequently overshadowed by poor infection control (IPC) programs in most hospitals of the network, leading to many peaks in consumption of these antibiotics. Some studies of ATB consumption measuring DID have been done, but due to the lack of resources, these are not sustained. The last one performed in 2011 showed a slightly increase in drugs for treatment of MDR Gram-negative bacilli, linked to weak IPC programs.

On the other hand, many bigger private hospitals have well-mounted ASPs that include all core members. These institutions showed overall significant reductions of antimicrobials consumption and improvements in its use. Outside of Buenos Aires City, ASPs are actually scarce.

Main barriers and challenges to improve the use of antimicrobials in Argentina are the following:

a. *Health authorities support, at all levels (national, regional, and hospital).* Public campaigns should begin and be maintained with seasonal boosters. Legislation toward implementation of ASP, in at least bigger and more complex public hospitals, is necessary.
b. *Increase human resources for implementing effective and sustainable ASP in hospitals.* IDs, clinical pharmacists, infection control staff, and IT specialists are scarce in most public hospitals throughout the country. AMS activities should be stimulated and supported by health authorities.
c. *Material resources.* Funding from regional authorities to their dependent hospitals is unevenly distributed. The following essential items should always be present: stable drug supply, microbiology laboratory diagnosis tools, basic supplies for an effective IPC program, computer support and internet connectivity, updated data on bacterial resistance, and

antimicrobial consumption. Ideally, pharmacokinetic/pharmacodynamic tools should also be available.

d. *Suboptimal undergraduate training on microbiological, ecological, and pharmacological aspects of antibiotic resistance*. In this regard, some interventions have been attempted, although not sustained due to the lack of support from the authorities.

REFERENCE

[1] Howard P, Pulcini C, Levy Hara G, West RM, Gould IM, Harbarth S, Nathwani D. An international cross-sectional survey of antimicrobial stewardship programmes in hospitals. J Antimicrob Chemother 2015;70(4):1245–55.

Chapter 19.22

Antimicrobial Stewardship in Spain

José R. Paño-Pardo* and Jesús Rodríguez-Baño**

*Hospital Clínico Universitario "Lozano Blesa", Instituto de Investigación Sanitaria Aragón, Zaragoza, Spain

**Unidad Clínica de Enfermedades Infecciosas y Microbiología, Hospital Universitario Virgen Macarena and Departamento de Medicina, Universidad de Sevilla - Instituto de Biomedicina de Sevilla (IBiS), Seville, Spain

Spain is administratively divided in 17 regions (autonomous communities), each of which has its regional public health-care system, funded and managed by the governments of the autonomous communities. The same principles apply to all regional public health-care systems as belonging to a single national public health-care system. Before 2013, health care was universally provided by the national public health care system. Since then, the national health-care system provides health care to every legally established citizen in Spain and to EU citizens visiting the country. Pensioners do not have to pay out of pocket for any health care provided; nevertheless, nonpensioners pay a proportional fee for outpatient prescription drugs, namely, antibiotics.

One of the missions of central government's Ministry of Health is to minimize disparities and inequities among regional public health-care system service portfolio. Spain is one of the very few countries in the EU that does not officially recognize the specialty of infectious diseases. Although the absence of the specialty of infectious diseases has limited citizens' access to expertise in this field, it has not prevented the development of a strong and prolific community of microbiologist and infectious disease experts who have worked in the field of antimicrobial stewardship (AMS). In 2006, in the setting of high rates of over-the-counter antimicrobial prescribing, the Spanish Ministry of Health released a population-targeted, mass-media campaign aiming to increase antimicrobial resistance awareness [1]. A comprehensive assessment of the impact of this campaign on antimicrobial use was not conducted. AMS programs began to be implemented in Spanish hospitals in the mid-2000s, some of them starting in the setting of multicenter research projects [2–4] as described in a national survey on the implementation of antibiotic stewardship programs [5].

Antimicrobial Stewardship. http://dx.doi.org/10.1016/B978-0-12-810477-4.00056-8

In 2012, the Spanish Society of Clinical Microbiology and Infectious Diseases (SEIMC), jointly with the Spanish Society of Hospital Pharmacists (SEFH) and the Spanish Society of Preventive Medicine, Public Health and Hygiene (SEMPSMH), promoted a consensus document on AMS programs, named *programas de optimización de uso de antibióticos (PROA)*, an acronym that has spread widely [6]. This document provided comprehensive information on the aims, structure, governance, activities, and indicators for AMS programs in the country. SEIMC also started an ambitious e-learning educational program aimed at training AMS teams. The Spanish consensus document on AMS, despite not being endorsed by any official institution, served as a practical guiding tool for all those interested in implementing AMS in hospitals and fueled the surge of multiple bottom-to-top initiatives all over the country. It also pushed some regions, such as Andalusia, to implement a region-wide institutional program (available at: http://ws140.juntadeandalucia.es/pirasoa/). In other regions, many of these initiatives were not generalized due to the lack of institutional support.

In 2014, the Spanish Ministry of Health, on the mandate of EU institutions, designated the Spanish Agency of Medicines (AEMPS) to design and implement a National Task Force Against Antimicrobial Resistance (PNRAN) that included AMS as a core strategy. Currently, the PNRAN is at the end of its design phase.

The main challenges to implement AMS programs in Spain are the lack of institutional support (and therefore the lack of specific human resources), the unavailability of ready-to-use tools for AMS programs, and the absence of officially recognized infectious disease specialists. PNRAN is trying to address the lack of institutional support by promoting the introduction of AMS as a necessary standard for hospitals' accreditations, but this is part of regional governments' competencies in health care and has still to be adopted and agreed upon. Currently, antimicrobial consumption and resistance monitoring has not been standardized, and comparisons and benchmarking cannot be established between regions and institutions. Spain would benefit from a wider surveillance network (i.e., European), but monitoring tools must be rapid and flexible enough for sites to provide useful input on these topics. A repository of tools that could be used by antimicrobial stewards is desirable; however, development is dependent on a funding, and no budgetary resource has been allotted for this work. Finally, the fact that AMS in hospitals and in primary care is expanding, makes other types of health-care institutions, such as nursing homes, one of the main challenges to address.

REFERENCES

[1] Campañas 2006—Uso responsable de antibióticos. Usándolos bien hoy, mañana nos protegerán 2016:1–2. Available at: http://www.msssi.gob.es/campannas/campanas06/Antibioticos.htm. Last accessed, 29 April 2106.

[2] López-Medrano F, San Juan R, Serrano O, Chaves F, Lumbreras C, Lizasoain M, *et al.* PACTA: efecto de un programa no impositivo de control y asesoramiento del tratamiento antibiótico sobre la disminución de los costes y el descenso de ciertas infecciones nosocomiales. Enferm Infecc Microbiol Clin 2005;23:186–90.

[3] García-San Miguel L, Cobo J, Martínez JA, Arnau JM, Murillas J, Peña C, *et al.* La «intervención del tercer día»: análisis de los factores asociados al seguimiento de recomendaciones sobre la prescripción de antibióticos. Enferm Infecc Microbiol Clin 2014;32:654–61. http://dx.doi.org/10.1016/j.eimc.2013.09.021.

[4] Cisneros JM, Neth O, Gil Navarro MV, Lepe JA, Jiménez-Parrilla F, Cordero E, *et al.* Global impact of an educational antimicrobial stewardship programme on prescribing practice in a tertiary hospital centre. Clin Microbiol Infect 2014;20:82–8. http://dx.doi.org/10.1111/1469-0691.12191.

[5] Paño-Pardo JR, Padilla B, Romero-Gómez MP, Moreno-Ramos F, Rico-Nieto A, Mora-Rillo-M, *et al.* Monitoring activities and improvement in the use of antibiotics in Spanish hospitals: results of a national survey. Enferm Infecc Microbiol Clin 2011;29:19–25. http://dx.doi.org/10.1016/j.eimc.2010.05.005.

[6] Rodriguez-Bano J, Paño-Pardo JR, Alvarez-Rocha L, Asensio Á, Calbo E, Cercenado E, *et al.* Programs for optimizing the use of antibiotics (PROA) in Spanish hospitals: GEIH-SEIMC, SEFH and SEMPSPH consensus document. Enferm Infecc Microbiol Clin 2012;30:22.e1–22.e23. http://dx.doi.org/10.1016/j.eimc.2011.09.018.

Chapter 19.23

Antibiotic Stewardship in Sweden

Cecilia Stålsby Lundborg*, Mats Erntell** and Katarina Hedin‡
*Karolinska Institutet, Stockholm, Sweden
**Strama Halland, Halmstad, Sweden
‡Region Kronoberg, Växjö, Sweden

BACKGROUND

In Sweden, a high-income country, responsibility for health care is shared by the government and national authorities, together with regional county councils and their local municipalities. The county councils and municipalities are independently responsible for financing and delivering health care, while the government and national authorities are responsible for policies and regulations. According to Swedish statistics, about 7.5% of GDP was spent on health care in 2012; most of this expenditure is financed through taxes, with patient fees covering a small part. Public providers are the dominant providers of health care in Sweden, with private providers accounting for about 12% of health care; this private provision is still publicly financed, and patients are covered by the same regulations and fees that apply to public facilities [1].

Antibiotic use is comparatively low in Sweden, and there is a strictly followed prescription-only policy. Antibiotic resistance incidences are also comparatively low [2].

FORMATION OF STRAMA: THE SWEDISH STRATEGIC PROGRAMME AGAINST ANTIBIOTIC RESISTANCE

Increased antibiotic prescribing and resistance among pneumococci were detected in Sweden in the late 1980s and early 1990s [3]. As a response, the Swedish strategic program against antibiotic resistance (Strama) was

Antimicrobial Stewardship. http://dx.doi.org/10.1016/B978-0-12-810477-4.00057-X

created in 1995 on a voluntary and informal basis by professionals and the authorities, with the aims of improving the use of antibiotics and containing antibiotic resistance. Local Strama groups supported by the county councils were formed soon afterward. In 2010, the national Strama structure was incorporated into what is now the Public Health Agency of Sweden.

There has been a long-term political commitment toward containment of antibiotic resistance, irrespective of government. Examples of this include that Strama received financial support from the government for its national work in 2000, only 5 years after its inception. Furthermore, a national target was set in 2009 for antibiotics of 250 prescriptions per 1000 inhabitants and year; this target was included as an indicator of rational prescribing in a government initiative to improve patient safety during 2011–14.

The main characteristic of the Strama work is the combination of national and local multiprofessional groups, including authorities, with close communication and personal contacts between the two levels [4]. At both levels, various professional groups with specific competence and engagement in treatment of infectious diseases are represented, mainly specialists in family medicine, microbiology, infectious diseases, ear, nose, throat, pediatrics, and pharmacists. The interactions and networking between the local and national level have always been intense and were formalized into a Strama program council (http:/strama.se/) in 2015, established by the counties via the Swedish Association of Local Authorities and Regions.

Activities at both national and local levels include (i) monitoring of antibiotic prescribing and sales, monitoring of antibiotic resistance, and feedback of these data to prescribers; (ii) development and educational outreach for new national guidelines; (iii) participation in various types of national and local studies (see Box 1 for some examples); (iv) communication with media; and (v) participation in international networks and collaborations. Specific activities at the local level include educational activities for health-care personnel, staff at preschools, nursing homes, and parents.

Fig. 1 shows that the total number of antibiotic prescriptions in outpatient care in Sweden has decreased over the past 20 years, including a reduction by 71% in prescriptions for children aged 0–4 years. The reduction in DDDs has not been as large due to increased dosage for common infections. About 80% of all antibiotics are prescribed in the outpatient setting; 60% of these are prescribed in primary care, making it an important target for efforts to improve antibiotic use.

MAIN SUCCESSES OF THE STRAMA WORK

1. High awareness and acceptance among health-care professionals and the general public of the risks of overprescription of antibiotics and its connection to antibiotic resistance.

BOX 1 Examples of Published Studies Conducted Within the Strama Framework

Mölstad S, Stålsby Lundborg C, Karlsson A-K, Cars O. Large variation in antibiotic prescribing comparing 13 European countries. Scan J Infect Dis 2002; 34:366–371.

Stålsby Lundborg C, Olsson E, Mölstad S, and the Swedish Study Group on Antibiotic Use. Antibiotic prescribing in outpatients—a one-week diagnosis–prescribing study in five counties in Sweden. Scan J Infect Dis 2002;34: 442–448.

Ganestam F, Stålsby Lundborg C, Grabowska K, Cars O, Linde A. Weekly Antibiotic Prescribing and Influenza Activity in Sweden—A Study throughout Five Influenza Seasons. Scan J Infect Dis 2003;35:836–842.

Andre M, Mölstad S, Stålsby Lundborg C, Odenholt I and the Swedish Study Group on Antibiotic Use. Management of urinary tract infections in primary care: A repeated 1-week diagnosis-prescribing study in 5 counties in Sweden in the years 2000 and 2002. Scan J Infect Dis 2004;36:134–138.

Hedin K, Andre M, Håkansson A, Mölstad S, Rodhe N, Petersson C. A population-based study of different antibiotic prescribing in different areas. Br J Gen Pract 2006;56:680–685.

Hedin K, Andre M, Håkansson A, Mölstad S, Rodhe N, Petersson C. Physician consultation and antibiotic prescription in Swedish infants: population-based comparison of group daycare and home care. Acta Paediatr. 2007;96 (7):1059–1063

André M, Vernby Å, Odenholt, Stålsby Lundborg C, Axelsson I, Eriksson M, Runehagen A, Schwan Å, Mölstad S. Diagnosis-prescribing surveys in 2000, 2002 and 2005 in Swedish general practice - consultations, diagnoses, diagnostics and treatment choices. Scan J Inf Dis 2008;40(8):648–654.

Lennell A, Kühlmann-Berenzon S, Geli P, Hedin K, Petersson C, Cars O, Mannerquist K, Burman LG, Fredlund H; Study Group. Alcohol-based hand-disinfection reduced children's absence from Swedish day care centers. Acta Paediatr 2008;97(12):1672–1680.

Ansari F, Erntell M, Goossens H, Davey P. The European surveillance of antimicrobial consumption (ESAC) point-prevalence survey of antibacterial use in 20 European hospitals in 2006. Clin Infect Dis. 2009;49(10):1496–1504

Andre M, Vernby A, Berg J, Lundborg CS. A survey of public knowledge and awareness related to antibiotic use and resistance in Sweden. J Antimicrob Chemother 2010;65(6):1292–1296

Björkman I, Berg J, Röing M, Erntell M, Stålsby Lundborg C. Perceptions among Swedish hospital physicians on prescribing of antibiotics and antibiotic resistance. Qual Saf Healthcare. Qual Saf Healthcare 2010;19(6):e8. Epub 2010 Jul 1.

Hedin K, Andre M, Håkansson A, Mölstad S, Rodhe N, Petersson C.Infectious morbidity in 18-month-old children with and without older siblings. Fam Pract 2010;27(5):507–512.

Soderblom T, Aspevall O, Erntell M, Hedin G, Heimer D, Hokeberg I, Kidd-Ljunggren K, Melhus A, Olsson-Liljequist B, Sjogren I, Smedjegard J,

Continued

BOX 1 Examples of Published Studies Conducted Within the Strama Framework—cont'd

Struwe J, Sylvan S, Tegmark-Wisell K, Thore M. Alarming spread of vancomycin resistant enterococci in Sweden since 2007. Euro Surveill 2010;15(29).
Bjorkman I, Erntell M, Roing M, Stalsby Lundborg C. Infectious disease management in primary care: perceptions of GPs. BMC Family Pract 2011, 12:1
Pettersson E, Vernby Å, Mölstad, Stålsby Lundborg C. Can a multifaceted educational intervention targeting both nurses and physicians change the prescribing of antibiotics to nursing home residents? A cluster randomized controlled trial. J Antimicrob Chemother 2011;66(11):2659–2666.
Kallstrom AP, Bjorkman I, Stålsby Lundborg C. The sterile doctor: doctors' perceptions on infection control in hospitals in Sweden. J Hosp Infect 2011;79:187–188.
Björkman I, Berg J, Viberg N, Stålsby Lundborg C. Awareness of antibiotic resistance a prerequisite for restrictive antibiotic prescribing in UTI treatment. A qualitative study among primary care physicians in southern Sweden. Scan J Prim Healthcare 2013;31(1):50–55. doi: 10.3109/02813432.2012.751695.
Strandberg EL, Brorsson A, Hagstam C, Troein M, Hedin K. I'm Dr Jekyll and Mr Hyde": are GPs' antibiotic prescribing patterns contextually dependent? A qualitative focus group study. Scand J Prim Healthcare 2013;31(3):158–165.
Vallin M, Polyzoi M, Marrone G, Rosales-Klintz S, Tegmark Wisell K, Stålsby Lundborg C. Knowledge and Attitudes Towards Antibiotic Use and Resistance—A Latent Class Analysis of a Swedish Population-based Sample. PLoS One 2016;11(4):e0152160
Andre M, Gröndal H, Strandberg EL, Brorsson A, Hedin K. Uncertainty in clinical practice - an interview study with Swedish GPs on patients with sore throat. BMC Fam Pract 2016;17(1):56.
Skoog G, Struwe J, Cars O, Hanberger H, Odenholt I, Prag M, *et al.* Repeated Nationwide Point Prevalence Surveys of Antimicrobial Use in Swedish Hospitals—data for actions over eight years (2003-2010). EuroSurveillance 2016;21(25).
Strandberg EL, Brorsson A, André M, Gröndal H, Mölstad S, Hedin K. Interacting factors associated with Low antibiotic prescribing for respiratory tract infections in primary health care - a mixed methods study in Sweden. BMC Fam Pract 201618;17(1):78.

Antibiotic resistance has been portrayed as something that can affect everyone. The use of specific messages on reducing *unnecessary* use by following evidence-based guidelines led to antibiotic sales decreasing substantially, especially for children. This was apparently achieved without negative effects, as demonstrated through follow-up analyses for complications using population registries.

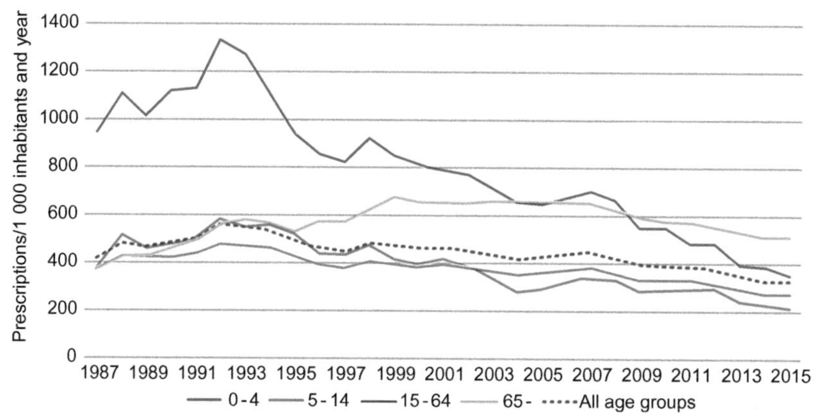

FIG 1 The sales of antibiotics for systemic use in outpatient care (all antibiotic sales on prescriptions) in 1987–2015, prescriptions/1000 inhabitants and year, both sexes, and different age groups. Source: *Public Health Agency of Sweden and National Veterinary Institute. Swedres-Svarm 2015. Consumption of antibiotics and occurrence of antibiotic resistance in Sweden. Public Health Agency of Sweden 2016.*

2. Political commitment and incentives secured visibility and funding for the challenge of antibiotic resistance. This was due in part to strategic media coverage and a strong well-reputed national leader, Professor Otto Cars.
3. The development and maintenance of the systematic and long-term strategy for the multistakeholder and multiprofessional cooperation at national and local levels and between the levels has been important for the success of Strama.
4. The use of educational outreach to many professional groups and especially to those responsible for telephone advice, which has led to fewer health-care visits for minor infections and thus less probability of receiving antibiotics.

The Swedish health-care system has structural aspects that probably facilitated the formation, development, and success of Strama:

1. A system of County Medical Officers for communicable disease control according to the Communicable Diseases Act. The County Medical Officers for communicable disease control has public authority status and liaises with all other parts of the health-care system including primary care. County Medical Officers for communicable disease controls have taken on great responsibility for the rational use of antibiotics.
2. Departments of infectious diseases and clinical microbiology are present in all counties. Early on, these specialists advocated the use of narrow-spectrum antibiotics.
3. Analysis of local resistance data has been used within the health-care system and by Strama as a basis for the strategies of recommending antibiotics.

4. Antibiotics are prescription-only drugs, and this is strictly followed. Sweden has had sales statistics on medicines since the 1970s by the parastatal company Apoteksbolaget AB. Information about all prescribed drugs has been collected in the prescribed drugs registry by the National Board of Health and Welfare since 2005. Pharmacists have been active since the beginning in the Strama work especially for analyses and feedback of prescribing data.

CURRENT CHALLENGES

To maintain enthusiasm and commitment at local, regional, and national levels, keeping up the activities of prescription feedback, guideline implementation, and awareness activities to various target groups.

To establish structures and secure permanent funding for local Strama groups throughout the country, to address the still unexplained variations in antibiotic use between different parts of the country.

To implement antibiotic stewardship programs in all hospitals.

To ensure that a relatively high number of both health-care providers and patients who have experiences from other countries with a different disease panorama and with potentially higher expectations for antibiotic prescribing will not increase antibiotic prescribing.

To limit internet sales and direct imports of antibiotics, which allow access to antibiotics without a proper diagnosis.

To limit increasing incidences of resistance and import of bacterial resistance through leisure travelers and those who have sought health care in countries with higher prevalence of resistance.

WHAT IS NEEDED NOW?

Continued commitment and further establishment of Strama within the knowledge management structure built up between the authorities and caregivers.

Increased awareness campaigns that reach all inhabitants in Sweden.

Simpler access to diagnosis-related prescribing data for all health-care levels, to provide improved feedback and "quality accounting" (not only economic figures) of antibiotic prescribing data.

Continued international cooperation, Sweden, should continue to contribute internationally with our experience and at the same time acknowledge that the situation in Sweden with universal access to health care and a strict prescription-only policy is relatively unique; therefore, transferring our experiences to other settings requires adapting them to drastically different economic and social contexts.

REFERENCES

[1] https://sweden.se/society/health-care-in-sweden/.

[2] Public Health Agency of Sweden and National Veterinary Institute. Swedres-Svarm 2015. Consumption of antibiotics and occurrence of antibiotic resistance in Sweden. Solna/Uppsala, 2016.

[3] Public Health Agency of Sweden. Swedish work on containment of antibiotic resistance. 2014. https://www.folkhalsomyndigheten.se/pagefiles/17351/Swedish-work-on-containment-of-antibiotic-resistance.pdf.

[4] Mölstad S, Erntell M, Hanberger H, Melander E, Norman C, Skoog G, *et al.* Sustained reduction of antibiotic use and low bacterial resistance. A ten year follow-up of the Swedish STRAMA programme. Lancet Infect Dis 2008;8:125–32.

Chapter 19.24

Antimicrobial Stewardship in Switzerland

Catherine Pluss-Suard, Laurence Senn and Giorgio Zanetti
Lausanne University Hospital, Lausanne, Switzerland

The Swiss health system is considered as excellent but is costly and complex. Switzerland devotes a large share of gross domestic product for health expenditure compared with countries involved in the Organization for Economic Cooperation and Development (OECD) [1]. Half of health expenditure is dedicated to hospitals, which can be public, private, or subsidized private. Funding sources for hospitals come from private health insurance companies, the cantons, and the cities. A quarter of health expenditure is devoted to outpatient care. Swiss citizens have to be covered by compulsory individual health insurance. They pay insurance premiums that are not based on incomes and only marginally covered by employers. The health system is managed at the region level. Each of the 26 cantons has its own organization not only offering local competencies and flexibility but also representing challenges for national public health actions.

A national surveillance and research program for antibiotic resistance and antibiotic consumption in Switzerland has been successfully setup (www.anresis.ch) [2]. The correlation of total antibiotic consumption between out- and inpatient settings seen in European countries is not observed in Switzerland [3]. The total consumption of antibiotics in the inpatient setting (data since 2004) is close to the median in comparison with countries participating in the European Surveillance of Antimicrobial Consumption Network (ESAC-Net) [2], showing that there may be room for improvement. However, antibacterial consumption in the outpatient setting (data since 2013) is relatively low compared with other European countries [2]. This can be partly explained by the fact that all antibacterials have always been prescription-only medicines. Second, Swiss citizens generally consume less drugs than OECD countries. Third, the franchise thresholds paid by the patients may lead people to wait or to use symptomatic treatments before consultation; it may also decrease willingness to be prescribed by drugs. Other sociocultural factors

Antimicrobial Stewardship. http://dx.doi.org/10.1016/B978-0-12-810477-4.00058-1

certainly contribute to a low consumption but have been poorly studied until now. Additional needs would include development of quality indicators at the patient level for which reporting could be centralized. Monitoring of antibacterial consumption should also be setup in long-term care facilities. Surveillance data of antibiotic resistance show that methicillin-resistant *Staphylococcus aureus* (MRSA) rates have decreased significantly since 2004. Vancomycin resistance in *Enterococci* is very low and has remained stable over the last 10 years. In contrast, quinolone resistance and third-generation cephalosporin resistance in *Escherichia coli* and *Klebsiella pneumoniae* has gradually increased.

Joint antibiotic stewardship initiatives other than surveillance have not been done so far, except for local projects. Most acute care hospitals have their own antibiotic guidelines, a list of restricted antibiotics, infectious diseases, and infection control specialists. A national strategy on antibiotic resistance involving all stakeholders only became available in 2015 [4]. This strategy following the One Health approach is a cooperation between the Federal Office of Public Health (FOPH), Food Safety and Veterinary Office (FSVO), Agriculture (FOAG), Environment (FOEN), and the cantons. It emphasizes the need to raise awareness of the problem of antibacterial resistance among public/patients, health-care providers, and stakeholders. The population will be informed about the importance of compliance, treatment alternatives, and preventive measures (e.g., vaccines). Undergraduate and continuing education of health-care providers will be enhanced. Responsible use of antibacterials will be promoted through antibiotic stewardship teams, development of national guidelines, and restriction of use of newly approved antibacterials. Rapid diagnostic tests will be recommended to guide prescribers for a more responsible use of antibacterials.

REFERENCES

[1] OECD. Health at a glance 2015: OECD indicators. Paris: OECD Publishing; 2015.

[2] Federal Office of Public Health. Joint report 2013. Usage of Antibiotics and Occurrence of Antibiotic Resistance in Bacteria from Humans and Animals in Switzerland. November 2015. FOPH publication number: 2015-OEG-17.

[3] Vander Stichele RH, Elseviers MM, Ferech M, Blot S, Goossens H. European Surveillance of Antibiotic Consumption Project G. Hospital consumption of antibiotics in 15 European countries: results of the ESAC Retrospective Data Collection (1997-2002). J Antimicrob Chemother 2006;58(1):159–67.

[4] The Federal Council. Strategy on Antibiotic Resistance—Switzerland, 2015/11/18. www.bag.admin.ch/fr/star.

Chapter 19.25

Antimicrobial Stewardship in Turkey

Önder Ergönül*, Füsun Can* and Murat Akova**
*Koç University, Istanbul, Turkey
**Hacettepe University, Ankara, Turkey

Turkey is a midincome country with 79 million inhabitants, located between Europe and Asia. The proportion of the health expenditures in gross national product was reported as 6.1% in 2010. Public sector funding of total health expenditures in 2010 was 75% [1]. The population is covered for free access to health care and has access to a family physician. Besides universal coverage, private insurance or self-paying is also common.

By 2010, the Social Security Institution had contracted with 421 private hospitals (90% of large hospitals) to provide care and complex emergency services such as burn care, intensive care, cardiovascular surgery, and neonatal care. The average length of stay was 4.1 days in 2010, with 65% occupancy.

Antibiotics are the most commonly consumed drugs in Turkey [2]. According to the 2011 data, Turkey has the highest antibiotic consumption rate among eastern European, non-EU countries, as 42 DID/1000 inhabitants [3]. This is 3.5 times higher than Netherlands, which has the lowest DID value. The most commonly used antibiotics among outpatients were penicillins and other beta-lactams with extended spectrum (31.4 DID), macrolides (3.9 DID), and fluoroquinolones (3.6 DID). Strict price regulations for antibiotics and also any other drugs for human use are enforced by the government, and prices per unit are comparatively cheaper than European countries. Generic antibiotics are used preferentially both in and out of hospital settings due to cost concerns. Antibiotic consumption is significantly different between the regions in Turkey and the most common at the southeast regions of Turkey [2].

In 2003, Ministry of Health of Turkey started a nationwide antibiotic restriction program in hospitals. According to this program, carbapenems, glycopeptides, piperacillin/tazobactam, and ticarcillin/clavulanate are defined as

Antimicrobial Stewardship. http://dx.doi.org/10.1016/B978-0-12-810477-4.00059-3

restricted antibiotics that can be used only with approval of infectious diseases specialists (IDSs). Parenteral fluoroquinolones, third- and fourth-generation cephalosporins, netilmicin, and amikacin can be prescribed by any doctor taking care of patients in the hospital within the first 72 h of the treatment, but infectious diseases specialist approval is required for the further utilization after 72 h. Short-term results of nationwide antibiotic restriction program were reported to be promising; the consumption of the restricted antibiotics decreased, the trend of resistance rates declined or stayed stable for two years, and significant saving of expenses occurred [4]. However, in the long term, substantial increase in resistance rates among Gram-negative bacteria has been reported. In a recent study, among the healthcare-associated Gram-negative bloodstream infections (GN-BSI), 38% of *Klebsiella pneumoniae* isolates were resistant to carbapenems, and 6% were resistant to colistin. Carbapenem resistance was significantly associated with mortality [5].

Overprescription of drugs is a national problem in Turkey, not limited to antibiotics, and patients' expectations for getting a prescription at each consultation are very high.

We can summarize the significant steps to be taken for AMS in Turkey, mainly under three headings:

1. Surveillance: Epidemiological studies should be performed to quantify antibiotic consumption and track antibiotic resistance. These data can be used to compare results with other countries.
   a. Effectiveness of national programs should be monitored, and relevant actions should be developed.
2. Health services:
   a. Sales of antibiotics over the counter (OTC) have been prohibited since 2015 in Turkey, but the prohibition of OTC sales should be monitored and sustained.
   b. The national antibiotic restriction program in hospitals is unique, and early reports were promising [4]; however, the program should be enhanced and revised based on new data and issues.
   c. Rapid diagnostic tests should be promoted and should be compensated by insurance systems.
   d. Implementation of antibiotic stewardship program should be an indicator of health-care quality control at local and national levels.
3. Education: The Ministry of Health of Turkey has defined a detailed and active education plan for rational use of drugs since 2011.
   a. Effective education programs for the rational use of antibiotics should be implemented at undergraduate level including medical schools, pharmacy, and nursing schools.
   b. Taking a "rational use of antibiotics course" should be mandatory in all medical residency programs.

c. The main targets of the national actions have so far been GPs and hospital doctors, but other health-care professionals, such as nurses, community pharmacists, and outpatient doctors (other than GPs), must also be involved.

d. Information documents on antibiotic stewardship should be available in long-term care facilities and day care centers.

e. National dissemination of information to the public is important. The Ministry of Health of Turkey and professional societies should enhance and sustain their work in this arena.

REFERENCES

[1] Atun R. Transforming Turkey's Health System—lessons for universal coverage. N Engl J Med 2015;373:1285–9.

[2] Akıcı A, Şahin A, Melik B, Aksoy M. Akıcı A, Alkan A, editors. National drug consumption surveillance-2011. Ankara: Ministry of Health; 2015.

[3] Versporten A, Bolokhovets G, Ghazaryan L, Abilova V, Pyshnik G, Spasojevic T, *et al.* Antibiotic use in eastern Europe: a cross-national database study in coordination with the WHO Regional Office for Europe. Lancet Infect Dis 2014;14:381–7.

[4] Altunsoy A, Aypak C, Azap A, Ergonul O, Balik I. The impact of a nationwide antibiotic restriction program on antibiotic usage and resistance against nosocomial pathogens in Turkey. Int J Med Sci 2011;8:339–44.

[5] Ergönül Ö, Aydin M, Azap A, Başaran S, Tekin S, Kaya S, *et al.* Health-care associated gram negative bloodstream infections: emergence of resistance and predictors of fatality. J Hosp Infect 2016;94(4):381–5.

Chapter 19.26

Antimicrobial Stewardship in the United States

Theodore I. Markou and Sara E. Cosgrove
Johns Hopkins University School of Medicine, Baltimore, MD, United States

The United States does not have a national health system with the exception of the Veterans Health Administration that serves military veterans. There are over 5500 registered hospitals of which the majority (>85%) are community hospitals; these are a mix of not for profit, for profit, and state and local government facilities [1]. The major payers are private insurance companies and the Centers for Medicare and Medicaid Services (CMS), a federal agency that covers adults aged 65 and older, those with permanent disabilities, and low-income persons.

Several federal initiatives are ongoing in the United States in the area of antimicrobial stewardship. The Centers for Disease Control and Prevention (CDC)'s Get Smart for Antibiotics Program was initially directed toward improving antibiotic prescribing in the outpatient setting but has been expanded to include the inpatient and long-term care settings [2]. Initiatives of these programs have included development of *Core Elements* documents describing the crucial components of antibiotic stewardship programs in a variety of healthcare settings [3,4]. The uptake of these components in acute care hospitals in the United States was assessed in a national survey in 2014: 39% of the 4184 hospital surveyed reporting having all core elements of an antibiotic stewardship program in place [5]. In parallel to the WHO's World Antibiotic Awareness Week, the CDC holds Get Smart for Antibiotics Week annually in November to bring public awareness to the issues of antibiotic resistance and overuse [2]. The CDC also has developed a mechanism for electronic reporting of antibiotic use and resistance data from acute care hospitals [6,7]. Reporting to this antibiotic use and resistance module of the National Healthcare Safety Network is currently voluntary, which has slowed large-scale, representative submission of data, although Centers for Medicare and Medicaid Services are exploring making such reporting a requirement for reimbursement [8].

Antimicrobial Stewardship. http://dx.doi.org/10.1016/B978-0-12-810477-4.00060-X

Efforts to combat antibiotic resistance and curb overuse in both animal and human populations was a focus of the Obama administration. In 2014, a report from the President's Council of Advisors on Science and Technology described a variety of needed initiatives to slow the spread of antibiotic resistance and ensure long-term availability of effective antibiotic therapy that included improved surveillance, wider adoption of antimicrobial stewardship programs across all healthcare settings, and novel approaches to incentivize new antibiotic development [9]. The release of this report was followed by the National Action Plan to Combat Antibiotic-Resistant Bacteria and formation of the Presidential Advisory Council on Combating Antibiotic-Resistant Bacteria [10,11] to advise on progress on the national plan.

Centers for Medicare and Medicaid Services has health and safety requirements known as conditions of participation that institutions must meet to be reimbursed by Centers for Medicare and Medicaid Services. Compliance is assessed for the most part by accreditation organizations, the largest of which is The Joint Commission (TJC). Requirements from Centers for Medicare and Medicaid Services and The Joint Commission are significant drivers for establishment of quality improvement efforts in hospitals and other settings. The Joint Commission has recently added a new requirement that hospitals and nursing care centers have antibiotic stewardship programs [12]. In addition, Centers for Medicare and Medicaid Services has released regulations requiring antibiotic stewardship programs in nursing homes and is expected to release a conditions of participation requiring antibiotic stewardship in acute care hospitals [13]. Challenges moving forward include insuring that stewardship programs that are established to fulfill these requirements implement and sustain approaches to optimize antibiotic prescribing across the United States. This includes the development of metrics that can be standardized and collected to evaluate the processes and outcomes of stewardship programs. Further work is needed to impact antibiotic prescribing in the outpatient arena where no regulatory levers exist currently.

REFERENCES

[1] American Hospital Association. http://www.aha.org/research/rc/stat-studies/fast-facts.shtml. Accessed 22 July 2016.

[2] CDC Get Smart Program. https://www.cdc.gov/getsmart/index.html. Accessed 22 July 2016.

[3] CDC, Core Elements of Hospital Antibiotic Stewardship Programs. Available at http://www.cdc.gov/getsmart/healthcare/pdfs/core-elements.pdf. Accessed 22 July 2016.

[4] CDC, The Core Elements of Antibiotic Stewardship for Nursing Homes. Available at http://www.cdc.gov/longtermcare/pdfs/core-elements-antibiotic-stewardship.pdf. Accessed 22 July 2016.

[5] Pollack LA, van Santen KL, Weiner LM, Dudeck MA, Edwards JR, Srinivasan A. Antibiotic Stewardship Programs in U.S. Acute Care Hospitals: findings from the 2014 National Healthcare Safety Network Annual Hospital Survey. Clin Infect Dis 2016;63:443–9.

[6] CDC NHSN Antibiotic Use and Resistance Module. Available at http://www.cdc.gov/nhsn/pdfs/pscmanual/11pscaurcurrent.pdf. Accessed 22 July 2016.

[7] Fridkin SK, Srinivasan A. Implementing a strategy for monitoring inpatient antimicrobial use among hospitals in the United States. Clin Infect Dis 2014;58:401–6.

[8] Federal Register. 2016;81(81):253. Available at https://www.gpo.gov/fdsys/pkg/FR-2016-04-27/pdf/2016-09120.pdf. Accessed 22 July 2016.

[9] President's Council of Advisors on Science and Technology. Report to the President on Combatting Antibiotic Resistance. September 2014. Available at https://www.whitehouse.gov/sites/default/files/microsites/ostp/PCAST/pcast_carb_report_sept2014.pdf. Accessed 22 July 2016.

[10] National Action Plan for Combating Antibiotic Resistant Bacteria. Available at https://www.whitehouse.gov/sites/default/files/docs/national_action_plan_for_combating_antibotic-resistant_bacteria.pdf. Accessed 22 July 2016.

[11] Presidential Advisory Council on Combating Antibiotic Resistant Bacteria. http://www.hhs.gov/ash/advisory-committees/paccarb/about-paccarb/index.html#. Accessed 22 July 2016.

[12] The Joint Commission Antimicrobial Stewardship Standard. Available at https://www.jointcommission.org/assets/1/6/New_Antimicrobial_Stewardship_Standard.pdf. Accessed 14 February 2017.

[13] https://www.cms.gov/Newsroom/MediaReleaseDatabase/Press-releases/2016-Press-releases-items/2016-09-28.html?DLPage=1&DLEntries=10&DLSort=0&DLSortDir=descending. Accessed 14 February 2017.

Section E

Research and Perspectives

Chapter 20

Methodological Challenges in Evaluating Antimicrobial Stewardship Programs: "Through Measuring to Knowledge"

Marlieke E.A. de Kraker and Stephan Harbarth
Infection Control Program, Geneva University Hospitals and Faculty of Medicine, Geneva, Switzerland

KEY POINTS

- Adequate estimation of the impact of antimicrobial stewardship programs (ASPs) is complicated by bias, confounding, and random time effects including concurrent changes in patient mix, care practices, and bacterial clones.
- Nonrandom intervention allocation and lack of blinding increases the risk of selection, detection, and performance bias in ASP evaluations.
- ASPs influence infectious disease transmission dynamics between individual patients, and therefore, cluster randomized controlled trials should be considered as the gold standard for ASP evaluations.
- Interrupted time series (ITS) design should be the design of choice if cluster-randomized intervention allocation is not feasible.
- Cluster randomized controlled trials should be analyzed by multilevel regression techniques acknowledging cluster-level and individual-level information.
- Time series design should record outcome data for at least one year before and after the intervention and should be analyzed by segmented regression methods or ARIMA models.

INTRODUCTION

"Through measuring to knowledge" was the motto of the Dutch Nobel laureate and professor in physics, H. K. Onnes (1853–1926). Although this was

Antimicrobial Stewardship. http://dx.doi.org/10.1016/B978-0-12-810477-4.00020-9

brought into play by the natural sciences, it can be applied to many topics in hospital epidemiology as well and is of paramount importance when talking about the evaluation of ASPs. While ASPs have been around for quite a while, the quantification of the impact of such programs remains a challenge, and the components critical for success remain to be discerned.

Scientific evidence, or knowledge, can only be derived from counterfactual reasoning: "What would have happened if we would not have implemented this ASP?" Scientists need to be able to create two parallel worlds, one in which a certain ASP is implemented and one where no intervention took place. Unfortunately, in real-life, we cannot observe the counterfactual and we have to makeshift with scientific solutions. For valid reasoning, all study components should adhere to state-of-the-art scientific standards: study design, setting, exposure and outcome measures, statistical analyses, and reporting of results.

There has been a lack of high-quality studies in the field of ASPs. Studies assessing the impact of interventions have often applied "before-after" designs [1,2], which inherently suffer from bias, confounding, and random time effects. In this chapter, we will discuss different designs to provide a framework for assessing the quality of evidence provided by studies investigating antimicrobial stewardship (AMS) interventions.

The choice of statistical methods depends on the design of the study and can range from simple Chi-squared tests for "before-after" designs to complex generalized linear mixed models for stepped-wedge designs. We will shortly summarize different methods, including their pros and cons.

Hopefully, the suggestions in this chapter will optimize "measuring" in future studies, paving the road to more "knowledge" about the critical components of successful ASPs.

STUDY DESIGN

In the undertaking of ASP evaluations, the first critical step is to choose the most appropriate design in order to minimize threats to the validity of inference. The most appropriate design will eliminate the influence of concurrent changes (confounding, maturation, and regression to the mean) and prevent the emergence of systematic errors during the study process (bias) [3].

Confounding, Maturation, and Regression to the Mean

Each time we observe a relationship between exposure (i.e. ASP) and outcome, a third variable could be obscuring, that is, *confounding* the causal relationship. Suppose a hospital has restricted the use of carbapenems by means of authorization requirements. After a few weeks, the number of carbapenem-resistant *Klebsiella pneumoniae* infections has decreased significantly. This could be due to the intervention, or due to the fact that the hospital initiated screening on admission for multidrug-resistant Gram-negative pathogens and intensified infection control measures (confounder) at the same time. Depending on the primary outcome measure, many confounding factors can jeopardize the validity of study results [4]. ASP outcome measures can be roughly divided into process outcome measures (*P*), like defined daily doses (DDDs) of antibiotics per 1000 patient days; clinical outcome measures (*C*), like median length of stay or mortality; or microbiological outcome measures (*M*), like the proportion of resistant isolates [5].

In Table 1, we have summarized the main confounding factors, we will exemplify the latter two. Seasonality could confound the relationship between an intervention and the outcome in many ways. For example, higher temperatures and humidity in summer in the Northern Hemisphere have been associated with increased susceptibility to colonization and infection by *Staphylococcus aureus* [6], *Escherichia coli*, and *Acinetobacter* species [7], so higher rates of infections (M) and antimicrobial prescriptions (P) are more likely in summer than in winter. The true effect of interventions implemented just before or right after summer will therefore be concealed by these innate phenomena, if they are not properly taken into account in the study design and analysis phase.

In the same way, natural wave trajectories of bacterial clones can obscure the efficacy of an intervention. The recent decreasing trends in methicillin resistant *Staphylococcus aureus* (MRSA) in parts of Europe, Australia, and the United States may reflect improved infection control practices and enhanced antibiotic stewardship. However, some argue that widespread

TABLE 1 Possible Confounding Factors in Evaluation Studies of Antimicrobial Stewardship Programs and the Category of Outcome Measures They Could Distort [5]

	P	*C*	*M*
Change in patient demographics	X	X	
Change in case mix	X	X	
Change in infection control measures	X	X	
Change in care practice (e.g., increased ratio of ambulatory care)	X	X	
Seasonality of quality of care, type of hospital admissions, infections, or pathogens	X	X	X
Natural life cycle and survival fitness of specific bacterial clones			X

P, process outcome measures; C, clinical outcome measures; M, microbiological outcome measures.

declining MRSA rates are possibly attributable to changes in the organism itself including shifts in circulating clones [8]. In the United Kingdom, EMRSA-16 already appeared to decline before any control measures were put in place, following a natural trajectory of initial expansion, stasis, and decline [9].

Fortunately, there are ways to disentangle the effects of confounders and the intervention of interest. First, one could include a control arm, which is exposed to all the same external factors except for the intervention. Second, the outcome measure of interest could be measured for an extended period before and after the intervention; this will expose existing trends associated with changes other than the intervention. Then, the possible confounding factors can be registered and/or measured and included in a final multivariate analysis. Finally, if the confounder is not so easily measured or quantified, like quality of care, one might use a "nonequivalent dependent variable" or "negative controls" [10]. A nonequivalent dependent variable is an outcome that has similar confounding and causal factors as the outcome of interest, but will not be impacted by the current intervention. Selection of an appropriate nonequivalent dependent variable can be a difficult task as ASP interventions are often very broad and can impact a wide variety of outcomes. In the case of evaluating the number of DDDs after restriction of antibiotics active against Gram-negative pathogens, one could measure the changes in antibiotics solely active against Gram-positive pathogens. If only the first group of antibiotics showed a decrease and the latter showed a stable trend, one can be more convinced that the intervention caused the reduction.

Next to confounding factors, two other methodological principles could distort the relationship between intervention and outcome: *maturation effects* and *regression to the mean.*

Maturation effects refer to natural changes that patients can experience over time. For example, patients may become decolonized spontaneously independent of their earlier antibiotic exposure. So if time between exposure and outcome is long, depending on clone characteristics, microbiological outcomes may become false negative [11]. If one is aware of this principle, the influence could be minimized beforehand by including a control group, and/or limiting the duration of data collection, or afterward by adjusting for a proxy for maturation in the final analytical model.

Regression to the mean is a statistical phenomenon that dictates that if the first measurement is extreme, the follow-up measurement will tend to be closer to the overall mean value [12,13], independent of any intervention. Studies about the impact of ASPs have often failed to explicitly document the reasons behind their decision to intervene [14], but are frequently a response to unusual levels of antibiotic use or resistance. Therefore, there is a high risk of regression to the mean in this field of study [3]. The impact of regression to the mean can be reduced by including a control group and by accurate randomization of the intervention. In uncontrolled before-after studies, the design should include multiple outcome assessments over time before and after the intervention, so preintervention trends can be compared with postintervention trends.

Bias

Contrary to the above-described effects, bias is not a natural phenomenon, but a systematic error or artifact in the study process introduced by the researcher. This is a bigger threat to validity than confounding. If systematic differences arise between the intervention arm (or before period) and control arm (or after period), adjustment in the analysis phase is impossible. According to the Cochrane methodology, five types of biases can be identified: *selection bias*, *performance bias*, *detection bias*, *attrition bias*, and *reporting bias*. All refer to different stages of the research process [15].

Selection bias indicates the emergence of systematic differences between participants in the intervention and control arm through differential selection. This improper assignment can be caused by nonrandom allocation sequence generation and/or inadequate concealment of the allocation. Proper randomization should make use of a centralized randomization scheme using opaque envelopes or an on-site computer. Invalid randomization methods include the use of an open list of random numbers, allocation based on patient record number, birth dates, days of the week or admission order, because they have a high risk of creating unbalanced study groups [16].

Performance bias refers to systematic differences in exposure between study arms. This can include quality of care, infection control, antibiotic policy, and blood culture rates. This often arises due to lack of blinding of caregivers or patients for intervention allocation. In ASP evaluations, blinding is often difficult, and, unintentionally, caregivers in the intervention arm may improve on other care practices as well. Patients may show different behaviors as well; they may search additional treatment if they know they are receiving a possibly less effective treatment or withdraw earlier if their treatment protocol may have a higher risk of side effects.

Detection bias is also associated with blinding, blinding for intervention allocation when assessing the outcome measure of interest. Dranitsaris *et al.* [17] measured differences in appropriateness of cefotaxime prescribing in patient groups with and without pharmacist-enforced guideline adherence. In this particular study, the nonblinded clinical pharmacists themselves collected all the outcome data. Unconsciously, their level of approval may have been higher for cases in the intervention arm, especially since guideline adherence can be quite subjective in complex clinical cases. Therefore, there was a high risk of detection bias in this study.

Finally, *attrition bias* and *reporting bias* are concerned with incomplete outcome data. This can be due to withdrawals or selective reporting, respectively, and should be systematically different for the intervention and the control arm to be able to cause bias. This can easily be detected by scrutinizing the study's flowchart. If patients drop out, this should be made explicit and reasons should be discussed.

Design

The choice for a certain study design will, in general, already determine how well one can challenge confounding, bias, maturation, and regression to the mean. In theory, the more control a researcher can exert over study processes, the higher the likelihood of appropriate causal inference.

Observational studies are highly vulnerable to methodological fallacies, since exposure is not controlled, but rather follows a natural process. Experimental designs have a higher internal validity because intervention allocation is controlled. The purest experimental design in medicine is the randomized controlled trial (RCT), where the intervention is randomly allocated. A second category is the quasiexperimental designs, where the intervention is allocated in a quasirandom way, like by consecutive order, or date of birth.

The Effective Practice and Organization of Care (EPOC) group, a Cochrane review group, has approved the following study designs to evaluate interventions in the field of hospital epidemiology [16]: RCT (experimental design), nonrandomized controlled trial (CCT; quasiexperimental design), controlled before-after design (CBA; quasiexperimental design), and interrupted time series (ITS; quasiexperimental design). RCTs can be further subdivided in: classical individual RCTs, cluster randomized controlled trials, cluster randomized controlled trials with crossover, and the stepped-wedge design. This selection of study designs was adopted in the ORION statement, which was developed to raise standards in the field of hospital epidemiology [3]. The seven different designs are summarized in Fig. 1, and below we will discuss the advantages and challenges of each design in more detail. Design-appropriate reporting guidelines for the different study designs can be found at www.equator-network.org, like STROBE-AMS [18], for observational studies focusing on antimicrobial resistance and AMS.

Experimental Designs: Randomized Controlled Trials

In many fields of research, RCTs are considered the golden standard for causal inference. It is required to have objective documentation of the study protocol and statistical analysis plan prior to initiation of the trial (with prior publication, e.g., at http://clinicaltrials.gov). Intervention and control arms are generated by randomization, which, in theory, creates groups that are, on average, similar to each other. This will prevent the possible distorting effect of confounding factors, maturation, or regression to the mean. In addition, blinding can be used to mask intervention allocation among participants, clinicians, data collectors, those evaluating the outcomes, and data analysts. This will enhance the objective evaluation of intervention effectiveness.

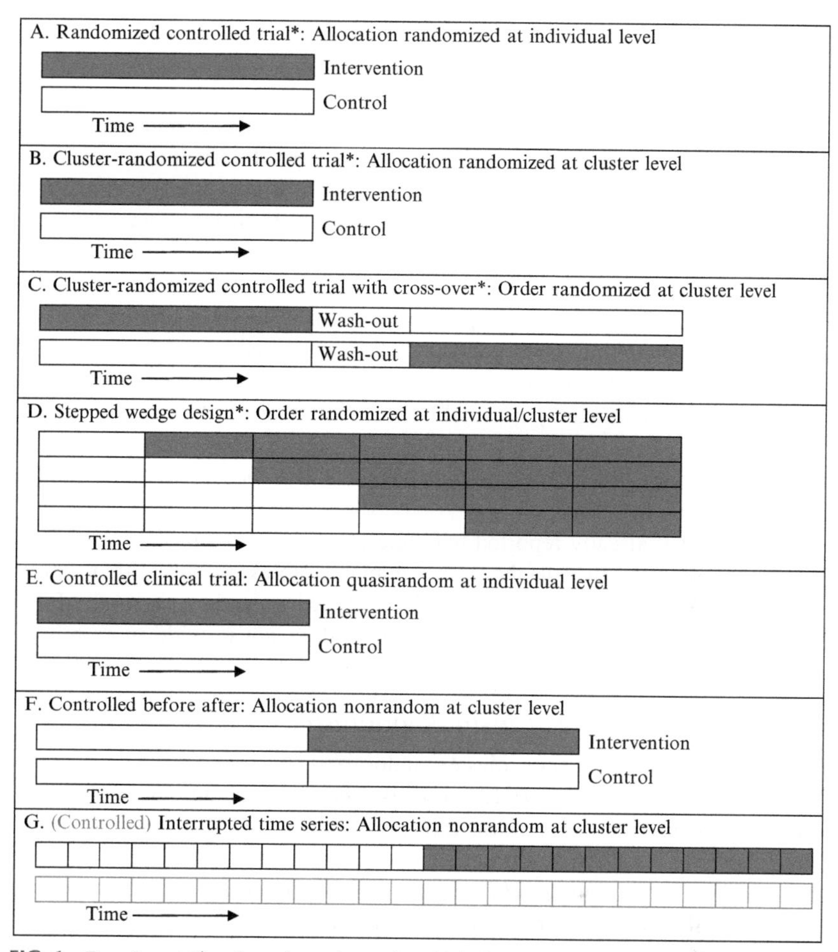

FIG. 1 Experimental* and quasiexperimental study designs appropriate for evaluation of antimicrobial stewardship programs.

Consequently, RCTs can provide strong evidence for the impact of ASP interventions.

Nevertheless, in the Cochrane review of Davey *et al.* [2], it is shown that many RCTs studying interventions to improve antibiotic prescribing have high risk of selection bias due to improper allocation methods. Then, it is often not feasible to apply blinding for all parties involved in ASP evaluation trials. Participants can be blinded to the drugs they are receiving, but it is often not feasible to blind clinicians, as ASP programs often entail behavioral change. Blinding those involved in data processing and coding is often more feasible, but Davey *et al.* [2] found that very few RCTs applied blinding for these parties. In many RCTs, none of the parties were blinded, blinding measures were unclear or not explicitly described. As a consequence, there was a high risk of performance and detection bias.

Most importantly, contamination jeopardizes valid inference from RCTs evaluating the effectiveness of ASPs. Contamination means that intervention effects spillover to participants in the control arm. Interventions related to dynamics of infectious diseases in a hospital setting, and/or behavioral changes, are especially sensitive to contamination. If an intervention reduces the incidence of infections due to a highly transmissible pathogen, this can have an impact on the risk of cross-transmission to patients in the control arm as well. If the aim of a program is to reduce duration of treatment via computer-based reminders, caregivers in the control arm may become aware of the intervention through discussions on the work floor or because of overlap/rotations in healthcare personnel between study arms. As a result, patients in the control arm may experience earlier treatment cessation as well. Contamination will result in bias toward zero for the measured impact of the intervention.

Shadish [19] already reported that "the experiment is a profoundly human endeavor, affected by all the same human foibles as any other human endeavor." So although there are well-developed procedures to make RCTs less susceptible to causal fallacy, this design does not provide more reliable results by definition, especially in the field of ASP evaluations.

Experimental Designs: Cluster-Randomized Controlled Trials

Despite their low occurrence, cluster randomized controlled trials have a number of advantages over RCTs [2]. The likelihood of contamination is greatly reduced, because the intervention allocation is randomized at group level, like ward, hospital, or district. Consequently, participants experiencing the intervention, or the clinicians who are part of the intervention, are separated geographically. Hallsworth *et al.* [20] carried out a cluster-RCT to reduce unnecessary antimicrobial prescriptions among general practitioners (GP). They randomized receipt of a feedback letter about excessive antibiotic prescriptions among GP practices in the United Kingdom. By randomizing at practice level instead of GP level, contamination of caregivers was prevented.

Logistically, a cluster-RCT is also easier to implement, particularly if it concerns a structural intervention. Meeker *et al.* [21] used electronic health records to generate alternatives for not-guideline-concordant antibiotic prescribing among American clinicians in primary care. By applying a cluster-RCT, the entire computer system within each practice could be adjusted, instead of having to create a personalized system. Finally, cluster randomized controlled trials can enhance compliance, as all professionals implementing the intervention will work within the same cluster.

Unfortunately, this method does also introduce some methodological and logistical challenges. Heterogeneity between clusters can arise more easily than between individuals, as the number of clusters is often relatively low.

This can result in flawed conclusions, because clusters are not comparable. And it can reduce the generalizability of findings, because the included clusters had particular features, like a high baseline incidence rate of infections or short length of stay. Heterogeneity in implementation of the intervention across clusters is another common challenge, especially since ASPs often contain many qualitative elements related to culture change.

Since randomization is at group level, cluster randomized controlled trials can also suffer from postrandomization selection bias at the individual-level. If it is known beforehand which hospital unit will apply rapid testing of blood cultures, physicians might be inclined to send the more severely ill cases to the intervention unit. This can be prevented by identifying eligible patients before randomization or by involving a third party for patient recruitment.

From a statistical point of view, cluster randomized controlled trials are less efficient than the classical individual RCT; they need to include more patients to achieve similar power. This is due to high correlation in outcomes between individuals within a cluster compared with the correlation between clusters. Efficiency is reduced even further if the number of patients included in each cluster is different. The most efficient method to increase the power of a cluster-RCT is to include a higher number of clusters, which can be quite challenging. Rules of thumb state that the absolute minimum of included clusters per arm is four to be able to yield a statistically significant result at the 5% level [22].

Experimental Designs: Cluster-Randomized Controlled Trials with Cross-Over

Cluster-randomized controlled trials with cross-overs are very similar to cluster randomized controlled trials, but now all clusters will experience the intervention. Each cluster will have a period of intervention, a washout period, to avoid carryover effects, and a period without any intervention or vice versa. The order of these sequences will be randomly allocated. In this way, the influence of cluster heterogeneity is partly eliminated, as all clusters will be part of the control and intervention arm, although in a different order. This results in cluster-randomized controlled trials with cross-overs being the most efficient cluster-RCT design from a statistical point of view. It requires the lowest number of clusters for an equally powered study compared with cluster randomized controlled trials without crossover, or stepped-wedge designs. However, if the number of clusters is greatly reduced, this does increase the probability of chance imbalances and reduces the level of generalizability [22]. Furthermore, since all clusters experience the control and intervention period in a serial fashion, study duration is more than doubled. Very few cluster-randomized controlled trials with cross-overs can be found in the field of ASP [2], as the design is often not suitable; if the intervention has a lasting or delayed effect, like an educational campaign, washout is not realistic.

Experimental Designs: Stepped-Wedged Design

Stepped-wedge design, experimentally staged introduction, delayed intervention, and phased implementation trials are all different names for the same design. In a stepped-wedge design, all trial participants (individuals or clusters) will experience the intervention, but there is a phased introduction. Participants will start the intervention sequentially, in a unidirectional fashion; in the end, all will be in the intervention arm. The time point (step) during which the intervention will be assigned is determined at random, and data are collected at each step until all groups are in the intervention arm. For design efficiency, only one cluster should crossover per time unit [23]. In Table 2, a summary of the advantages and constraints of stepped-wedge design can be found.

A famous example of an applied stepped-wedge design is the Matching Michigan study, where 223 intensive care units (ICUs) from the United Kingdom implemented a successful intervention bundle from the United States to reduce catheter-related bloodstream infections [26]. Although the popularity of the stepped-wedge design has increased, not many studies in the field of ASP have embraced this design [27].

Quasiexperimental Designs

In the field of health system intervention research, RCTs are often not feasible to address effectiveness questions. Trials are expensive and time-consuming; they require ethical approval and a reasonable number of participants or

TABLE 2 Advantages and Constraints of the Stepped-Wedge Design [22,24,25]

Pros	Cons
A proven effective measure can be implemented and evaluated at a large scale without the need for withdrawal or withholding	Duration of implementation is greatly prolonged
It is easier to implement from a logistical, practical, and financial view, since the intervention is implemented in a sequential order	Blinding of participants or care providers is very difficult, increasing the risk for performance and detection bias
Temporal changes in effectiveness of the intervention can be modeled	It has a high intrinsic risk of attrition bias, due to the required repeated measurements after each step
Statistically more efficient than cluster-randomized controlled trials	There is still a strong need for further guidance on sample size calculations, analytical methods, and reporting

hospital units to be able to adequately randomize intervention allocation. Therefore, EPOC has also identified acceptable quasiexperimental designs. Campbell and Stanley [28] and later Harris *et al.* [29] have popularized the name quasiexperiments for studies that evaluate the effect of a nonrandomized group level intervention. In literature, quasiexperiments are often referred to as before-after or pre-post-test studies, descriptive study designs, or controlled intervention trials. These studies can utilize observational data that are readily available and are therefore more time and money efficient and encompass less ethical difficulties.

The internal validity of quasiexperimental designs can, however, be lower than their experimental counterparts. Since they lack randomization, the risk of selection bias is higher. Often, perceived low-risk patients or patients with a good prognosis will be more likely to be subjected to interventions aimed at reducing antibiotic exposure than high risk patients and vice versa. These systematic differences between study arms will result in an upward or downward bias, respectively, for intervention effectiveness. Internal validity can be improved by increasing the homogeneity of the patients in the intervention and the control arm. This can be achieved through restriction of the type of patients to be included, through matching of intervention and control patients based on important confounders, or through multivariate regression in the analysis phase.

Quasiexperimental designs are often (partially) retrospective, and therefore allocation concealment and blinding of participants or clinicians are rare in these types of studies. As a result, the risk of performance and detection bias is also higher compared with RCTs. Since attrition bias and reporting bias depend on the researchers rather than the study design, these biases are not necessarily higher for quasi- vs experimental designs.

The advantage of quasiexperiments is that the external validity, that is, the generalizability of the conclusions, is often higher, because data are more representative of routine clinical care, antibiotic prescribing practices, and real-world patient heterogeneity.

Here we will only discuss quasiexperimental designs where the likelihood of causation is strengthened by inclusion of a control group, as approved by the Cochrane EPOC Review Group for evaluation of interventions [16]. This includes nonrandomized controlled trial, CBA, and ITS.

Quasiexperimental Designs: Nonrandomized Controlled Trial

In this design, an intervention is assigned by the researcher in a quasirandom manner at the individual or group level, by methods like alternation, date of birth, and patient identifier. This does not assure homogeneity between the intervention and control arms and greatly increases the risk of selection bias. This should not be preferred over proper randomization, however, when only few participants or units are available, randomization would only add complexity and not improve homogeneity between the study arms. One of the few nonrandomized controlled trials in ASP research is the study by Toltzis *et al.* [30] studying the impact of antibiotic cycling on colonization of patients at the neonatal ICU (NICU). Since their hospital only had two NICUs, and the intervention was introduced at unit level, they could only include two study clusters and randomization would not have made a difference. In this setup, baseline differences between the two study units are very likely (experience of attending physicians and baseline disease severity). Furthermore, the nurse who assigned patients to the wards was not blinded, so there was a high risk of selection bias due to nonrandom allocation and lack of allocation concealment.

There can also be pragmatic reasons to apply a nonrandomized controlled trial. Lee *et al.* [31] compared strategies to reduce MRSA rates in surgical patients in ten different hospitals across Europe and Israel. In this case, the choice of allocation was restricted due to factors such as cost and personnel, capacity of the microbiology laboratories, and mandatory local or national interventions.

A nonrandomized controlled trial should only be considered for hypothesis-generating research for which participant selection or randomization is complex or expensive.

Quasiexperimental Designs: CBA Studies

After RCTs, this is the most popular design for evaluation of ASPs [2]. An intervention is implemented at group level, often initiated by a third party, and outcomes before and after the intervention period are contrasted to outcomes in the same time periods in a comparable control arm without the intervention under study. So in contrast with nonrandomized controlled trial, heterogeneity between the intervention and control arm has less impact on the evaluation of effectiveness as the preintervention period in the intervention arm serves as an additional control for the postintervention period.

Nevertheless, inclusion of a comparable control group, with regard to baseline values of exposure and outcome variables, is important to discern regression to the mean and other temporal changes. The quality of CBA designs also depends on the timeliness of data collection; preferably, data collection for the intervention and control arm should be carried out within the same time period, preferably in a prospective fashion. Furthermore, the Cochrane EPOC group suggests that at least two intervention and two control units should be included [16].

Quasiexperimental Designs: Interrupted Time Series

ITS is considered the strongest quasiexperimental design [4], if analyzed in the correct way. In a review by Ramsay *et al.* [32], they used real-life data [33] to show how simple pre-post comparisons (change in mean) can overestimate the effect compared with proper time series analyses (change in level and

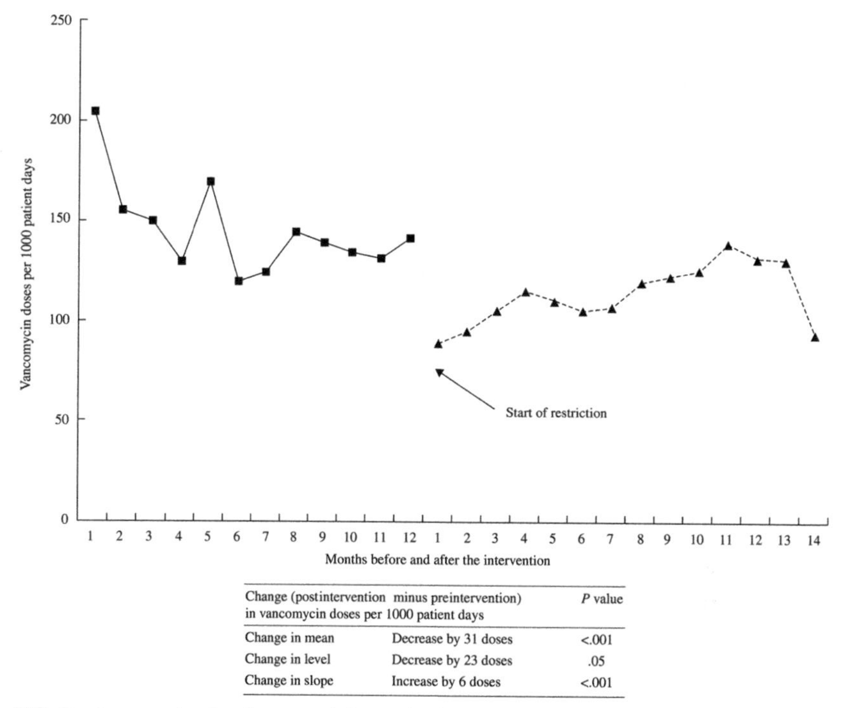

Change (postintervention minus preintervention) in vancomycin doses per 1000 patient days		*P* value
Change in mean	Decrease by 31 doses	<.001
Change in level	Decrease by 23 doses	.05
Change in slope	Increase by 6 doses	<.001

FIG. 2 An example of an interrupted time series in which the effect of the interventions is overestimated by analysis of mean data before and after the intervention [32].

change in slope); see Fig. 2. Although ITS, by definition, has a higher risk of bias than well-performed RCTs, Davey *et al.* [2] found that, for studies in the field of ASP evaluations, ITS had lower risk of bias than RCTs, nonrandomized controlled trials, and CBAs. This was associated with the fact that many of the RCTs lacked appropriate randomization methods, allocation concealment, and blinding and suffered from contamination bias. The ITS, however, provides information about preintervention trends and assesses the extent to which the effect of an intervention was sustained.

A time series consists of multiple assessments of a specific outcome measure, at group level, at regularly spaced time intervals. The "interruption" or "change point" of the time series is an identifiable real-world event, like the start of an AMS intervention. The advantage of this design is that the preintervention segment can serve as a control for the postintervention segment. Due to the large number of sequential data points, ITS can determine trends before and after intervention, which enables discrimination of the impact of the intervention from regression to the mean, seasonality, or other preexisting trends. It also indicates whether an intervention had an abrupt effect (abrupt change in level) or a gradual effect (change in trends). For example, the number of vancomycin DDDs may drop straight after initiation of the ASP, or an increasing trend may be turned into a decreasing trend. Time series can also be used to identify whether an intervention effect is immediate or delayed.

Confounding in ITS is limited to factors related to the outcome of interest that changed simultaneously with the intervention. There is also a risk of performance and reporting bias, since ITS lacks allocation concealment and this could influence behavior of physicians and researchers involved. This type of confounding or bias can be overcome by extending ITS with a control group or by including a nonequivalent (not affected by the intervention) outcome variable.

The Cochrane EPOC group quality criteria [16] for ITS are twofold: a clearly defined point in time when the intervention occurred and at least three data points before and three after the intervention. However, 12 monthly data points before and 12 monthly points after the intervention are advisable to be able to rule out seasonality [4,32].

Wagner *et al.* [4] extended the quality criteria of ITS by underlining the need for a sufficient number of observations/patients (minimum number of 100) at each time point to achieve an acceptable level of variability of each estimate. A final important quality criterion is whether the researchers prespecified the shape of the effect they were expecting to find; without such a hypothesis, an ITS design could become a fishing expedition (by increasing the number of comparisons, one increases the chance of finding a significant finding that is false positive) [2].

STATISTICAL METHODS

The choice of statistical methods clearly depends on the chosen design, but among the experimental designs and quasiexperimental designs, there is some overlap in methods. Here, we will first discuss a general methodological issue and basic statistical tests, followed by design-specific statistical methods for experimental designs and quasiexperimental designs. Methodological and statistical advice should always be sought from the outset of each evaluation study.

A general issue in studies focusing on clinical or microbiological endpoints, like colonization, infection, or hospital mortality is competing events. For all these outcomes, there are other possible competing endpoints like hospital discharge. As soon as patients are discharged, they are no longer at risk for the outcome of interest and should no longer be included in the risk set. Ignoring competing events can result in an overestimation of the cumulative probability for the outcome of interest and thus underestimate the effect of the intervention. Therefore, it should be assessed whether the intervention has an impact on the competing event by event-specific analysis. See Wolkewitz *et al.* [34] for more details on this specific topic.

If only one outcome measurement before and after intervention is available, basic statistical test to compare groups can be utilized, like *t*-tests

(2 groups), ANOVA (3 groups), and MANOVA (3 or more groups) for continuous data and Pearson χ^2 or Fisher's exact tests (if counts are low) for categorical data. If the outcome variable has a skewed distribution, Mann-Whitney *U*, for continuous data, or the Wilcoxon test, for categorical data, would be appropriate. These methods can be applied for many of the experimental designs and quasiexperimental designs, but require balanced intervention and control arms and a single center setting without clustering. If any of these requirements are not met, or multiple measurements over time before and after intervention are available, more complicated analyses should be applied, as will be discussed in the following section.

Statistical Methods: Experimental Designs

There are two different general approaches for analyzing RCTs: "intention to treat" analysis (ITT) and the "per-protocol" analysis (PP). In the first method, participants are analyzed according to their allocation regardless of adherence to the intervention, and randomization is kept intact. PP only includes participants who completed the study protocol and therefore disturbs randomization. ITT can suffer from misclassification bias: If there is a high level of noncompliance, patients in the intervention arm are actually mislabeled controls. If the noncompliant patients drop out, and the endpoint cannot be ascertained, attrition bias is another risk. PP provides a more accurate estimate of intervention effectiveness, but can suffer from loss of power and selection bias [35]. Fortunately, interventions related to ASP often involve minimization of treatment by deescalation, or reduced duration of treatment, or replacing one treatment by another treatment, and consequently systematic noncompliance and dropouts are not very likely, and ITT and PP analyses will often provide similar results.

Statistical analyses for RCTs can be very straightforward, but they can become more complicated in multicenter trials, if one assumes that the effect of the intervention is different for each center. In that case, multilevel models, also called hierarchical, mixed-effects, or random-effects models, or generalized estimating equations (GEE) can be applied. Further reading is available in the paper by Chu *et al.* [36] and Mody *et al.* [37]

In case of cluster randomized controlled trials, one needs to address the fact that observations are more strongly correlated within each cluster than between clusters. There are three different ways of doing that. First, means or proportions can be determined at cluster-level and then basic statistical tests can be used to compare these cluster-level outcome measures. Second, patient-level analyses can be undertaken by applying adjustments for the clustering effect. If clustering is ignored, standard errors will be underestimated, and statistical tests will be invalid. In most statistical packages, regular procedures are available to inflate the standard errors by accounting for the intracluster correlation (so-called Huber-White, robust, or Sandwich standard errors). However, none of these methods allows for adjustment of patient or cluster characteristics. To this effect, multilevel analysis comes into play again, whereby covariates measured at all the different levels can be incorporated. These methods are very flexible but require extensive statistical expertise. Peters *et al.* [38] provide a helpful overview of the different applicable models. The most appropriate analytical choice depends on the research question, presence of confounders, the number and size of clusters, and the variability of cluster size [39]. However, the ORION guidelines state that analysis of aggregated data should be avoided when disaggregated data are available [3].

In cluster-CO-RCTs and the stepped-wedge design, each cluster will be part of the control and intervention arm; thus, the most straightforward effect measure is a within cluster comparison using a paired *t*-test. However, stepped-wedge designs are implemented over a long time, and intervention effectiveness may vary over time. If time effects are present, linear mixed models for continuous and normally distributed data should be applied. If cluster sizes vary, or data are not normally distributed, generalized linear mixed models or GEE are warranted. They should, however, be used with care in case of low number of clusters or low number of time points [23,24]. An example of an applied generalized linear mixed models can be found in the paper of Fuller *et al.* [40].

Statistical Methods: Quasiexperimental Designs

In quasiexperimental designs, the intervention is not randomly allocated, and consequently, the intervention and control arm are not balanced. This warrants stratified or multivariate analytical techniques. If multiple confounders may play a role, propensity scores can be calculated, which account for the covariates that predict receiving the treatment, and this score can then be used to match similar patients in the intervention and control arm. See, for example, the study by Chi *et al.* [41], where they retrospectively matched high risk preterm infants based on a propensity score to evaluate the effectiveness of a six monthly intramuscular administration of palivizumab to prevent RSV hospitalizations. Use of the propensity score can result in exclusion of outliers and will reduce the power of the study. Another method to adjust for multiple confounders is multivariate generalized linear models, like linear regression for continuous outcomes, logistic regression for dichotomous outcomes, or Poisson regression for counts or rates. A rule of thumb in regression models is that there should be a minimum of 10 events for each additional explanatory variable, but this number increases if the probable effect size of the explanatory variable is low or if correlation with other explanatory variables is high [42].

The advantage of the CBA design is that the outcome of interest is also measured preintervention in both study arms, and one can apply the double-difference method. The change in outcome between the before and after intervention period can be determined in both study arms, and these change measures can be compared between the intervention and control arm with basic statistical tests. This method removes the difference in outcome between treatment and comparison group at baseline, however, it does assume that the trends in both study arms were similar before intervention. If required, stratification, propensity score matching, and generalized linear models can also be applied.

The preferred method of analyses for ASP evaluations is time series analyses, whereby aggregated outcome measures over time are used to calculate trends before and after intervention. Segmented regression can assess changes in mean outcome levels (intercepts) and trends (slopes) before and after intervention. Multiple change points can be included, if multiple, sequential interventions are implemented. Lags can also be included, if one assumes the intervention takes some time to have an impact. The data from the "lag" period can either be excluded or one can model it as a separate segment [4].

If outcome measures are highly autocorrelated, as is often the case for ASP evaluation studies, regular segmented regression may not be the optimal solution. Autocorrelation implies that outcomes at two time points close together are more similar than outcomes further apart in time. Correlation between adjacent data points is termed first-order autocorrelation, and correlation between the current point and 2 months before or after is termed second-order autocorrelation. Seasonal patterns in monthly time series are called higher-order autocorrelation. Failing to correct for autocorrelation may lead to underestimated standard errors and overestimated significance of the effects of an intervention. Time series methodologies can take this autocorrelation into account, for example, through autoregressive integrated moving average (ARIMA) models. To detect autocorrelation, one can visually inspect a plot of residuals against time and conduct statistical tests (Durbin-Watson statistic). The challenge is to correctly model the data; autocorrelation, moving averages, and trends can be expressed within one model. Confounding in ARIMA models is limited to factors that are related to the outcome of interest and occur at the same time as the intervention, which makes this type of analyses much more robust than any of the other methods. A limitation of this method is the complexity of building the model and interpreting the results.

Control groups can make time series analysis even stronger, by building ARIMA models for both study arms and comparing these qualitatively. The same can be done for a nonequivalent outcome measure. The effect of possible explanatory time series like antibiotic usage or bed occupation rate, including a possible time delay, can be assessed by the use of transfer functions or cross-correlation functions. A nice example can be found in the paper of Vernaz *et al.* [43] where the authors studied the impact of antibiotic use and hand rub consumption on the incidence of MRSA and *Clostridium difficile*.

CONCLUSIONS

The field of ASP evaluations is still dominated by low-quality studies, which do not adhere to current methodological standards. The biggest threats to causal inference are related to contamination, including cross-transmission, and external secular trends, like regression to the mean. Although RCTs are considered the golden standard in clinical research, cluster randomized controlled trials should be considered the golden standard in the field of ASP evaluations, where infectious disease dynamics and overlap in personnel can easily bias study outcomes. Thereby, multilevel analysis based on individual-level data should be preferred over aggregated data analyses. While stepped-wedge design may seem interesting, implementation is challenging and costly, and sample size calculations and analytical methods still lack a strong theoretical basis.

If only few clusters are available or ASPs are implemented in a single center setting, cluster randomization will not result in balanced groups, and controlled ITS should be the design of choice. Ideally, the outcome of interest should be measured at equally spaced time points for at least one year before and after the intervention, and analyses should acknowledge the data structure: Plot the dots! Segmented regression analysis is a straightforward analytical method to provide insight in ITS data. Time series modeling, like ARIMA models, can provide even stronger evidence, but requires solid statistical expertise.

We are optimistic that the quality of future endeavors will benefit from these recent insights and are confident that our understanding of the critical components of ASP will continue to grow.

REFERENCES AND FURTHER READING*

[1] Akpan MR, Ahmad R, Shebl NA, Ashiru-Oredope D. A review of quality measures for assessing the impact of antimicrobial stewardship programs in hospitals. Antibiotics 2016;5. http://dx.doi.org/10.3390/antibiotics5010005.

[2*] Davey P, Brown E, Charani E, *et al.* Interventions to improve antibiotic prescribing practices for hospital inpatients. Cochrane Database Syst Rev 2013;4:CD003543.

[3] Stone SP, Cooper BS, Kibbler CC, *et al.* The ORION statement: guidelines for transparent reporting of outbreak reports and intervention studies of nosocomial infection. Lancet Infect Dis 2007;7:282–8.

[4*] Wagner AK, Soumerai SB, Zhang F, Ross-Degnan D. Segmented regression analysis of interrupted time series studies in medication use research. J Clin Pharm Ther 2002;27:299–309.

[5] McGowan JE. Antimicrobial stewardship—the state of the art in 2011: focus on outcome and methods. Infect Control Hosp Epidemiol 2012;33:331–7.

[6] Leekha S, Diekema DJ, Perencevich EN. Seasonality of staphylococcal infections. Clin Microbiol Infect 2012;18:927–33.

[7] Eber MR, Shardell M, Schweizer ML, Laxminarayan R, Perencevich EN. Seasonal and temperature-associated increases in gram-negative bacterial bloodstream infections among hospitalized patients. PLoS One 2011;6:e25298.

[8] Livermore DM. Fourteen years in resistance. Int J Antimicrob Agents 2012;39:283–94.
[9] Wyllie DH, Walker AS, Miller R, *et al.* Decline of meticillin-resistant *Staphylococcus aureus* in Oxfordshire hospitals is strain-specific and preceded infection-control intensification. BMJ Open 2011;1:e000160.
[10] Lipsitch M, Tchetgen Tchetgen E, Cohen T. Negative controls: a tool for detecting confounding and bias in observational studies. Epidemiology 2010;21:383–8.
[11] Miller RR, Walker AS, Godwin H, *et al.* Dynamics of acquisition and loss of carriage of staphylococcus aureus strains in the community: the effect of clonal complex. J Infect 2014;68:426–39.
[12] Barnett AG, van der Pols JC, Dobson AJ. Regression to the mean: what it is and how to deal with it. Int J Epidemiol 2005;34:215–20.
[13] Morton V, Torgerson DJ. Regression to the mean: treatment effect without the intervention. J Eval Clin Pract 2005;11:59–65.
[14] Cooper BS, Stone SP, Kibbler CC, *et al.* Systematic review of isolation policies in the hospital management of methicillin-resistant *Staphylococcus aureus*: a review of the literature with epidemiological and economic modelling. Health Technol Assess 2003;7:1–194.
[15] The Cochrane Collaboration. Cochrane handbook for systematic reviews of interventions, www.cochrane-handbook.org; 2011.
[16*] Effective Practice Organisation of Care (EPOC). EPOC resources for review authors. Oslo: Norwegian Knowledge Centre for the Health Services; 2015. http://epoc.cochrane.org/epoc-specific-resources-review-authors.
[17] Dranitsaris G, Spizzirri D, Pitre M, McGeer A. A randomized trial to measure the optimal role of the pharmacist in promoting evidence-based antibiotic use in acute care hospitals. Int J Technol Assess Health Care 2001;17:171–80.
[18] Tacconelli E, Cataldo MA, Paul M, *et al.* STROBE-AMS: recommendations to optimise reporting of epidemiological studies on antimicrobial resistance and informing improvement in antimicrobial stewardship. BMJ Open 2016;6:e010134.
[19*] Shadish W, Cook T, Campbell D. Experimental and quasi-experimental signs for generalized causal inference. Boston, MA: Houghton Mifflin Torgerson; 2002.
[20] Hallsworth M, Chadborn T, Sallis A, *et al.* Provision of social norm feedback to high prescribers of antibiotics in general practice: a pragmatic national randomised controlled trial. Lancet 2016;387:1743–52.
[21] Meeker D, Linder JA, Fox CR, *et al.* Effect of behavioral interventions on inappropriate antibiotic prescribing among primary care practices: a randomized clinical trial. JAMA 2016;315:562–70.
[22] Taljaard M, Teerenstra S, Ivers NM, Fergusson DA. Substantial risks associated with few clusters in cluster randomized and stepped wedge designs. Clin Trials 2016;13:459–63.
[23] Hussey MA, Hughes JP. Design and analysis of stepped wedge cluster randomized trials. Contemp Clin Trials 2007;28:182–91.
[24] Brown CA, Lilford RJ. The stepped wedge trial design: a systematic review. BMC Med Res Methodol 2006;6:54.
[25*] Beard E, Lewis JJ, Copas A, *et al.* Stepped wedge randomised controlled trials: systematic review of studies published between 2010 and 2014. Trials 2015;16:353.
[26] Bion J, Richardson A, Hibbert P, *et al.* 'Matching Michigan': a 2-year stepped interventional programme to minimise central venous catheter-blood stream infections in intensive care units in England. BMJ Qual Saf 2013;22:110–23.
[27] Palmay L, Elligsen M, Walker SA, *et al.* Hospital-wide rollout of antimicrobial stewardship: a stepped-wedge randomized trial. Clin Infect Dis 2014;59:867–74.

[28] Campbell DT, Stanley JC. Experimental and quasi-experimental designs for research on teaching. In: Handbook of research on teaching. Chicago IL: Rand McNally; 1963. p. 171–246.

[29] Harris AD, Bradham DD, Baumgarten M, Zuckerman IH, Fink JC, Perencevich EN. The use and interpretation of quasi-experimental studies in infectious diseases. Clin Infect Dis 2004;38:1586–91.

[30] Toltzis P, Dul MJ, Hoyen C, *et al.* The effect of antibiotic rotation on colonization with antibiotic-resistant bacilli in a neonatal intensive care unit. Pediatrics 2002;110:707–11.

[31] Lee AS, Cooper BS, Malhotra-Kumar S, *et al.* Comparison of strategies to reduce meticillin-resistant *Staphylococcus aureus* rates in surgical patients: a controlled multicentre intervention trial. BMJ Open 2013;3:e003126.

[32*] Ramsay C, Brown E, Hartman G, Davey P. Room for improvement: a systematic review of the quality of evaluations of interventions to improve hospital antibiotic prescribing. J Antimicrob Chemother 2003;52:764–71.

[33] Belliveau PP, Rothman AL, Maday CE. Limiting vancomycin use to combat vancomycin-resistant *Enterococcus faecium.* Am J Health Syst Pharm 1996;53:1570–5.

[34*] Wolkewitz M, Vonberg RP, Grundmann H, *et al.* Risk factors for the development of nosocomial pneumonia and mortality on intensive care units: application of competing risks models. Crit Care 2008;12:R44.

[35*] Anderson DJ, Juthani-Mehta M, Morgan DJ. Research methods in healthcare epidemiology and antimicrobial stewardship: randomized controlled trials. Infect Control Hosp Epidemiol 2016;37:629–34.

[36*] Chu R, Thabane L, Ma J, Holbrook A, Pullenayegum E, Devereaux PJ. Comparing methods to estimate treatment effects on a continuous outcome in multicentre randomized controlled trials: a simulation study. BMC Med Res Methodol 2011;11:21.

[37] Mody L, Krein SL, Saint S, *et al.* A targeted infection prevention intervention in nursing home residents with indwelling devices: a randomized clinical trial. JAMA Intern Med 2015;175:714–23.

[38] Peters TJ, Richards SH, Bankhead CR, Ades AE, Sterne JA. Comparison of methods for analysing cluster randomized trials: an example involving a factorial design. Int J Epidemiol 2003;32:840–6.

[39] Eccles M, Grimshaw J, Campbell M, Ramsay C. Research designs for studies evaluating the effectiveness of change and improvement strategies. Qual Saf Health Care 2003;12:47–52.

[40] Fuller C, Michie S, Savage J, *et al.* The feedback intervention trial (fit)—improving hand-hygiene compliance in UK healthcare workers: a stepped wedge cluster randomised controlled trial. PLoS One 2012;7:e41617.

[41] Chi H, Hsu CH, Chang JH, *et al.* A novel six consecutive monthly doses of palivizumab prophylaxis protocol for the prevention of respiratory syncytial virus infection in high-risk preterm infants in Taiwan. PLoS One 2014;9:e100981.

[42] Courvoisier DS, Combescure C, Agoritsas T, Gayet-Ageron A, Perneger TV. Performance of logistic regression modeling: beyond the number of events per variable, the role of data structure. J Clin Epidemiol 2011;64:993–1000.

[43] Vernaz N, Sax H, Pittet D, Bonnabry P, Schrenzel J, Harbarth S. Temporal effects of antibiotic use and hand rub consumption on the incidence of MRSA and *Clostridium difficile.* J Antimicrob Chemother 2008;62:601–7.

Index

Note: Page numbers followed by "*t*" indicate tables.

M

N

O

P

Q

R

S

T

U

List of Abbreviations

ABC	ATP binding cassette
ABS	antibiotic stewardship
ACSQHC	Australian Commission for Safety and Quality in Health Care
ADKA	Association of Hospital Pharmacists
AEMPS	Spanish Agency of Medicines
AFS	antifungal stewardship
AGES	Austrian Agency for Health and Food Safety
allo-HSCT	Allogeneic hematopoietic stem cell transplantation
AML	acute myeloid leukemia
AMR	antimicrobial resistance
AMS	antimicrobial stewardship
AMTs	antimicrobial management teams
ANC	current absolute neutrophil count
APUA	Alliance for the Prudent Use of Antibiotics
ARHAI	antimicrobial resistance and healthcare associated infection
ARIMA	autoregressive integrated moving average models
ARTI	acute respiratory tract infections
ASP	antimicrobial stewardship program
ATC	anatomical therapeutic chemical
$AUC_{0\text{-}24}$	area under the concentration-time curve over 24 h
AURA	antimicrobial use and resistance in Australia
BAL	bronchoalveolar lavage
BAPCOC	Belgian Antibiotic Policy Coordination Committee
BHI	brain heart infusion
BLBLI	beta-lactam/beta-lactamase inhibitor combinations
BP	blood pressure
BSIs	bloodstream infections
CAESAR	Central Asian and Eastern European Surveillance of Antimicrobial Resistance
CAP	community-acquired pneumonia
CBA	controlled before-after design
CC258	clonal complex 258
CDC	Centers for Disease Control and Prevention

CDDEP Center for Disease Dynamics, Economics, and Policy
CDI *Clostridium difficile* infection
CDSS computerized decision support system
CEO chief executive officer
CFU colony-forming units
CIDEIM Centro Internacional de Entrenamiento e Investigaciones Médicas
CLSI Clinical and Laboratory Standards Institute
C_{max} maximum concentration
CME continuing medical education
C_{min} minimum concentration
CNS central nervous system
CPD continuing professional development
CPOE computerized physician order entry
CR carbapenem-resistant
CRP C-reactive protein
CSF cerebrospinal fluid
CVCs central venous catheters
DART German Antimicrobial Resistance Strategy
DDD defined daily dose
Defra Department for Environment, Food, and Rural Affairs
DGI German Infectious Diseases Society
DOH Department of Health
DOTs days of therapy
DRIVE-AB driving re-investment in Research & Development (R&D) and responsible antibiotic use
DTC Drug and Therapeutic Committee
EAAD European Antibiotic Awareness Day
EARS-Net European Antimicrobial Resistance Surveillance Network
EARSS European Antimicrobial Resistance Surveillance System
ECDC European Center for Disease Prevention and Control
ECIL European conference on infections in leukemia
ED emergency department
EMR electronic medical records
EORTC-MSG European Organization for Research and Treatment of Cancer-Mycoses Study Group
EPOC effective practice and organisation of care
ESAC European Surveillance of Antimicrobial Consumption
ESAC-Net Surveillance of Antimicrobial Consumption Network
ESBL extended-spectrum β-lactamase
ESCMID European Society of Clinical Microbiology and Infectious Diseases
ESGAP ESCMID Study Group for Antibiotic Policies

ESKAPE	*Enterococcus faecium*, *S. aureus*, *Klebsiella pneumoniae*, *Acinetobacter baumannii*, *Pseudomonas aeruginosa*, and *Enterobacter* spp.
ESR	erythrocyte sedimentation rate
ESVAC	European Surveillance of Veterinary Antimicrobial Consumption
EU	European Union
EUCAST	European Committee on Antimicrobial Susceptibility Testing
EUCIC	European Committee on Infection Control
FDA	Food and Drug Administration
FOAG	Federal Offices of Agriculture
FOEN	Federal Offices of Environment
FOPH	Federal Offices of Public Health
FSVO	Food Safety and Veterinary Office
FTE	full-time equivalent
FUO	fever of unknown origin
GARP	global antimicrobial resistance partnership
GEE	generalized estimating equations
GLASS	global antimicrobial resistance surveillance system
GM	galactomannan
GMS	Gomori's Methenamine-Silver stain
GNB	gram-negative bacteria
GN-BSI	gram-negative bloodstream infections
GP	general practitioner
HAIs	healthcare-acquired infections
HCDCP	Hellenic Center for Disease Control and Prevention
HE	hematoxylin and eosin
HMO	healthcare maintenance organization
HSCT	hematopoietic stem cell transplantation
ICU	intensive care unit
ID	infectious diseases
IDSA	Infectious Diseases Society of America
IDSA/SHEA	Infectious Diseases Society of America and the Society for Healthcare Epidemiology of America
IFD	invasive fungal disease
IMI	innovative medicines initiative
IPC	infection prevention and control
ISC	International Society of Chemotherapy
ISKRA	Interdisciplinary Section for Antimicrobial Resistance Control
IT	information technology
ITS	interrupted time series

JANIS	Japan Nosocomial Infections Surveillance
LOS	length of stay
LRTI	lower respiratory tract infections
LTCFs	long-term care facilities
MALDI-TOF	Matrix Assisted Laser Desorption Ionization Time of Flight
MASCC	multinational association for supportive care in cancer
MBLs	metallo-β-lactamases
MDR	multidrug resistant
MHLW	Ministry of Health, Labor, and Welfare
MIC	minimum inhibitory concentration
MOH	Ministry of Health
mprF	multipeptide resistance factor
MRSA	methicillin-resistant *Staphylococcus aureus*
NAAT	nucleic acid amplification tests
ND4BB	new drugs for bad bugs
NDM-1	New Delhi metallo-β-lactamase 1
NHS	National Health Service
NICE	National Institute for Clinical Excellence
NIICAR	National Institute for Infection Control and Antibiotic Resistance
NSQHC	National Safety and Quality in Health Care
OD	optical density
OECD	Organization for Economic Cooperation and Development
OPAT	outpatient parenteral antibiotic therapy
OTC	over-the-counter
OVLC	open virtual learning community
PACCARB	Presidential Advisory Council on Combating Antibiotic Resistant Bacteria
PAHO	Pan-American Health Organization
PAS	Periodic Acid-Schiff
PBPs	penicillin-binding protein
PCR	polymerase chain reaction
PCT	procalcitonin
PD	pharmacodynamics
PDA	Potato Dextrose Agar
PDD	prescribed daily dose
PDSA	PLAN-DO-STUDY-ACT
PK	pharmacokinetics
PK/PD	Pharmacokinetics/pharmacodynamics
PNA-FISH	peptide nucleic acid fluorescence in situ hybridization
PNSP	penicillin-nonsusceptible *S. pneumoniae*
POC	point of care
PPS	point prevalence survey
PROA	Programas de optimización de uso de antibióticos

Pro-ADM	proadrenomedullin
QIs	quality indicators
qPCR	quantitative polymerase chain reaction
RCT	randomized controlled trial
RDTs	rapid diagnostic tests
RIVM	National Institute for Public Health and the Environment
RTI	respiratory tract infections
RT-PCR	reverse transcriptase PCR
SAASP	South African Antibiotic Stewardship Programme
SACAR	Standing Advisory Committee on AMR
SAPG	Scottish Antimicrobial Prescribing Group
SCCmec	staphylococcal cassette chromosome mec
ScotMARAP	Scottish Management of Antimicrobial Resistance Action Plan
SDA	Sabouraud Dextrose Agar
SEFH	Spanish Society of Hospital Pharmacists
SEIMC	Spanish Society of Clinical Microbiology and Infectious Diseases
SEMPSMH	Spanish Society of Preventive Medicine, Public Health and Hygiene
SSTI	skin and soft tissue infections
ST	sequence type
Strama	Swedish strategic programme against antibiotic resistance
suPAR	Soluble urokinase-type plasminogen activator receptor
TATFAR	Transatlantic Task Force on AMR
TDM	therapeutic drug monitoring
TDS	three times daily administration
TID	1000 inhabitants per day
URTI	upper respiratory tract infections
UTI	urinary tract infections
VAP	ventilator-associated pneumonia
VIM	Verona integron–encoded metallo-β-lactamase
VISA	vancomycin intermediate *S. aureus*
VRE	vancomycin-resistant enterococci
WHO	World Health Organization

Edwards Brothers Malloy
Ann Arbor MI. USA
May 30, 2017